Microbes for Humanity and its Applications

THE EDITORS

Dr. Deepak Kumar Malik is Faculty, Department of Biotechnology, UIET, Kurukshetra University, Kurukshetra. He has a PhD degree in Microbiology from Choudhary Charan Singh Haryana Agriculture University, Hisar in the area of PGPR. He is having 15 year of teaching and 20 years of research experience in the field of Microbiology. He has published research paper in various disciplines of Agriculture Microbiology and Nanotechnology.

Dr. Meenu Rathi is Assistant Professor at G.M.N College Ambala Cantt. She has a PhD degree in Botany from Choudhary Charan Singh University Meerut in the area of phosphate solubilization by bacteria. She is having 10 year of teaching experience and 4 years research experience in the field of plant microbe interaction.

Dr. Divya Bhatia is presently working as Assistant Professor at UIET, K.U.K. She has done M. Tech and Ph.D. in Biotechnology from Kurukshetra University, Kurukshetra. She has expertise in Microbial Biotechnology and Nanobiotechnology. She has more than six-years teaching experience and has contributed 25 research papers and 3 book chapters.

Dr. Shikha Jaggi presently working as Head Department of Zoology at G.M.N College Ambala Cantt. She has a PhD degree in the field of Cytogenetics from department of Zoology from Kurukshetra University Kurukshetra. She has published more than 8 research papers and two edited books.

Microbes for Humanity and its Applications

– Editors –

Deepak Kumar Malik

Meenu Rathi

Divya Bhatia

Shikha Jaggi

2022

Daya Publishing House®

A Division of

Astral International Pvt. Ltd.

New Delhi – 110 002

ISBN : 9789354614941

Published by : **Daya Publishing House®**
A Division of
Astral International Pvt. Ltd.
– ISO 9001:2015 Certified Company –
4736/23, Ansari Road, Darya Ganj,
New Delhi-110 002
Ph. 011-43549197, 8130496929
E-mail: info@astralint.com
Website: www.astralint.com

Preface

Microorganisms are the most abundant and diverse form of life on Earth. Microbiota, microorganisms residing inside the human body, shares a symbiotic relationship with humans. Microorganisms have diverse range of application in different fields like human welfare, agriculture, textile, detergent, food and beverages, bioremediation and effluent treatment. Microbes has been exploited for various applications *viz.* pest resistance, plant growth promotion, waste treatment, therapeutic proteins and vaccine production, biosensor applications, fermented products, nanoparticles synthesis. New generation sequencing technology has helped to decipher microbial genetic constitution with accuracy and speed. Metabolomics studies and simulation approaches have assisted to understand the complex regulatory pathways. Application of Genetic engineering and editing tools has revolutionized their application potential. The overall purpose of the book is to provide deep insights into the microbial applications in food, agriculture, pharmaceuticals, brewery, textile industry and environment. The basic idea of book is to compile and to make available the different aspects of microbial applications for leading researcher. This book highlights the significant role of microorganisms in different fields *viz.* bioremediation, pest control, fermented food production, nanoparticles synthesis, health industry (as probiotic), antimicrobial synthesis and intervention of sequencing techniques.

Deepak Kumar Malik

Meenu Rathi

Divya Bhatia

Shikha Jaggi

Contents

Chapter 1

Microbial Degradation of Organophosphate Compounds: A Review

*Yashika Sood, Divya Bhatia, Harnek Singh Saini and Deepak Kumar Malik**

Department of Biotechnology, University Institute of Engineering and Technology, Kurukshetra University, Kurukshetra – 136 119, Haryana

**e-mail: dmalik2015@kuk.ac.in*

ABSTRACT

Organophosphate compounds are used globally as insecticides, as pesticides and chemical warfare agents. Contamination caused by the organophosphate compounds is one of a major reason of soil and water pollution as organophosphates are toxic to flora and fauna. Decontamination done by physical and chemical ways is costly, consume time, and not able to provide an apt solution in most of the cases. Therefore, bioremediation is an appropriate method for contamination removal from environment with the help of certain microorganisms. There are reports of several isolates of bacteria, fungi, cyanobacteria which show potential ability to degrade organophosphates into non-toxic compounds. This review article discusses different microorganisms which are reported till date to degrade different organophosphates like diazinon, chlorpyrifos, monocrotophos, glyphosate etc.

Introduction

Organophosphates (OP) with structure $O{=}P(OR)_3$, known as esters of phosphoric acid, are the important class of organophosphorus compounds.

The structure consists of a central phosphate molecule with alkyl or aromatic substituents. R1 and R2 in general structure are aryl or alkyl groups that are bonded to a Phosphate atom. R1 is attached to the phosphate atom directly whereas R2 is bounded with either to an oxygen or sulphur atom in case of phosphonates (phosphonothioates) (Figure 1.1). At the minimum, one of the R groups in phosphoramidates is connected to –NH2. The X group is also said "leaving group" because it is released when hydrolysis of OP compounds is done. The X group may be a heterocyclic group, aromatic, aliphatic, or halogen.

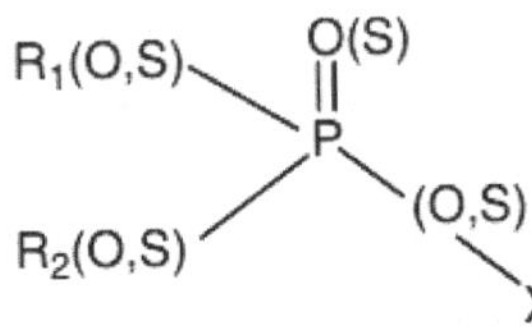

Figure 1.1: General Structure of Organophosphate.

Organophosphate insecticides, such as diazinon, chlorpyrifos, monocrotophos, disulfoton, azinphos-methyl, dimethoate, *etc.*, have extensive applications in agriculture and in households as pesticides.

Table 1.1 : Different OP Compounds Used as Insecticides

Insecticides	***Structure***
Diazinon	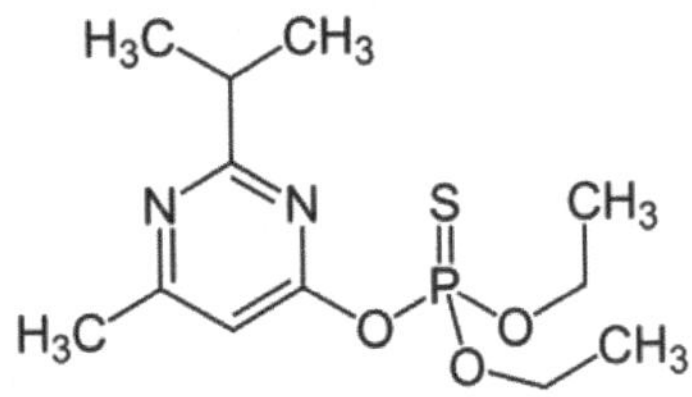
Chlorpyrifos	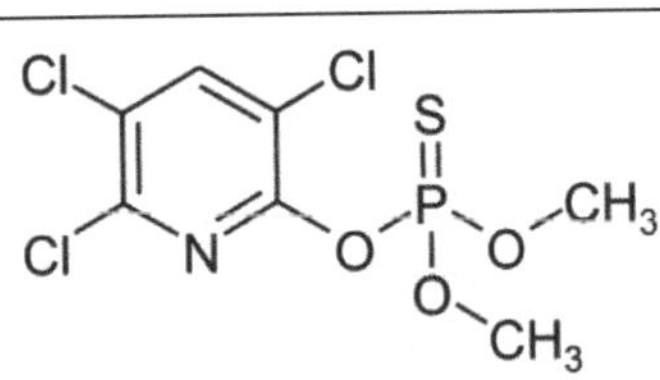
Dimethoate	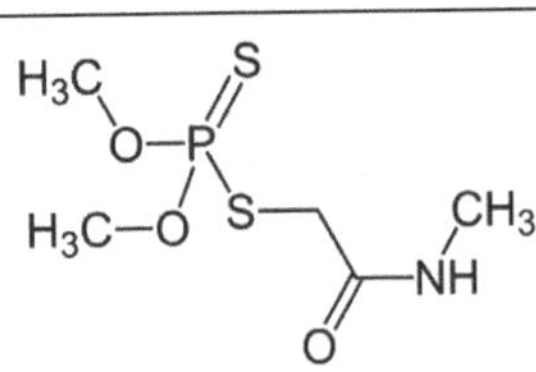

Insecticides	*Structure*
Malathion	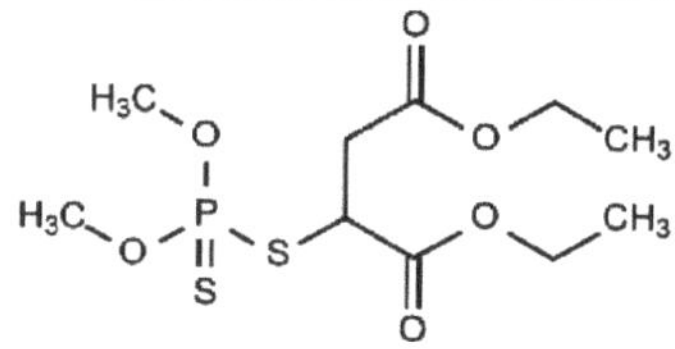
Monocrotophos	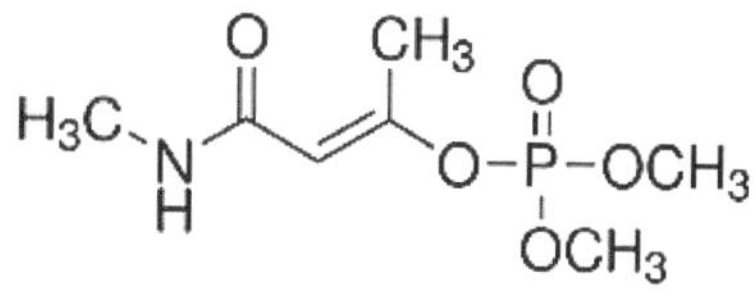
Parathion	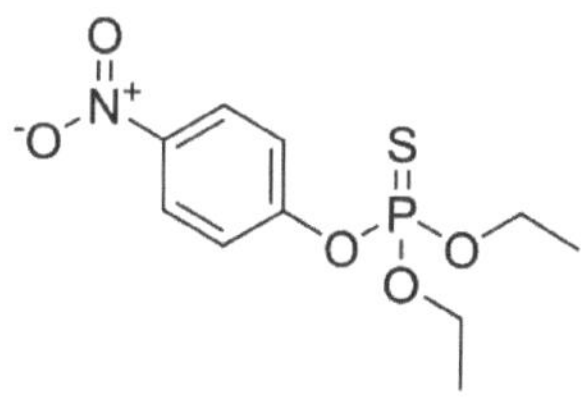
Glyphosate	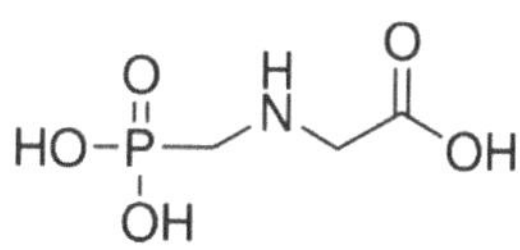
Ethoprophos	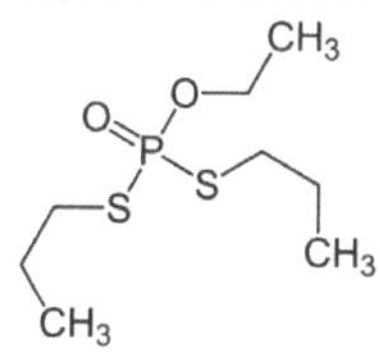

Tetraethyl pyrophosphate, the first organophosphorus insecticide, was made in use in 1937. Severe or persistent exposure to OPs can be the cause of vast amount of toxicity in mammals, plants, and insects. The bulk handling of OPs at the farm fields or unwanted release may be the cause of soil surface and ground water contamination. A number of reports have concluded that organophosphorus compounds can contaminate a large amount of water and terrestrial ecosystems.

Mode of Action

Organophosphate compounds work by their action in inhibition of acetylcholine breakdown. Nerve impulses are transmitted to the brain skeletal and muscular systems which are accomplished by acetylcholine where it plays an important role (Toole and Toole, 1995). Acetylcholine, after transmission, should be hydrolysed in order to neglect overstimulation of the nervous systems.

Acetylcholinesterase (AChE), is an esterase enzyme which catalyse the hydrolysis breakdown and results in choline and acetyl CoA formation. The AChE helps in forming an enzyme-substrate complex by the substrate binding at its active site at serine 203. The normal activity of AChE is disturbed by organophosphates as it binds to the AChE's active site followed by change in its structure and function. The leaving group present in OPs removed, hence AChE phophorylates. Phosphorylated AChE hydrolysis is extremely slow process, could take up to hours or days and causes acetylcholine accumulation at the synapses which thereby over stimulate the nervous system (Manahan,1992). The end result of over stimulation results in agitation, paralysis, excess salivation, confusion, convulsion, failure of respiratory system and fatality of insects and mammals.

Microbial Degradation of Organophosphate Compounds

The organophosphorus is present either as a phosphonate or as a phosphate ester. They consist of number of sites that can easily undergo hydrolysis as they are esters. The principle reactions associated with the microbial degradation are oxidation, hydrolysis, dealkylation and alkylation(Singh *et al.,* 1999). Notable step in decontamination of organophosphate is the microbial degradation causing hydrolysing of the bonds P-O-alkyl and P-O-aryl. There have been several research reported where isolated bacterial species helps in co-metabolism and bio-mineralization of organophosphate compounds.

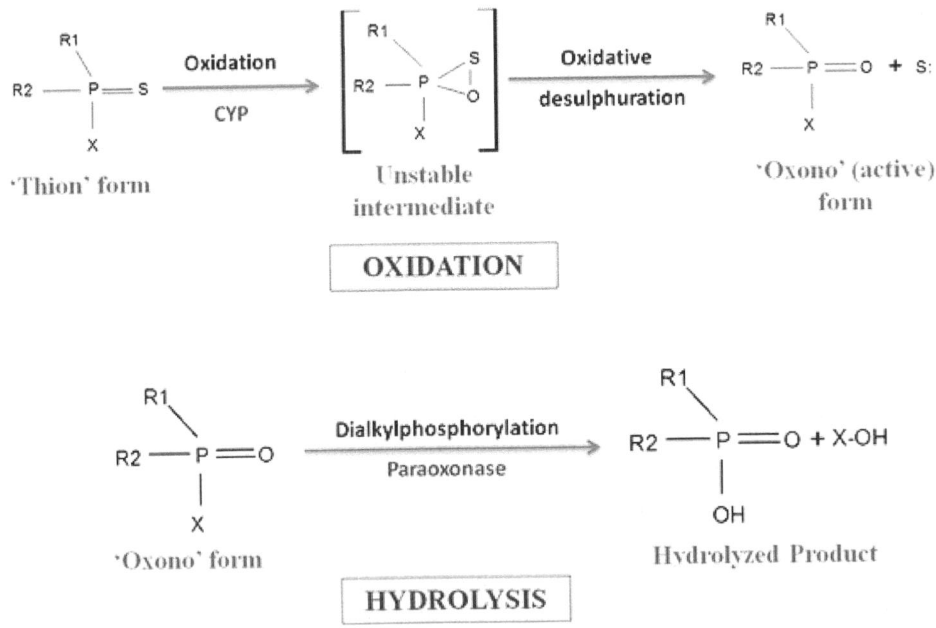

Figure 1.2: Degradation of Organophosphate Compounds by Mechanism of Oxidation and Hydrolysis (L.S.B. Upadhyay, 2017).

Oxidation is processed by the mode of action of cytochrome P450 (CYP) enzymes. Oxidation of the thion group's sulphur atom is carried out by CYP. This oxidation forms an unstable intermediate. This intermediate disintegrates and oxono metabolite is formed which has activated sulphur atom (oxidative desulphuration). "Oxono" forms are strong inhibitors of acetylcholinesterase. After oxidation process of hydrolysis takes place. Enzyme esterase A which is paraoxonase, is used which helps in cleavage of the organophosphates and form dialkylphosphate and leaving group.

Several reports have shown different microorganisms function in degradation of organophosphates as shown in table. Soils which show biodegradation of organophosphorus compounds in enhanced way, or great degree of contaminated soils with excessive concentrations of organophosphate compounds, is considered for the isolating the microorganisms which are having ability in degradation of organophosphates.

The concept of different aspects like physiology, microbiology, ecology, molecular and biochemical study of microorganisms which is being used for the degradation/bioremediation is needed in order to attain pollutant transformation (Iranzo *et al.,* 2001). Isolating organophosphate degrading microorganisms is important step for: (i) Determining the microbial metabolomics mechanism of natural processes (ii) Understanding gene or enzyme evolution mechanisms (iii) Using microorganisms selected in process of bioremediation in order to detoxify and decontaminate polluted aquatic and terrestrial ecosystem.

For isolation of these microorganisms, techniques like enrichment culture in selective mineral salts media has been used in the many cases. Media is prepared in such a way in which the organophosphate compounds are consumed by microorganisms as the sole carbon, nitrogen, sulphur or phosphorus source.

Table 1.2: List of Microorganisms Reported for Degradation of Organophosphate Compounds

Organophosphate Compounds	*Microorganisms Reported*	*References*
Diazinon	*Flavobacterium* sp.	Sethunathan *et al.* (1973)
	Pseudomonas spp.	Rosenberg *et al.*(1979)
	Arthrobacter spp.	Barik *et al.* (1979)
	Pseudomonas putida	Rosenberg *et al.*(1979)
	Bacillus sp.	Sharmila (1989)
	Agrobacterium sp.	Yasouri(2006)
	Serratia marcescens sDI101	Abo-Amer, Aly E (2010)
	Burkholderia caryophylli	Mahiudddin (2014)
	P. fluorescens	Mehdi Hassanshahian (2016)

Organophosphate Compounds	*Microorganisms Reported*	*References*
Parathion	*Flavo bacterium* spp.	Sethunathan *et al.*(1973)
	Pseudomonas spp.	Siddaramappa *et al.* (1973)
	Xanthomonas spp., *Pseudomonas* spp.	Rosenberg and Alexander (1979)
	Pseudomonas stutzeri	Daughton *et al.* (1977)
	Bacillus spp.	Nelson *et al.* (1982)
	Plesiomonas spp.	Cui zhongli *et al.* (2001)
	Agrobacterium radiobacter	Horne (2002)
	Sphingobium fuliginis	Kawahara, Tanaka, Yoon, and Yokota (2010)
Methylparathion	*Pseudomonas* spp.	Chaudry *et al.* (1988)
	Bacillus spp.	Sharmila *et al.* (1989)
	Pseudomonas putida	Rani and Lalitha-kumari (1994)
	Flavobacterium balustinum	Somara and Siddavattam (1995)
	Burkholderiaspp	Hayatsu *et al.* (2000)
	Plesimonas spM6	Zhongli *et al.* (2001)
	Burkholderia cepacia, Bacilluss spp.	Keprasertsup *et al.* (2001)
	Pseudomonas spp. *A3*	Zhongli *et al.* (2002)
	Pseudomonas spp. *WBC*	Yali (2002)
	Burkholderia cepacia	Ekkhunnatham, Jongsareejit (2012)
	Scytonema hofmanni	Balkrishna Tiwari (2017)
	Pseudomonas aeruginosa	Archana *et al.* (2020)
	Azohydromonas australica	Sumao Zhao (2020)
Chlorpyrifos	*Enterobacter* spp.	Singh *et al.* (2003)
	Trichoderma harzianum, Pencillium brevicompactum.	Omar (1998)
	Flavobacterium spp. ATCC27551	Mallick *et al.* (1999)
	Pseudomonas diminuta	Serdar *et al.* (1982)
	Micrococcus spp.	Guha *et al.* (1997)
	Phanerochaete chrysosporium	Bumpus *et al.* (1993)
	Hypholama fascicularae,	Bending *et al.* (2002)
	Coriolus versicolor	Obojska *et al.* (2002)
	Aspergillus spp.	Mukherjee *et al.* (2004)
	Chlorella vulgaris	Yang (2006)
	Stenotrophomonas spp.	Xu *et al.* (2007)
	Serratia spp.	Li *et al.* (2007)
	Sphingomonas spp.	Ghanem *et al.* (2007)

Organophosphate Compounds	Microorganisms Reported	References
	Klebsiella spp.	Fang *et al.* (2008)
	Verticillium spp. DSP *Pseudomonas nitroreducens*	Korade *et al.*(2009)
	Spirulina platensis ARM 728	Thengodkar *et al.* (2010)
	Pseudomonas stutzeri	Awad *et al.* (2011)
	Acremonium spp. (*GFRC-1*).	Kulshrestha and Kumari (2011)
	Cladosporium cladosporioides.	Gao *et al.* (2012)
	Pseudomonas nitroreducens AR-3	Aswathi *et al.* (2019)
Monocrotophos	*Chlorella vulgaris, Nostoclinkia, Synechococcus elongates, Phormidium tenue, Scenedescus bijugatus.*	Megharaj *et al.* (1987)
	Bacillus spp.	Rangaswamy and Venkateswaralu (1992)
	Aulosira fertilissima, Nostoc Muscorum.	Subramaniam *et al.* (1994)
	Pseudomonas spp., *Arthrobacter* spp., *Pseudomonas mendocina, Arthrobactera trocyaneu, Bacillus megaterium*	Bhadbhade *et al.* (2002 a, b, c)
	Clavibacter michiganense SBL1, *Pseudomonas aeruginosa* F10B.	Singh *et al.* (2003)
	Paracoccus sp. M-1	Jia *et al.* (2006)
	Starkeya novella YW6	Lina Sun (2015)
Glyphosate	*Alcaligene* sp.	Tolbot *et al.* (1984)
	Flavobacterium sp.	Balthazor and Hallas (1986)
	Agrobacterium sp.	Wacket *et al.* (1987)
	Arthrobacter sp. *GLP*	Pipke *et al.* (1987)
	Arthrobactera trocyaneus, Arthrobacter spp. *GLP-1/Nit-1.*	Pike and Amrhein (1988)
	Pseudomonas sp. *Lbr*	Jacob *et al.* (1988)
	Bacillus megaterium 2BLW	Quinn *et al.* (1989)
	Agrobacterium radiobacter	Mcauliffe *et al.* (1990)
	Rhizobium sp.	Liu *et al.* (1991)
	Pencillium notatum	Pothuluri *et al.* (1992)
	Trichoderma viridae, Scopulariopsis spand, Aspergillus niger.	Zboinska *et al.* (1992b)
	Pseudomonas ssp.	Kertesz *et al.* (1994a)
	Pseudomonas pseudomallei	*Penaloza-Vazquez et al. (1995)*
	Penicillium citrium	Pothuluri *et al.* (1998)

Organophosphate Compounds	***Microorganisms Reported***	***References***
	Penicillium chrysogenum	Klimek *et al.* (2001)
	Geobacillus caldoxylosilyticus T20	Obojska *et al.* (2002)
	Alternaria alternata	Lipok *et al.* (2003)
	Ochrobactrum anthropi GPK3	Sviridov *et al.*(2012)
	Comamona sodontotermitis strain *P2*	Ermakova *et al.*(2017)
	Ochrobactrum intermedium strain Sq20, *Pseudomonas* sp. strain 4ASW	Firdous *et al.* (2017a, 2017b)
	Aspergillus oryzae A-F02	Fu *et al.* (2017)
Dimethoate	*A. sydowii*	Hasan (1999)
	Pseudomonas aeruginosa	Deshpande, Dhakephalkar, Kanekar (2001)
	Aspergillus niger ZHY256	Liu (2001)
	Pseudomonas putida	Nazarian A. (2007)
	Lactobacillus plantarum	Yuan *et al.* (2021)

Chlorpyrifos

Chlorpyrifos [O,O-diethyl O-(3,5,6-trichloro-2-pyridinyl)- phosphorothioate] considered as extensively utilized organophosphate in form of pesticides and broad-spectrum chlorinated organophosphate insecticides specially in agricultural fields and used to save economically important crops from pest attack. They have shown effective results against taxonomically diverse insect pests (Fang *et al.*, 2006). The route of administration or absorption of chlorpyrifos can be by dermal absorption, inhalation and ingestion (Yurumez *et al.*, 2007). Chlorpyrifos affects the nervous system of the targeted insects. Phosphorylation is done at the neurons synapse and in the plasma in order to inhibit the activity of acetylcholinesterase (Hui *et al.*, 2010). The chlorpyrifos fate is destined by owns physico-chemical properties and also by soils attributes in which it is present, how it is managed and environmental factors (Halimah *et al.*, 2010). Chlorpyrifos is able to bind toplants, particles of soil (Gebremariam *et al.*, 2012). Aftera some time, majority of chlorpyrifos is either biodegraded, volatilized, or hydrolyzed. Different elements like soil characteristics, temperature and concentration influence the volatilization of chlorpyrifos from soil. Several cases are studied and reported regarding the volatilization. Racke *et al.* (1993) reported that in the time period of 30 days of its application, volatilization of 2.6 per cent and 9.3 per cent of the applied chlorpyrifos taken from sand and a silt loam soil respectively was observed.

Chlorpyrifos biodegradation in water is highly significant process as chlorpyrifos contamination in water is highly toxic to aquatic life; the persistent metabolites are complex in nature. The reports regarding the chlorpyrifos biodegradation in water are very less (Lu *et al.*, 2006). Still there have been many researches reporting total *in vitro* chlorpyrifos biodegradation in liquid media (Yang *et al.*, 2005; Xu *et al.*, 2008; Li *et al.*, 2008; Li *et al.*,2011). *Acremonium* spp. strain GFRC-1 degraded 83.9 per cent of chlorpyrifos when they were grown in the complete medium (Kulshrestha and Kumari, 2011). Bacterial species for *e.g.*, *Enterobacter, Klebsiella, Alcaligenes, Pseudomonas, Providencia, Lactobacillus, Leuconostoc, Serratia* and *Synechocystis* have been researched which showed degradation capacity of chlorpyrifos in liquid media (Xu *et al.*, 2007; Ghanem *et al.*, 2007; Surekha *et al.*, 2008; Madhuri and Rangaswamy, 2009; Singh *et al.*, 2011; Awad *et al.*, 2011). Microbes (bacterial species and fungal) can degrade chlorpyrifos in two ways; one in which chlorpyrifos are utilized directly as carbon, nitrogen and phosphorus source (Singh *et al.*, 2004; Awad *et al.*, 2011) or second by co-metabolizing the chlorpyrifos. Certain bacterial isolates are reported with ability to utilize chlorpyrifos as only source of carbon, nitrogen or phosphorus (Ghanem *et al.*, 2007; Zhu *et al.*, 2010). *Paracoccus TRP* in liquid culture was reported to utilize chlorpyrifos completely (Xu *et al.*, 2008; Li *et al.*, 2011). Experiments were performed where activated sludge samples was taken from pesticide manufacturing unit from which isolates were taken. Results of experiments with activated sludge samples takenfrom pesticide manufacturing unit show that isolates utilize chlorpyrifos as only source of nitrogen, carbon or phosphorus. TCP and diethylthiophosphoric acid (DETP) were formed by hydrolyzing chlorpyrifos. The intermediates formed were used as nitrogen, carbon or phosphorus source, therefore performing complete mineralization of chlorpyrifos.

Soil can also said to be purified by inoculating it with precisely adapted microorganisms (Diez, 2010; Abo-Amer, 2011). The soils, already been contaminated with chlorpyrifos, involve different microorganisms containing organophosphate degrading enzymes (Sasikala *et al.*,2012). Singh *et al.* (2003) was the first researcher who studied soils of Australia and reported presence of amplified chlorpyrifos biodegradation in soils and identified the isolates as strain of *Pseudomonas* spp. The report said that soils with pH >6.7 were able to degrade for 90 days after inoculation. In 2004, they reported biodegradation of chlorpyrifos (35 mg) in soil by adding strain *Enterobacter* B-14 having a small native population (Singh *et al.*, 2004).

Chlorpyrifos degradation produce two vital metabolites TCP and DETP (Bootharaju *et al.*, 2012), while there are reports that additional metabolites, like desethyl chlorpyrifos-oxon, chlorpyrifosoxon,3,5,6-trichloro-2 methoxypyrimidine and desethyl chlorpyrifos have their production in comparatively little amount, which depends upon the microorganisms and environmental factors.

Isolated novel strain of *Pseudomonas nitroreducens* AR-3 from agricultural soil contaminated with pesticide was reported to remove 97 per cent of chlorpyrifos in time period of 8 hours at 30 °C. It was done in a mineral salt medium (MSM) which contained glucose (1.0 g/L) and yeast extract (0.5 g/L) and 2 per cent (v/v) inoculum along with chlorpyrifos (Awasthi *et al.*, 2019). Glucose addition showed to enhance the degradation of chlorpyrifos. Organophosphate hydrolase (OPH) of *P. nitroreducens AR-3* was inducible and intracellular. *Pseudomonas nitroreducens AR-3* exhibit potential to bioremediate chlorpyrifos due to its properties like high degrading ability, minimum time of incubation and ease to maintain in minimal media. The strain AR-3 is said to have rapid rate of degradation of organophosphorous compound reported up to now in *Pseudomonads.*

The reports of fungi with organophosphate compounds hydrolysing enzymes are less. White-rot fungus *Pleurotuso streatus* consists of enzyme laccase (phenol oxidase) which was extracted by Amitai *et al.* (1998). The enzyme showed activity of hydrolysing phosphorothiolates by attacking the P-S bond. More fungal enzymes like A-OPH from *Aspergillus niger* ZHY256, OPH (P-OPH) enzyme from *P. lilacinum* BP303, chlorpyrifos hydrolase from a fungus *Cladosporium cladosporioides* (Liu *et al.*, 2001; Liu *et al.*, 2004; Gao *et al.*, 2012 respectively) are also reported.

Parathion and Methylparathion

One of the most toxic insecticides known is *parathion* [O,O-diethyl-O-p-nitrophenyl phosphorothioate]. There are reports of fatal cases to several human and non-target species in many developing countries caused due to extreme toxicity of parathion (McConnell *et al.*, 1999). Parathion is reported to degrade in biologically active soil. As the concentration of parathion increases, the population of bacteria in soils was observed to expand (Nelson, 1982). There are reports of isolation of several species of parathion hydrolysing bacteria from parathion enriched or environments with a high concentration of other organophosphates. First organophosphorus degrading bacterium was isolated by Sethunathan and Yoshida (1973), which was identified as *Flavobacterium* spp. It was reported that they could degrade parathion and diazinon. *Pseudomonas* spp. was isolated which show ability to hydrolyse parathion and product p-nitrophenolwas used as a C or N source (Siddaramappa *et al.*,1973). Rani and Lalithakumari, (1994) reported metabolization of p-nitrophenol to 1,2,4-benzenetrioland hydroquinone by *Pseudomonas putida* strain. The product formed was further cleaved to produce maleylacetate by the help of enzyme benzenetriol oxygenase.

In 2010, parathion degrading bacteria *Flavobacterium sp.* ATCC 27551was re-evaluated by examining its chemotaxonomical and phylogenetical data. The phylogenetic analysis showed the strain to belong to family *Sphingomonadaceae.* The data therefore reclassified *Flavobacterium* sp. ATCC 27551 as *Sphingobium fuliginis* ATCC 27551 is identified as *a Sphingobium fuliginis* strain, not a *Flavobacterium sp.*(Kawahara *et al.*, 2010).

Methylparathion [O, O-dimethyl O-4-nitrophenyl Phosphorothioate] has shown neurotoxicological properties. They are named a constant natural toxin because of their toxic nature, are reported to cause chromosomal damage and disrupt the endocrine (Ang *et al.*, 2005; Webster *et al.*, 2002). Microbial hydrolysis of methylparathion results in production of principle product *p*-nitrophenol

(4-nitrophenol). *p*-nitrophenol (PNP) has shown less toxicity in animals, but high toxic effects are reported in microflora of the soil (Errampalli, 1999). Reportedly *Burkholderiacepacia*, isolated from site in Thailand which was methyl parathion treated (Keprasertsup *et al.*, 2001), showed ability of degradation of methyl parathion and p-nitrophenol and utilized them as sole C source. Isolation of *Plesiomonas* sp. was done which has also shown potential for hydrolysing 200 mg of methyl parathion into p-nitrophenolin time period of 15 minutes but was not able to convert p-nitrophenol which assembled in the medium (Zhongli *et al.*, 2001). These researchers also identified a *Pseudomonas* sp. strain p3 that utilized methyl parathion for its sole source of C and N for growth (Zhongli *et al.*, 2002). Bacterial consortium was obtained capable of degrading methyl parathion and p-nitrophenol utilizing them as the only source of C (Nancy *et al.*,2011). The consortium obtained contained *Bacillus* sp., *Acinetobacter* sp., *Flavobacterium* sp., *Pseudomonas aeruginosa, Pseudomonas putida, Citrobacter freundii, Stenotrophomonas* sp., *Proteus vulgaris, Acinetobacter* sp., *Klebsiella* sp. and *Proteus* sp. The consortium reported to exhibit ability for degradation of p-nitrophenol and methyl-parathion in concentrations of 150 mgL^{-1} within 120 hours. However, when glucose or peptone is added to the nutrient medium, it results in reduction of degradation time to 24 hours.

Diazinon

Diazinon [O, O -diethyl O -2-isopropyl-6-methylpyrimidin-4-yl phosphorothioate] is a broad-spectrum organophosphate contact insecticide. It is used as an insecticide, acaricide, and nematicide and extensively used for controlling soil and foliage bugs on crops like rice, fruits, sugarcane, wine grapes, potatoes, and corn (Aggarwal *et al.*, 2012).

First two strains of diazinon-degrading microorganisms were reported by Gunner and Zuckerman (1968) which were an *Arthrobacter* sp. and a *Streptomyces*. They exhibit their ability to consume diazinon as sole carbon source and when cultured separately, were capable of attacking the diazinon's ethyl ester moiety. However, they failed to degrade the pyrimidinyl ring of diazinon. *Pseudomonas putida* strain was isolated from sewage and soil samples diazinon contaminated, and shows potential for utilizing diazinon and additional organophosphate compounds like malathion, parathion and dimethoate for source of P (Rosenberg and Alexander, 1979). *Pseudomonas putida* showed ability for hydrolysing diazinon to produce 2- isopropyl-4-methyl-6 hydroxypyrimidine (IMHP) and diethylthiophosphate (DETP). DEPT further degraded by enzymes phosphomonoesterases and phosphodiesterases and releases inorganic phosphate.

Three strains of bacteria were identified from samples of soil and screening and identification was done morphologically and biochemically. They were identified as *Burkholderia caryophylli, Pseudomonas peli* and *Brevundimonas diminuta*. The study concluded that all these isolates showed ability for completely degrading 20 mg/l diazinon in mineral salt medium (MSM) within incubation period of 12 days and utilized diazinon as carbon only source (Mahiudddin *et al.,* 2014). Mehdi Hassanshahian, 2016 reported the presence of degradation of diazinon bacteria in agricultural land of Kerman pistachio orchards, Southeastern Iran. They secluded and characterised *Achromobacter piechaudii* strain D8, *Pseudomonas putida* strain D3 and *Pseudomonas fluorescens* strain D1 out of which finest degrading strain was *Pseudomonas fluorescens* strain D1 was able to potentially grow in concentration up to 100 ppm. Gas chromatography (GC) was done for confirmation of organic pesticides decomposition by isolates.

Monocrotophos

Monocrotophos (3-hydroxy N-methyl-cis-crotonamide dimethyl phosphate) is a water-soluble organophosphate insecticide, first manufactured in 1965 by Ciba AG and Shell Development Co., which is extensively utilized because of affordable prices and its efficacy (Sha, 1999). Monocrotophos have shown potential where toxicity levels in the oral cavity are extremely high, while dermal toxicity levels are moderate. (Lee *et al.,* 1990). The mode of action applicable to the toxicity is that they inhibit the cholinesterase activities (Skripsky and Loosli, 1994). Because, they are hydrophilic in nature, they are weakly adsorbed by soil particles and have the ability for contaminating the groundwater (Singh and Singh 2003). Characterization of Monocrotophos can be done by a P–O–C linkage and amide bond. Singh and Singh (2003) notified monocrotophos degradation and utilization as a P source in liquid medium by *Pseudomonas aeruginosa* sp., *Clavibacter michiganense*s sp. insidiosum. Bhadbhade *et al.* (2002) reported isolation of a *Pseudomonas mendocina* MCM B-424 strain consists of 7.5 kb plasmid which was the reason of its ability to metabolize monocrotophos. It utilized monocrophos as carbon source.

An isolated *Paracoccus* sp. strain was said to be obtained from residue sample collected from the bottom of a chemical factory's wastewater treatment system. Identification was done based on morphological analysis of cells, physiological analysis, chemo-taxonomical characterisation and phylogenetical similarity of 16S rDNA gene sequences. The study showed potential mineralization of monocrotophos by the strain and use them as a source of C under conditions including aerobic and anaerobic (Kai-zhiJia, 2006). *Starkeya novella YW6* isolated from paddy soil showed potential for degrading monocrotophos, reported by Lina Sun, 2015. It utilized monocrotophos exclusive sources for C or N for development. The study showed that monocrotophos is hydrolysed at the P–O–C bond and result in formation of N-methylacetoacetamide and dimethyl phosphate. N-methylacetoacetamide is further transformed to N-methyl-4-oxopentanamide, that later transforms to 5-(methylamino)- 5- oxo-pentanoic acid. This 5-(methylamino)-5-oxo-pentanoic acid is broken down into methylamine andglutaric acid.

Glyphosate

Glyphosate [N-(phosphonomethyl)glycine], is used as herbicides to control yearly and eternal unwanted plants in agricultural fields, for forest development, metropolitan and in home gardens (Zhan *et al.*, 2018). Glyphosate tends to bind with particles of soil and results in its deposition in the soils upper layer. Glyphosate is reported to target 5-enolpyruvylshikimate-3-phosphate synthase (EPSPS) and causes obstruction of aromatic amino acid biosynthesis in shikimate pathway (Haslam, 2014).The glyphosate inhibition of EPSPS further suppress the production of required proteins and important secondary metabolites, and hampers the essential kinetic pathways in plants and soil microorganisms (Sviridov *et al.*, 2015; Bai and Ogbourne 2016).

Several glyphosate-degrading microbes have been isolated from contaminated areas using enrichment cultivation technique. They are *Achromobacter* sp. strain MPK 7A, *Ochrobactrum intermedium* strain Sq20, *Comamonas odontotermitis* strain P2and *Pseudomonas* sp. 4ASW (Dick and Quinn 1995b; Firdous *et al.*, 2017a; Ermakova *et al.*, 2017; Firdous *et al.*, 2017b). Firdous *et al.*, 2017 has reported *Comamonaso dontotermitis* strain, glyphosate-adulterated Australian soil was the source of this isolate which used glyphosate as only source of C or P. The absolute breakdown of glyphosate (1.5 g/L) was reported to accomplish within104 h. In same year, Firdous *et al.*, 2017b reported identification of *Ochrobactrum intermedium* Sq20as glyphosate degrading microorganism which uses the organophosphate compound as sole source of carbon.

There are two major glyphosate-degrading microorganisms' degradation pathways which include conversions of glyphosate to amino-methylphosphonic acid (AMPA) and sarcosine. Firstly, the formation of AMPA and glyoxylate takes place by the catalysis of glyphosate by glyphosate oxidoreductase. Secondly, glyphosate degradation pathway is catalyzed by C-P lyase result in production of sarcosine which is converted to glycine and formaldehyde by sarcosine oxidase (Kishore and Jacob, 1987; Shinabarger and Braymer, 1986).

There are reports of a distinct AMPA degrading pathway in *Ochrobactrum anthropi* GPK3, where phosphonoformaldehyde is formed as a result of AMPA metabolization with the help of transaminase and was further catabolized using phosphonatase to produce formaldehyde (Sviridov *et al.*, 2014).

Different fungi have shown potential for degradation of glyphosate. The first fungal strain isolated to be studied was of strain *Penicillium citrinum,* which has the potential to metabolise glyphosate (Zboinska *et al.*, 1992). Later reports identified a *P. Notatum* isolate which showed the ability to metabolize glyphosate via the AMPA pathway (Bujacz *et al.*, 1995). *Trichoderma viride, T. harzianum, Scopulariopsis* sp., *Alternaria* sp. and *Aspergillus niger*, are some of the soil-isolated fungal strains demonstrated their potential to grow on a variety of organophosphonate compounds, including glyphosate and employ it as phosphorus source (Krzysko-Lupicka *et al.*, 1997). Klimek *et al.*, 2001 isolated a non-nitrate-using strain of *Penicillium chromogenum* using glyphosate for nitrogen source for the first time. Lipok *et al.*, 2003 identified variety of fungi from non-infected carrot seeds, which includes *Penicillium janthinellum, Penicillium simplicissimum, Mucor* sp. and *Alternaria alternata,* which signify their propensity to utilize glyphosate as source of phosphorus.

Immobilized whole cells of six wild type microorganisms—*Streptomyces phaeochromogenes, Streptomyces setonii, Nocardiacoryne bacterioides, Nocardiaa steroides* and two *Arthrobacter oxydans* as well as enzyme extracts potentially useful biocatalysts used for water decontamination (Santillan, 2020). The adaptability of these investigated biocatalysts allows them to be used to select, determined by the circumstances (environmental conditions and type: water or soil), whole cells (to avoid cellular disruption and centrifugation) or enzyme extracts (which is simple to manage), free or immobilised. To attain the required end product, the usage of this novel technique of whole cell being directly introduced into the polluted surroundings or using consortium having distinct enzymes in free or immobilised state can be researched and used.

Conclusion and Future Perspective

Bioremediation is the transformation of pollutants and converting it into less toxic compounds with the help of microorganisms and enzymes present in them for degradation. This method tends to be safe, efficient and cost friendly. The negative effects of various organophosphates on flora and fauna have been reported, highlighting the importance of removing organophosphates from polluted environments. Reports show that many organophosphates are biodegradable. Using reported different taxonomically diverse range of microorganisms to degrade different organophosphate compounds can be beneficial for environment.

There are possibilities to genetically modify these microorganisms in order to detoxify organophosphate compounds. This modification can result in increased production of the organophosphorus hydrolase, which can be used to develop biosensors or immobilised onto various surfaces for efficient bioremediation.

REFERENCES

Abo-Amer, A.E., 2011. Biodegradation of diazinon by *Serratia marcescens* DI101 and its use in bioremediation of contaminated environment. J. Microbiol. Biotechnol. 21, 71-80.

Aggarwal, V., Deng, X., Tuli, A., and Goh, K. S. (2012). Diazinon—Chemistry and Environmental Fate: A California Perspective. Reviews of Environmental Contamination and Toxicology Volume 223, 107–140.

Alexander, M., 1999.Biodegradation and Bioremediation. Academic Press, San Diego, pp. 249-264.

Ang, L.; Zhao, M.; Obbard, P. Recent advances in the bioremediation of persistent organic pollutants via biomolecular engineering J. Enz. Microb. Technol. 2005, *37*, 487–496.

Archana, AppasamySurendran and Antony Joseph Thatheyus, 2020. Microbial Degradation of the Organophosphorus Insecticide, Methyl Parathion Using the Natural Bacterial Isolate, *Pseudomonas aeruginosa*. Asian Journal of Biotechnology, 12: 136-144.

Aswathi, A., Pandey, A., and Sukumaran, R. K. (2019). Rapid degradation of the organophosphate pesticide – Chlorpyrifos by a novel strain of *Pseudomonas nitroreducens AR-3*.Bioresource Technology, 292, 122025.

Awad, N.S., Sabit, H.H., Abo-Aba, S.E.M., Bayoumi, R.A., 2011. Isolation, characterization and fingerprinting of some chlorpyrifos-degrading bacterial strains isolated from Egyptian pesticides-polluted soils. Afr. J. Microbiol. Res. 5, 2855-2862

Bai SH, Ogbourne SM(2016) Glyphosate: environmental contamination, toxicity and potential risks to human health via food contamination. Environ SciPollut Res 23(19): 18988–19001

Bhadbhade, B.J., Dhakephalkar, P.K., Sarnaik, S.S. and Kanekar, P.P. (2002) Plasmid-associated biodegradation of an organophosphorus pesticide, monocrotophos, by *Pseudomonas mendocina*.Biotechnol.Lett. 24, 647–650.

Brajesh K. Singh and Allan Walker.Microbial degradation of organophosphorus compounds.FEMS Microbiol Rev 30 (2006) 428–471.

Chaudhry, R.; Ali N.; Wheeler, B. Isolation of a methyl parathiondegrading*Pseudomonas sp*. that possesses DNA homologous to the *opd*gene from a *Flavobacterium*sp. Appl Environ Microbiol. 1988, *54*, 288–293.

Chen S, Geng P, Xiao Y, HuM (2012) Bioremediation of β-cypermethrin and 3-phenoxybenzaldehyde contaminated soils using *Streptomyces aureus* HP-S-01. ApplMicrobiolBiotechnol. 94: 505–515

Cui zhongli, LI., Shunpeng, and Fu Guoping. 2001. Isolation of Methyl Parathion-Degrading Strain M6 and Cloning of the Methyl Parathion Hydrolase Gene. Applied and environmental microbiology, 67(10): 4922–4925.

Deshpande NM, Dhakephalkar PK, Kanekar PP. Plasmid-mediated dimethoate degradation in *Pseudomonas aeruginosa* MCMB-427.Lett ApplMicrobiol. 2001 Oct; 33(4): 275-9.

Diez, M.C., 2010. Biological aspects involved in the degradation of organic pollutants. J. Soil Sci. Plant Nutr. 10, 244-267.

Dimitrios G. Karpouzasand Brajesh K. Singh.Microbial Degradation of Organophosphorus Xenobiotics: Metabolic Pathways andMolecular Basis. Advances in Microbial Physiology Vol. 51.

Dragun J., Kuffner AC and Schneiter RW (1984) Groundwater contamination. Transport and transformation of organic chemicals. ChemEng 91: 65–70.

Ekkhunnatham, A., Jongsareejit, B., Yamkunthong, W. *et al.,* Purification and characterization of methyl parathion hydrolase from *Burkholderia cepacia* capable of degrading organophosphate insecticides. *World J Microbiol*

Biotechnol 28, 1739–1746 (2012). https: //doi.org/10.1007/s11274-011-0985-y

Errampalli, D.; Tresse, O.; Lee, H.; Trevors, J.T. Bacterial survival and mineralization of *p*-nitrophenol in soil by green fluorescent protein-marked *Moraxella sp.* G21 encapsulated cells. FEMS Microbiol. Ecol. 1999, *30*, 229–236.

Firdous S, Iqbal S, Anwar S (2017a) Optimization and modeling of glyphosate biodegradation by a novel *Comamonas odontotermitis* P2 through response surface methodology. Pedosphere (17)60381-3

Firdous S, Iqbal S, Anwar S, Jabeen H (2017b) Identification and analysis of 5-enolpyruvylshikimate-3-phosphate synthase (EPSPS) gene from glyphosate resistant *Ochrobactrum intermedium* Sq20. Pest Manage Sci 74: 1184–1196.

Fang, H., Yu, Y.L., Wang, X., Shan, M., Wu, X.M., Yu, J.Q., 2006.Dissipation of chlorpyrifos in pakchoi-vegetated soil in a greenhouse. J. Environ. Sci. 18, 60-764.

Gunner, H.B. and Zuckerman, B.M. (1968) Degradation of diazinon by synergistic microbial action. Nature 217, 1183–1184.

Gebremariam, S.Y., Beutel, M.W., Yonge, D.R., Flury, M., Harsh, J.B., 2012. Adsorption and desorption of chlorpyrifos to soils and sediments. Rev. Environ. Contam. Toxicol. 215, 123-175.

GAO, Y., Chen, S., Hu, M., Hu, Q., Luo, J., Li, Y., 2012.Purification and Characterization of a novel chlorpyrifos hydrolase from *Cladosporium cladosporioides* Hu-01. PLOS ONE 7 (6), -38137

Haslam, E. The Shikimate Pathway: Biosynthesis of Natural Products Series; Butterworths: London, UK, 2014.

Hui, T.J., Ariffin, M.M., Tahir, N.M., 2010.Hydrolysis of chlorpyrifos in aqueous solutions at different temperatures and pH.Malays. J. Anal. Sci. 14, 50-55.

Halimah, M., Tan, Y.A., Ismail, B.S., Nashriyah, M., 2010.Downward movement of chlorpyrifos in the soil of an oil palm plantation in Sepang, Selangor, Malaysia. J. Oil Palm Res. 22, 721-728.

Iranzo M, Sain-Pardo I, Boluda R, Sanchez J and Mormeneo S (2001).The use of microorganisms in environmental remediation. Annals Microbiol 51: 135–143.

J. Y. Santillan, A. Muzlera, M. Molina, E. S. Lewkowicz, A. M. Iribarren. (2020) Microbial degradation of organophosphorus pesticides using whole cells and enzyme extracts. Biodegradation.

Jia, K., Cui, Z., He, J., Guo, P., and Li, S. (2006). Isolation and characterization of a denitrifying monocrotophos-degrading*Paracoccussp.*M-1. FEMS Microbiology Letters, 263(2), 155–162.

Karpouzas, D.G., Morgan, J.A.W. and Walker, A. (2000) Isolation and characterization of ethoprophos-degrading bacteria.FEMS Microbiol. Ecol. 33, 209–218.

Klimek, M., Lejczak, B., Kafarski, P. and Forlani, G. (2001) Metabolism of the phosphonate herbicide glyphosate by a non-nitrate-utilizing strain of *Penicilliumchrysogenum.* Pest Manage. Sci. 57, 815–821.

Krzysko-Lupicka, T., Stof, W., Kubs, K., Skorupa, M., Wieczorek, P., Leijczak, B. and Kafarski, P. (1997).The ability of soil-borne fungi to degrade organophosphonate carbon-to-phosphorus bonds. Appl. Microbiol. Biotechnol. 48, 549–552.

Kawahara, K., Tanaka, A., Yoon, J., and Yokota, A. (2010). Reclassification of a parathione degrading*Flavobacterium sp. ATCC 27551 as Sphingobium fuliginis.* The Journal of General and Applied Microbiology, 56(3), 249–255.

Keprasertsup, C., Upatham, E.S., Sukhapanth, N. and Prempree, P. (2001) Degradation of methyl parathion in an aqueous medium by soil bacteria. Sci. Asia 27, 261–270.

Lata S.B. Upadhyay and Aditya Dutt.Microbial Detoxification of Residual Organophosphate Pesticides in Agricultural Practices. 2017

Lee PW, Fukuto JM, Hernandez H, Stearns SM (1990) Fate of monocrotophos in the environment. J Agric Food Chem 38: 567–573

Li, X., Jiang, J., Gu, L., Ali, S.W., He, J., Li, S., 2008. Diversity of chlorpyrifos-degrading bacteria isolated from chlorpyrifos-contaminated samples. Int. Biodeter. Biodegr. 62, 331-335.

Li, K., Wang, S., Shi, Y., Qu, J., Zhai, Y., Xu, L., Xu, Y., Song, J., Liu, L., Anisur Rahman, M., Yan, Y., 2011. Genome sequence of *Paracoccus sp.* strain TRP, a chlorpyrifos biodegrader. J. Bacteriol. 193, 1786-1787.

Lina Sun, Shijun Zhu, Zhengzhong Yang, Qing Chen1, Hongming Liu, Jun Zhang, Gang Hu, Shunpeng Li, Qing Hong. Degradation of monocrotophos by *Starkeya novella* YW6 isolated from paddy soil. Environ. Sci. Pollut. Res. 2015

Lu, J., Wu, L., Newman, J., Faber, B., Gan, J., 2006. Degradation of pesticides in nursery recycling pond waters.J. Agric. Food Chem. 54, 2658-2663.

Liu, B., McConnell, L.L., Torrents, A., 2001. Hydrolysis of chlorpyrifos in natural waters of the Chesapeake Bay. Chemosphere 44, 1315-1323.

Liu, Y.H., Liu, H., Chen, Z.H., Lian, J., Huang, X., Chung, Y.C., 2004. Purification and characterization of a novel organophosphorus pesticide hydrolase from *Penicilliumlilacinum* BP303.Enz.Microb. Technol. 34, 297-303.

Lupi L, Miglioranza KS, Aparicio VC, Marino D, Bedmar F, Wunderlin DA (2015) Occurrence of glyphosate and AMPA in an agricultural watershed from the southeastern region of Argentina. Sci Total Environ 536: 687–694

Liu, Z., Chen, X., Shi, Y., Su, Z., 2012. Bacterial degradation of chlorpyrifos by *Bacillus cereus.* Adv. Mater. Res. 356e360, 676-680.

Liu J, Chen S, Ding J, Xiao Y, Han H, Zhong G (2015) Sugarcane bagasse as support for immobilization of *Bacillus pumilus* HZ-2 and its use in bioremediation of mesotrione contaminated soils. ApplMicrobiolBiotechnol 99(24): 10839–10851

Mahiudddin, M and Fakhruddin, A. N. M. and Chowdhury, M. and Rahman, Mohammad and Alam, Mohammad. (2014). Degradation of the organophosphorus insecticide diazinon by soil bacterial isolate. International Journal of Biotechnology. 2014. 12-23.

Mallick, B.K., Banerji, A., Shakli, N.A., Sethunathan, N.N., 1999. Bacterial degradation of chlorpyrifos in pure culture and in soil. Bull. Environ. Contam. Toxicol. 62, 48 - 55.

McConnell R, Pacheoco F, Wahlberg K, Klein W, Malespin O, Magnotti R, Akerblorn M and Murray D (1999) Subclinical health effects of environmental pesticide contamination in a developing country: cholinesterase depression in children. Environ Res 81: 87–91.

Mehdi Hassanshahian. Isolation and Characterization of Diazinon Degrading Bacteria from Contaminated Agriculture Soils. Iranian Journal of Toxicology Vol 10, No 4, July-August 2016.

Mukherjee I, Gopal M, Dhar DW. (2004). Disappearance of chlorpyrifos from cultures of *Chlorella vulgaris*. Bull. Environ. Contam.Toxicol., 73, 358–63.

Nancy J. Pino, Maria C. DomiNguezAnd Gustavo A. Pen Uela. Isolation of a selected microbial consortium capable of degrading methyl parathion and *p*-nitrophenol from a contaminated soil site.Journal of Environmental Science and Health Part B (2011) 46, 173-180

Nazarian A. Utilization of dimethoate by wild type *Pseudomonas putida* from polluted sites in Iran. Pak J Biol Sci. 2007 Feb 15; 10(4): 664-7.

Nelson LM (1982) Biologically induced hydrolysis of parathion in soil: isolation of hydrolysing bacteria. Soil BiolBiochem 14: 223–229.

Newton M, Horner LM, Cowell JE, White DE, Cole EC (1994) Dissipation of glyphosate and aminomethylphosphonic acid in north American forests. J Agric Food Chem 42(8): 1795-1802

Ohshiro, K., Kakuta, T., Sakai, T., Hirota, H., Hoshino, T. and Uchiyama, T. (1996) Biodegradation of organophosphorus insecticides by bacteria isolated from turf green soil. J. Ferment. Bioeng. 82, 299–305.

Racke, K.D., 1993. Environmental fate of chlorpyrifos. Rev. Environ. Contam.Toxicol. 131, 1-154.

Ragnarsdottir, K.V. (2000) Environmental fate and toxicology of organophosphate pesticides. J. Geol. Soc. 157, 859–876.

Rani NL and Lalitha-kumari D (1994) Degradation of methyl parathion by *Pseudomonas putida*.Can J Microbiol 4: 1000–1004.

Rosenberg, A. and Alexander, M. (1979) Microbial cleavage of various organophosphorus insecticides. Appl. Environ. Microbiol. 37, 886–891.

Sasikala, C., Jiwal, S., Rout, P., Ramya, M., 2012. Biodegradation of chlorpyrifos by bacterial consortium isolated from agriculture soil. World J. Microbiol. Biotechnol.28, 1301-1308.

Shushkova T, Ermakova I, Leontievsky A (2010) Glyphosate bioavailability in soil. Biodegradation 21(3): 403–410

Simon, D., Helliwell, S., Robards, K., 1998.Analytical chemistry of chlorpyrifos and diuron in aquatic ecosystems. Anal.Chim.Acta 360, 1-16.

Siddaramappa R, Rajaram KP and Sethunathan NN (1973) Degradation of parathion by bacteria isolated from flooded soil. ApplMicrobiol 26: 846–849.

Singh S, Singh DK (2003) Utilization of monocrotophos as phosphorus source by *Pseudomonas aeruginosa F10B* and *Clavibactermichiganense* subsp. *insidiosum* SBL 11. Can J Microbiol 49: 101–109

Singh, B.K., Walker, A., Morgan, J.A., Wright, D.J., 2003. Effects of soil pH on the biodegradation of chlorpyrifos and isolation of a chlorpyrifos-degrading bacterium. Appl. Environ. Microbiol. 69, 5198-5206.

Singh, B.K., Walker, A., Morgan, A.W., Wright, D.J., 2004. Biodegradation of chlorpyrifos by *Enterobacter strain* B-14 and its use in bioremediation of contaminated soils. Appl. Environ. Microbiol. 70, 4855-4863.

Skripsky T, Loosli R (1994) Toxicology of monocrotophos. Rev Environ Contam T 139: 13 39

Subramanian, G., Sekar, S. and Sampoornam, S. (1994) Biodegradation and utilization of organophosphorus pesticides by cyanobacteria. Int. Biodeter. Biodegr. 33, 129–143.

SumaoZhaoa, Wei Xua Wen li Zhanga HaoWua Cuie Guanga Wanmeng Mu. In-depth biochemical identification of a novel methyl parathion hydrolase from *Azohydromonas australica* and its high effectiveness in the degradation of various organophosphorus pesticides. Bioresource Technology Volume 323, March 2021, 124641.

Sviridov A, Shushkova T, Ermakova I, Ivanova E, Leontievsky A (2014) Glyphosate: safety risks, biodegradation, and bioremediation. Current environmental issues and challenges. Springer, Dordrecht, pp 183–195.

Sviridov AV, Shushkova TV, Ermakova IT, Ivanova EV, Epiktetov DO, Leontievsky AA (2015) Microbial degradation of glyphosate herbicides (review). ApplBiochemMicrobiol 51(2): 188–195

Thengodkar RRM, Sivakami S. (2010). Degradation of chlorpyrifos by an alkaline phosphatase from the cyanobacterium *Spirulina platensis*. Biodegradation, 21, 637–44.

Toole G and Toole S (1995) Understanding Biology.3rd edn. Stanley Thornes, Cheltenham, UK.

Webster, R.; Mckenzie, G.H.; Moriarty, H.T. Organophosphate– based pesticides and genetic damage implicated in bladder cancer. Cancer Genet.Cytogen.2002, *133*, 112–117.

Xu, G., Zheng, W., Li, Y., Wang, S., Zhang, J., Yan, Y., 2008. Biodegradation of chlorpyrifos and 3,5,6-trichloro-2-pyridinol by a newly isolated *Paracoccus sp.* Strain TRP. Int. Biodeter. Biodegr. 62, 51-56.

Yang, L., Zhao, Y., Zhang, B., Yang, C., Zhang, X., 2005. Isolation and characterization of a chlorpyrifos and 3,5,6-trichloro-2-pyridinol degrading bacterium. FEMS Microbiol.Lett. 251, 67-73.

Yali, C., Ruifu, Z., Jian, H. and Shunpeng, L. (2002) Study on *Pseudomonas sp.* WBC-3 capable of complete degradation of methyl parathion. WeishengwuXuebao 42, 490–497.

Yasouri, F. (2006). Plasmid mediated degradation of Diazinon by three bacterial strains Pseudamonas sp., Flavobactetiumsp and Agrobacterium sp.Asian Journal of Chemistry. 18. 2437-2444.

Yuan S, Yang F, Yu H, Xie Y, Guo Y, Yao W. Biodegradation of the organophosphate dimethoate by *Lactobacillus plantarum* during milk fermentation. Food Chem. 2021 Oct 30; 360: 130042

Yurumez, Y., Cemek, M., Yavuz, Y., Birdane, Y.O., Buyukokuroglu, M.E., 2007.Beneficial Effect of N-Acetylcysteine against Organophosphate Toxicity in Mice. Biol. Pharm. Bull. 30, 490–494.

Zia Chishti,Sarfraz Hussain, Khaliq R. Arshad, Azeem Khalid, Muhammad Arshad. Microbial degradation of chlorpyrifos in liquid media and soil. Journal of Environmental Management 114 (2013) 372-380

Zhan H, Wang H, Liao L, Feng Y, Fan X, Zhang L, Chen S (2018) Kinetics and novel degradation pathway of permethrin in *Acinetobacter baumannii* ZH-14. Front Microbiol 9: 98

Zhongli, C., Shunpeng, L. and Guoping, F. (2001) Isolation of methyl parathion degradingstrain M6 and cloning of the methyl parathion hydrolase gene. Appl. Environ. Microbiol. 67, 4922–4925.

Zhongli, C., Shunpeng, L. and Guoping, F. (2002) Isolation and characterization of a p-nitrophenol degradation *Pseudomonas sp.* strain p3 and construction of a genetically engineered bacterium. WeishengwuXuebao 42, 19–26.

Chapter 2

Mycorrhizal Fungi: An Eco-friendly Input for Sustenance of Soil Fertility and Plant Health

Satish Kumar[1], Twinkle Chaudhary[2], Diksha[1], Satyavir S. Sindhu[1], Deepika Chaudhary[1] and Rakesh Kumar[1]*

[1]Department of Microbiology,
CCS Haryana Agricultural University, Hisar – 125 004, Haryana
[2]Enzyme Technology and Protein Bioinformatics Lab.,
Department of Microbiology, Maharshi Dayanand University, Rohtak, Haryana
**Corresponding author*

ABSTRACT

The soil microbiome has enormous potential in absorption of nutrients from soil and shields the plant against biotic and abiotic stresses. Various soil-inhabiting microorganisms that interact with plants include rhizobia, plant growth promoting rhizobacteria, pathogens, endophytes, epiphytes and mycorrhiza. Mycorrhizal fungi develop intricate symbiotic association with plant roots, which can be differentiated on the basis of morphology and the plant and fungal taxa so involved. Three primary types of mycorrhizal fungal association with plant roots include ectomycorrhiza, endomycorrhiza and ectendomycorrhizae. Arbuscular mycorrhizal symbiosis (endomycorrhizal type) is the most widespread mutualistic symbiosis among these associations. Plant responses to colonization by mycorrhizal fungi include effective exploitation of soil mineral resources and their defensive role against abiotic and biotic stresses leading to improved plant growth and reproduction. Due to better tolerance to environmental stresses, mycorrhizal fungi are involved in the persistence and fitness of many agricultural, horticultural and forestry plant species in diverse ecosystems. The aim of this review is to explore the current understanding of different types of mycorrhizae, signaling pathways involved in development of arbuscular mycorrhizal symbiosis and their agronomic significance.

Keywords: *Mycorrhizal fungi, Plant growth-promoting rhizobacteria, Symbiosis, Nutrient absorption, Abiotic stress, Biotic stress.*

Introduction

Current agriculture production strategies are associated with environmental challenges,which may cause loss of soil productivity, ferocious cycles of environment degradation and other harsh impacts on the mother earth (Duhamel and Vanderkoornhuyse, 2013; Garcia *et al.*, 2020; Wuepper *et al.*, 2020). Adoption of sustainable agriculture is based upon the efficient use of fertilizers in such a way that it could help in restoring soil quality and enhancing the agriculture productivity (Edwards, 2020). Also, strategies must be developed to meet the demands of increasing population, which must be practical in approach for increasing food supply with limited cultivable land (Godfray *et al.*, 2010). Sustainable strategies include no tillage agriculture, which is characterized by large density of fungal hyphae, enhanced soil aggregates and soil carbon (Simpson *et al.*, 2004; Paul *et al.*, 2013), intercropping (Brooker *et al.*, 2015) and conservation agriculture(Palm *et al.*, 2014). These strategies are known for establishing a balance in below ground microbial diversity along with detoxification of pollutants, carbon sequestration, nutrient retention and enhanced capacity of water storage (Bardgettand van der Putten, 2014; Bender and van der Heijden, 2015; Cavagnaro *et al.*, 2015; Stirling *et al.*, 2016).

Mycorrhiza, a symbiotic relationship between plant and fungi, could be counted as the foundation for beneficial effects observed in agriculture and can be considered as a prime component for sustainable intensification (Godfray *et al.*, 2010). Literal meaning of Mycorrhiza in Greek language is 'Fungus root' which is highly known for the nutrient translocation among soil and host plant. Fungi derive crucial nutrients from plant host in return of providing essential plant nutrients and mineral from the soil by increasing the root surface area. Other actions of symbiotic fungi include protection of host from various abiotic stresses and various plant diseases (Evelin *et al.*, 2009; Miransari, 2010, Hakeem *et al.*, 2016). Mycorrhizae has been classified basically in three groups: ectomycorrhiza, endomycorrhiza and ectendomycorrhizae. The ecologically and economically important classes are ecto-mycorrhizal and vasicular-arbuscular mycorrhizal associations, which are known for symbiosis with 75–80 per cent of aquatic and terrestrial plants including agricultural, horticultural and hardwood crop species (Christie *et al.*, 2004; Liu and Chen, 2007; Willis *et al.*, 2013).

Arbuscular mycorrhiza (AM) is the most common in ecosystem and are branched forms, which inhabit the cortex region of roots and can be observed in 80 per cent of the vasicular plants (Manchanda and Garg, 2007). Fungal-host association involves production of volatile organic compounds (VOCs), exopolysaccharides, microbe-associated molecular patterns (MAMPs), Myc factors and Nod factors (Goh *et al.,* 2014). Further, secreted volatile organic compounds help in root modification and Nod factor signaling pathway, which contribute towards fungal colonization and establishment of symbiosis (Maillet *et al.*, 2011). Plants secrete hormones and strigolactones for attracting fungi, while fungi secrete lipochito-oligosaccharides for initiating symbiotic relationship (Mohanta and Bae, 2015). Also, NSP1 and NSP2 and DELLA proteins like SLR1 (S-locus related gene 1) have a significant role in arbuscule development and infection of arbuscular mycorrhizal fungi (AMF) (Jin *et al.*, 2016). Once the fungi invade the cortical cells, it will produce haustoria-like structure, called as arbuscule for the transfer of nutrients among fungi and host cytoplasm (Oueslati, 2003).

Mycorrhizae improve plant growth by various mechanisms, including nutrient cycling (Rillig and Mummey, 2006; Shokri and Maadi, 2009; Wu *et al.,* 2011a, b; Yaseen *et al.,* 2012), improved plant growth and yield (Cavagnaro *et al.*, 2006; Nunes *et al.,* 2010) and enhanced absorption of slow diffusing ions for instance P, N, K, Mg, Cu and Zn (Meding and Zasoski, 2008; Smith and Read, 2008; Sharda and Koide, 2010). Besides this, hyphae of mycorrhizal fungi make complexes with heavy metals and helps in decontaminating the soil (Souza *et al.,* 2012). Moreover, mycorrhizal fungi inoculation has also been reported in amelioration of salt stress and drought stress (Sheng *et al.,* 2011; Ruiz-Lozano *et al.,* 2016; Santander *et al.,* 2019). Mycorrhizal colonization has also been reported to contribute towards plant protection from various diseases through systemic priming (Jung *et al.,* 2012) and by competing for space and resources (Linderman, 1994). Furthermore, combined application of mycorrhizae with other plant growth promoting rhizobacteria (PGPR) is a practical and better approach for sustainable agriculture (Denton, 2007; Ordookhani *et al.,* 2010; Najafi *et al.,* 2012). However, further characterization of efficient fungi, its molecular understanding along with genetic engineering may offer ways to harvest maximum benefits from this symbiotic association.

Types of Mycorrhizae

Mycorrhizae have been categorised into three distinct groups: ectomycorrhiza, endomycorrhiza and ectendomycorrhizae (with combined characters of ectomycorrhiza and endo-mycorrhizae). Ectomycorrhiza is constituted of obligatory symbiont from fungal classes Basidiomycetes and Ascomycetes, and plants of cold temperate regions such as pines and fir in subalpine forest and oaks (Read *et al.,* 1992) as well as poplars in deciduous forests. Ectomycorrhiza forms a mantle composed of hyphae covering the external area of root tip only (Allen, 1995). The second class endomycorrhiza has fascinated many scientists because of an intense invasive symbiosis. The fungi penetrate the root cells and colonise in internal regions of root including root cell wall and cell membrane (Bonfante and Genre, 2010). Endomycorrhiza further classified into: arbuscular, ericoid, arbutoid, monotropoid and orchid mycorrhizae (Peterson *et al.*, 2004). Arbuscular mycorrhiza is the most omnipresent group belonging to Glomeromycota and associated with 80 per cent of vascular plants (Simon *et al.*, 1993; Brundrett and Tedersoo, 2018). It penetrates through epidermis and colonize in cortical parenchyma (Powell and Bagyaraj, 1984). AM fungi are the obligatory symbionts involving a mutual exchange of carbon and nutrients between fungi and plant, respectively(Wipf *et al.*, 2019). Fungi take up 10-40 per cent of total produced carbon via photosynthesis from the host for the completion of its life cycle (Johns *et al.*, 2014). Moreover, extraradical mycelial networks seem to maintain their ability to establish mycorrhizal symbioses with plants as long as five months after shoot removal, adsorbing nutrients released from dead roots (Pepe *et al.*, 2018).

Ectomycorrhiza

The fungal mycelium ensheaths the plant root and forms tubercles around the rootlets with limited penetration of hyphae inside the root tissue. Infection restricts elongation of the absorbing roots but may stimulate branching. Short roots become swollen and may become variously coloured due to the colour of the fungal symbionts. Such roots become shrivelled and black when they are dead. In some cases, the surface of the plant roots may become enveloped with a mycelium of a saprophytic soil fungus. Ectomycorrhiza is found in all genera belonging to Abietaceae family among conifers and among Betulaceae, Fugaceae and some members in Soliaceae, Juglandaceae, Myrtaceae and Caesalpinaceae families among hard woods (Agerer, 1991). The predominant Basidiomycetes genera which form the ectomycorrhiza are *Amanita, Boletus, Cantharellus, Entoloma, Lactarius, Rhizopogon* and *Scleroderma etc.*

Ectomycorrhizal fungi are having no intracellular hyphae and form a mantle of hyphae over root surface while the network of hyphae established within the root surface is known as Harting net. Ectomycorrhizal fungi are a diverse group and the members of Ascomycota and Basidiomycota have been discovered to associate independently at least 80 times. (Tedersoo and Smith, 2013). Lately, based upon the molecular studies, it has been concluded that the ectomycorrhizal fungi arose from saprotrophs and plant endophytes via convergent evolution providing similar morphological and physiological features and certain shared genetic alteration resulting in mycorrhizal formation. However, independent rise of mycorrhiza is linked with some of the physiological changes. Ectomycorrhiza is somewhat moderately functional in host specialization in driving speciation on comparing with arbuscular and orchid mycorrhizal fungi.

Host specialization ranges from either species that from association with a gamut of host species or fungi that associates with a thin range of plant species. Though, some fungi may showmore host specialization than others. Ectomycorrhizal fungi associate with both autotrophic and mycoheterotrophic plants, although the former one is the most typical host. The effect of host interactions on fungal diversity and dissemination is still unrevealed. Also, this group has constrained geographic distribution than arbuscular mycorrhizal fungi (Allen *et al.,* 1995). Certain abiotic factors for instance, nitrogen and phosphorous level, and successional dynamics have been described to influence ectomycorrhizal fungal distributions. Hence, host specialization and environmental factors cause higher niche differentiation and specialization among ectomycorrhizal fungal species.

Endomycorrhiza

These endomycorrhiza are characterized by intracellular infection within the root. These associations produce fungal structure (vesicles and arbuscules) having extensive intracellular penetration in the cortex region of the root. The fungal symbionts give rise to different structure, extrametrical hyphae, globose or oval terminal swellings called vesicles and arbuscules. Arbuscular mycorrhizal (AM) fungi are geographically ubiquitous. They are commonly found in association with agricultural crops, most shrubs, most tropical and some temperate tree species. Arbuscular mycorrhizal fungi have been observed in 1000 genera of plants representing some 200 families (at least 300,000 receptive hosts in the world flora). Arbuscular mycorrhizae are formed by non-septate Zygomycetous fungi, belonging to the 6 genera *Glomus, Gigaspora, Acaulospora, Enterophospora, Sclerocystis* and *Scutellospora*. The fungi being obligate biotrophs do not grow on synthetic media and are classified according to the morphological characteristics of the spores formed in the soil. There are five major groups categorised under Endomycorrhiza: arbuscular, ericoid, arbutoid, monotropoid and orchid mycorrhizae.

Arbuscular Mycorrhiza

Arbuscular mycorrhizal association is the most usual interaction pervading in soil and can be observed for shrubs, tropical trees and few of the temperate species. Families including Brassicaceae, Chenopodiaceae, Polygonaceae and Cyperaceae are known to form ectomycorrhiza as well as arbuscular association. Families that form both ectomycorrhiza and AM include Juglandiaceae, Tiliaceae, Myrtaceae, Salicaceae, Fagaceae and Caesalpiniaceae. AM is highly distributed with in plant kingdom thus it is omnipresent and distribution ranges from aquatic to desert environment including tropical, temperate and arctic regions (Bagyaraj, 1992).

Arbuscules formation, which are tree shaped structure, is the characteristic feature of AM fungi and are formed inside the root after the fungal hyphae penetration in the roots. Arbuscules enhances the root surface area providing better nutrient exchange. Though the relationship is quite close but there is cytoplasmic separation due to fungal cell wall and the plasma membrane of host. AM association has been observed with roughly 300,000 species and the fossil study suggested the origin of AM fungi to be originated around 400 million years ago. Low species diversification of AM fungi in spite of ancient origin and wide host range reflects at least in part a higher range of genetic diversity among fungi coupled with minimum host specificity. Also, some of the endosymbionts and mycoplasma related endobacteria having high genetic plasticity associates with fungi and are well described for sustaining arbuscular mycorrhiza (Kuo, 2015; Naito *et al.*, 2015; Torres-Cortés *et al.*, 2015). However, some of the fungi like Glomeromycota are asexual with basic morphology making the morphological distinction very much complex. However, molecular techniques are used rapidly to characterize the diversity and evolution of Glomeromycota (Krüger *et al.*, 2012; Redecker *et al.*, 2013; Redecker and Schüßler, 2014; Hart *et al.*, 2015).

Orchid Mycorrhiza

Orchids are the members of Orchidaceae family constituted of 30,000 species. They are the obligatory associates of mycorrhizal fungi however the relationship is distinct in terms of mycohetertrophic nature of orchids during seed germination and initial stages of seedling growth, and count on fungi for carbohydrates (Dearnaley *et al.*, 2012). Orchids shifts from mycohetertrophy to autotrophy after the leaf's emergence leading to photosynthetic phase thus, they are mixotrophic. But some of the species of orchid remains mycoheterotrophic throughout their life cycle.

Around 25,000 species of Ascomycota and Basidiomycota are categorized as orchid mycorrhizal fungi (Dearnaley *et al.*, 2012; van der Heijden *et al.*, 2015) and interacts with orchids reflecting 10 per cent of total plant species (Brundrett, 2009). Earlier, the fungi were identified as *Rhizoctonia solani*, *R. repens* and other *Rhizoctonia* species but lately their prefect stage was characterised, and were described as members of genera *Thanatephorus, Ceratobasidium, Sebacina* and *Tulasnella*. *Tulasnella calospora* is an orchid mycorrhizal fungus, which is culturable and can lead its life cycle as saprophyte in host absence (Krishnamurthy and Senthilkumar, 2005).In contrast to the arbuscular mycorrhizal fungi, host specialization appears to have played a key role in the diversification of orchid mycorrhizal fungi. Species specificity among closely related orchids is based upon the association with particular combination of mycorrhizal fungi (Taylor *et al.*, 2004; Okayama *et al.*, 2012; Xing *et al.*, 2015). Even though, orchid mycorrhizal fungi show great host specificity, they had arisen independently from saprophytic and ectomycorrhizal fungi as evident from the phylogenetic analysis (Schüßler *et al.*, 2001), however there are certain reports suggesting them to originated from arbutoid (Okayama *et al.*, 2012) and endophytic fungi (Oberwinkler *et al.*, 2013).

Ericoid Mycorrhiza

Ericoid mycorrhizal fungi usually colonize plants belonging to the families Ericaceae, Empetraceae and Epacridaceae, which are commonly referred to as heath plants *e.g.*, azalea, rhododendron, blueberry, cranberry *etc.* These plants occur in temperate regions of the world. Most ericaceous species characteristically grow on nutrient-poor acidic soil, where ammonium predominates over nitrate. Hence ericoid fungi accelerate the Nitrogen and phosphorous uptake.Fungi of these group form hyphal coils in root cells of plants belonging to Ericaceae. Ericoid fungi are predominantly arisen from saprophytes but some are ascended from ectomycorrhizal as well as endophytic fungi. Host specialization within Ericoid mycorrhizal fungi ranges from narrow to broad host range with Ericaceae. Arbutoid and monotropoid mycorrhiza are the end result of host specialization among ericoid mycorrhizal fungi belonging to Ascomycota and Basidiomycota. The ericoid mycorrhizal fungicould be further explored for enhancing the productivity of these plants (Smith and Read, 2008).

Arbutoid Mycorrhiza

Arbutoid fungi associate with a slim range of host belonging only to subfamily Arbutoideae. Arbutoid has ascended a number of times mainly from taxa known for ectomycorrhizal association related to Ascomycota and Basidiomycota. Arbutoid fungi form hyphal sheath and grow intracellularly, thus sharing an in between space among endo- and ectomycorrhizal association hence termed as ectendomycorrhiza (Oberwinkler *et al.*, 2013). Arbutoid mycorrhizal fungi are the lineages of ericoid mycorrhizal fungi and endophytic fungi, which arose due to physiological and morphological convergent evolution among disparate lineages.

Monotropoid Mycorrhiza

Monotropoid and orchid mycorrhiza are similar in terms of mixotrophic nature of host plant. However, the difference lies in the fact that monotropoid fungi are morphologically distinct and from mantle and Harting net and they themselves have originated from orchid mycorrhiza. Also, montropoid mycorrhiza differs from arbutoid mycorrhiza and the former does not show hyphal coil formation and instead of cell wall penetration hyphae form pegs causing cell wall indentation without penetration in cell wall. The former has a narrow range of host and only associates with plants of subfamily Monotropoideae in the Ericaceae. Similar molecular evidences have been reported for the monotropoid mycorrhizal fungi evolution from Ascomycota and the Basidiomycota. Monotropoid mycorrhizal fungi have evolved from ectomycorrhizal fungi, ericoid mycorrhizal fungi and possibly endophytic fungi, followed by convergent evolution.

Piriformosporo indica

Piriformospora indica was first found in rhizosphere of *Prosopsis juliflora* (Swartz) DC and *Zizyphus nummularia* (Burm. fil.) Wt. and Arn. from the sandy soil of Rajasthan, India (Varma *et al.*, 1998; 1999). The fungus is easily cultivable, lacks host specificity and colonizes roots of many different plants, mostly in an endophytic fashion (Varma *et al.*, 2001). It is a member of order Sebacinales (recently formed) and is highly diversified in mycorrhizal association and plant growth promotion. *P. indica* can be considered as a asymptomatic root endophyte, known to inhabit members of gymnosperms, pteridophytes, bryophytes and a great number of monocots and dicots (Peškan-Berghöfer *et al.*, 2004; Fakhro *et al.*, 2009; Oelmuller *et al.*, 2009). The colonization of roots begins with a biotrophic growth phase, in which living cells are colonized and continues with a cell death-dependent phase, in which root cells are actively killed so that fungi can be established followed by proliferation (Varma *et al.,* 2013b). Colonization is restricted to some parts of root and it does not penetrate into aerial plant parts or endodermis. Thus, involved endophyte increases the nutrient uptake efficiency owing to the elaborated hyphal networks (Das *et al.,* 2013).

P. indica shows inter as well as intracellularly growth and forms fluorescent chlamydospores with pear shape appearance in the cortex region of roots. It increases nutrient and mineral uptake, better survival efficiency under abiotic stress like water, temperature and salt and provides systemic resistance against pathogens, insect, heavy metal and certain toxic compounds (Das *et al.,* 2012). *P. indica* positively influence the plant growth by promoting adaptation to biotic and abiotic stresses and early flowering, increasing seed yield and modifying the secondary metabolites (Verma *et al.,* 1998; Yadav *et al.,* 2010; Varma *et al.,* 2012b; 2014). Combined effect of micro- and macronutrients, phytohormones and mutualism with *P. indica* influences the plant metabolism (Danesh, 2015).*P. indica* can be cultured on a range of media constituted of starch and glucose as best carbon source (Serfling *et al.,* 2007; Varma *et al.,* 2014). *P.indica* is the model organism for the studies of plant-microbes or fungi relationship and allow a better evaluation of various aspects like fungal morphogenesis, stress tolerance and production of secondary metabolites in plants. Recent,genome sequencing of *P. indica* has provided a better insight into the mechanism behind the interaction (Zuccaro *et al.,* 2011).

Ectendomycorrhiza

Ectendomycorrhiza, earlier termed as 'ectendotrophic mycorrhiza' by Melin in 1923 are the least discussed mycorrhiza and are of limited occurrence (Egger and Fortin, 1988). Fungi in this group includes usually *Wilcoxina* (*W. mikolae* and *W. rehmii*), *Sphaerosporella brunnea* and, sometimes *Phialophora finlandia* and *Chloridium paucisporum* with host *Pinus* and *Larix* species. These are characterized with Hartig net, with presence or absence of a thin mantle and shows hyphal penetration into epidermis and cortex region of roots. The use of term ectendomycorrhiza is restricted for fungal species, which penetrate intracellularly and lead to Hartig net formation. This characteristic feature differentiates them from ECMs. However, more studies are required for elaborating nutrient exchange process between hyphae and root cells. Arbutoid and monotropoid mycorrhiza share some features with ectendomycorrhiza and thus denoted as ectendomycorrhiza (Zak, 1974, 1976). The hyphae are restricted to epidermis while in case of monotropoid mycorrhiza, pegs are formed in the epidermis; these gets enclosed by an intricated host derived cell wall (Smith and Read, 1997). Studies on ectendomycorrhiza associated with angiosperms like *Uapaca kirkana* (Högberg, 1982) andpetrocarp *Shorea parvifolia* (Louis, 1988) has revealed the occurrence of intracellular hyphae, which can be of senescing ectomycorrhizas. Fungi of these groups are phylogenetically related with Peizazle (operculate discomycetes) and

Leotialies (inoperculate discomycetes) order of Ascomycetes. The host generally is *Pinus* and *Larix*, but can be a range of conifer and angiosperm. Therefore, ectendomycorrhiza have a broad host range (Molina *et al.*, 1992).

Development of AM Symbiosis: Exchange of Signals between Fungus and the Plant

Arbuscular mycorrhiza has extremely wide host range (more than 80 per cent of species in plant kingdom depicts AM symbiosis) and they have ancient origins (origin lies even before 400 million years ago and first found in tissue of first land plants) (Remy *et al.*, 1994). However, only six genera including 10 species grouped under order Glomales (Zygomycetes family) can form AM association (Morton and Benny, 1990). However, there is insufficient information regarding physiology and genetic makeup of AM fungi, which could be due to an intricated biological system and obligatory nature of the symbiont. Recently, extensive studies have been made towards unravelling plant mechanism of AM symbiosis by the application of molecular cytology, use of mycorrhiza defective plant mutants and molecular biology techniques and comparative analysis with other plant microbe interactions.

AM fungi initiates the interaction with the plant by establishing a contact with plant roots via a hypha developed from a germinating soil borne spore. But before contact there is a mutual exchange of signals followed by recognition between fungi and plant (Figure 2.1). Root exudates secreted by host plant increases spore germination and fungal growth while the root exudates from non-host shows no such effect. Some of the active plant metabolites or exudates *i.e.* strigolactones and flavonoids have been identified (Phillips and Tsai, 1992; Smith *et al.*, 2006). However, flavonoids are essentially required in fungal interaction (Becard *et al.*, 1995)

Plant response to any infection is reliant on the type of organism and recognition of elicitors molecules secreted by pathogens, which triggers a signalling cascade leading to development of AM symbiosis (Figure 2.1). Fungi produce soluble compounds, which are known as Myc factors (mycorrhization factors) that results in mycorrhizal responsive genes expression leading to phenotypic changes in roots. SYM genes involved in recognition of Myc factors have also been characterized (produced by *myc* genes) that induce signaling cascade on the microbial colonization of roots. SYM genes are common in bacterial and fungal endosymbiosis (Parniske, 2004; Genre *et al.*, 2005). Symbiotic development includes various steps.

Initial Symbiotic Phase

Initiation of symbiosis includes cell to cell contact by the development of appressorium between plant, roots and fungus (García-Garrido and Ocampo, 2002). Appressoria are flattened, elliptical hyphal tips, which induces alteration in gene expression of fungi. After the appressoria has been established, the coordinated action of host cell helps in propagation and penetration of fungus (Paszkowski,

2006). Arbuscular mycorrhizal symbiosis is governed by signal transduction cascade resulting in an intricated signalling. For instance, strigolactones produced by plant aids in fungal growth during pre-symbiotic stage (Akiyama *et al.*, 2005; Besserer *et al.*, 2008; Kretzschmar *et al.*, 2012; Mori *et al.*, 2016). As a result, strigolactone production induces the activation of symbiotic signaling pathway by the secretion of short chain chitin oligomers (Lopez-Raez *et al.*, 2017). Functioning of strigolactones can be observed for *Arabidopsis thaliana* and the mold *Mucor* spp. symbiosis (Rozpadek *et al.*, 2018).

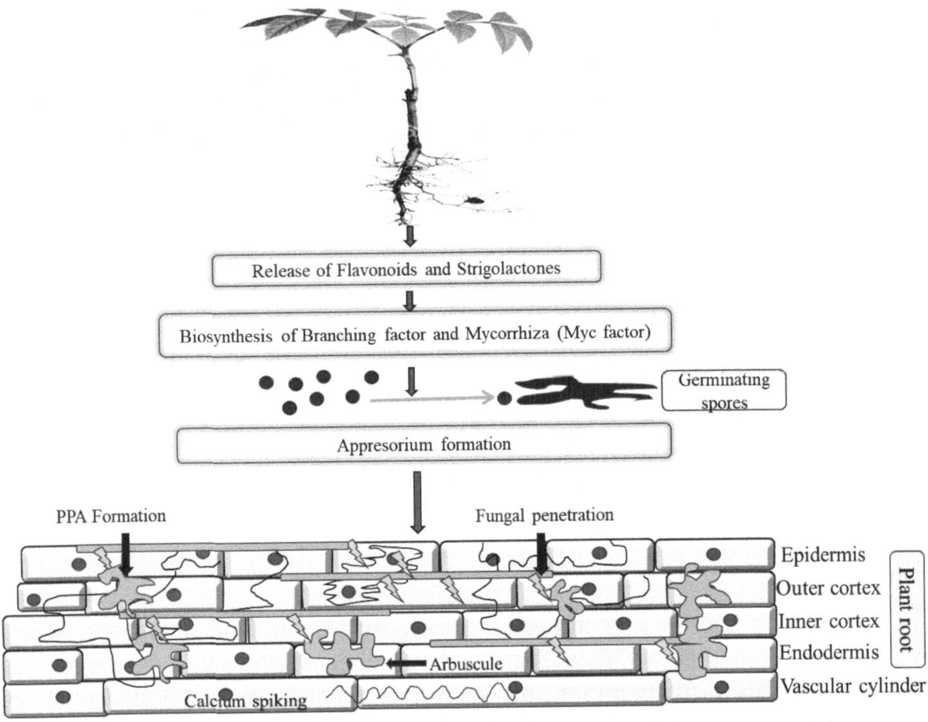

Figure 2.1: Release of Strigolactones and Myc Factors Leads to Development of Mycorrhiza.

Branching factors or BFs are another signaling molecules produced by host plant that are required for the morphogenesis and differentiation of fungal hypha (Giovannetti *et al.*, 1994). Root exudates influences spore germination, fungal growth and differentiation and root colonization (Nair *et al.*, 1991; Siqueira *et al.*, 1991; Kape *et al.*, 1993). Akiyama *et al.* (2005) extracted BFs produced by *Lotus japonicum* associated with *Gigaspora margarita* in P-deficient soil and were observed to induce hyphal branching. Symbiotic relationship is initiated when the fungal spore is germinated in soil. Myc factors or lipo-chitooligosaccharides (LCO) also function in plant fungi interaction. Once the fungi receive the signal, interaction beigns with spore germination and later with hyphal penetration via epidermis followed by either inter- or intracellularly colonization and growth of fungi in cortex. After penetrationin cortical region, fungal hyphae differentiate into branched structures called as arbuscles for the nutrient exchange and later establishe symbiotic matrixes of symbiosome. Similarly, AM fungi also produce some signaling molecules so to alter plant physiology and metabolism including Krebs cycle and plastid biosynthesis and induces starch accumulation in roots, calcium spiking and root formation. (Lohse *et al.*, 2005; Mukherjee and An´e, 2011; Maillet *et al.*, 2011). Vangelisti *et al.* (2018) studied gene expression of AM fungi in the rhizosphere of *Helianthus annus* and stated their role in nutrient exchange, plant defence, hormonal balance and signal transduction. In rice, an alpha/beta hydrolase is released by an intracellular receptor hydrolase DWARF 14 LIKE for the initiation of symbiosis (Gutjahr *et al.*, 2015). In *Rhizobium*–legume symbiosis also, fungal chitin derivatives were found to be efficient plant receptors in *Parasponia*, a non-leguminous plant (Den Camp *et al.*, 2011).

An increased concentration of cytokinin can be observed for mycorrhizal symbiosis (Shaul-Keinan *et al.*, 2002). Jin *et al.* (2016) identified DELLA proteins as prime regulators of gibberellic acid signalling in *Medicago truncatuala*. DELLA proteins regulate mycorrhizal symbiosis and arbuscular development as evident from the study of DELLA mutant (Floss *et al.*, 2013). Carbon dioxide alters the carbon, nitrogen and phosphorus cycle, thus induces strigolactone production in root, which activates redox-auxin-strigolactone systemic signaling cascade required for symbiosis (Zhou *et al.*, 2019).

AM fungal colonization in roots also requires Nod factors NFR1 and NFR5 also known as *LysM* receptor kinases. Further, myc-LCO receptors are analogous to *NFR1/LYK3* and *NFR5/NFP* in terms of structure but can possess similar functions (Sun *et al.*, 2015; Rasmussen *et al.*, 2016). During Nod signaling, symbiotic signaling pathway is linked with SYMRK or symbiosis receptor-like kinase by the action of receptor kinases. However, SRYM is also a coreceptor for some of the specific receptors. SYMRK interacts with MAPKK or mitogen-activated protein kinase and 3-hydroxy-3methylglutaryl-CoA reductase (HMGR) signifying their role in signal coordination between ion channels related with nucleus and certain membrane bound receptos. CYCLOPS, is another gene involved but exact functioning is still unclear, and is phosphorylated by CCaMK(calcium and calcium/calmodulin-dependent serine/threonine-protein kinase) (Oldroyd *et al.*, 2013). Calcium spiking is influence by nucleoporin protein like NUP133 and NUP85 in common symbiosis pathway (CMP) (Kanamori *et al.*, 2006). Thus, it is the coordinated network of genes behind such intricated physiological process of arbuscular formation and its accommodation.

Penetration of Fungal Hyphae

The process of penetration of fungi into host cell is mediated by one of the following mechanisms including formation of a penetration hypha, an infection peg developed from the appressorium and enzymatic action on host cell wall. Further, another mechanism for fungal invasion has been observed in *Lotus* involving an epidermal cleft formed midst of two rhizodermal cells allowing fungal to invade and later fungi gets translocated to exodermal cell in a slow manner(Paszkowski, 2006). However, fungi can directly invade the root cells as seen in *Medicago truncatula* (Genre *et al.,* 2005). Also, fungal invasion and colonization involves mechanical and enzymatic mechanism; the former includes enzymes for cell wall degradation, for instance, cellulases, xyloglucanases and proteolytic enzymes like polyglacturonase. Further, this process is totally based uponsynchronization with host machinery (García-Garrido and Ocampo, 2002). Mycorrhizal symbiosis is characterised by the production of MiSSP7 *i.e.* mycorrhizal induced small secreted protein, a crucial effector protein, which along with genes for cell wall remodelling and auxin homeostasis, establishes the Hartig net for nutrient exchange between the two symbiotic parteners. Gene XHT1 for endotransglucosylase/hydrolase identified from *Medicago truncatula* aids in fungal penetration by altering the structure of cell wall of plants so to establish a symbiotic relationship (Balestrini and Bonfante, 2014).

Mature Phase

AM symbiosis is bifurcated into Paris and Arum symbiosis, the former is characterized with fungal coils produced from the cell with no intracellular stage, while AM type is marked by arbuscules, which are tree-like structures (Dickson *et al.,* 2007). Arbuscules are the sites of nutrient exchange (Hughes *et al.,* 2008). Zhang *et al.* (2010) studied the symbiosis of *M. truncatula* mutants having stunted *str* gene, which normally recognised by *STR*2 gene induced on symbiosis. Hetero-dimerization of *STR*2 by the action of *STR*1 caused the formation of a complex in the peri-arbuscular membrane (PAM), suggesting its essential role in arbuscular growth and therefore establishment of AM association (Zhang *et al.,* 2010). Once the arbuscules has been developed, the host cell proceeds through enormous changes. However, arbuscules breakdowns after 2 to 4 days and prepares the cortical cells for another arbuscule (Paszkowski, 2006; Pumplin and Harrison, 2009). Prior to initiation of arbuscule, a sugar gradient arises among outer cell layer and vascular tissue, which is further perceived by host cell. Major proteins involved in arbuscule formation belong to SYM pathway and includes *LjCASTOR*, *LjSYM*15, *LjNup*133 and *LjSYM*24 (Kistner *et al.*, 2005).

Role of Mycorrhiza in Plant Growth Promotion

Mycorrhiza plays a vital role in acquisition of nutrients especially phosphorus and carbon. Mycorrhizal associations are also involved protection of host plants from various abiotic stresses and plant diseases (Evelin *et al.,* 2009; Hakeem *et al.,* 2016).

Nutrient Transport

The growth of AMF has been reported to enhance the acquisition of large number of nutrients including phosphorous, calcium, silicon, manganese, magnesium, aluminium, boron, iron, sodium, copper and zinc (Hua *et al.*, 2020). The accumulation of nutrients like magnesium, potassium and calcium enhanced the growth of host plants in acidic soil while the nutrients like nitrogen, zinc, copper and phosphorus stimulated plant growth in alkaline conditions. The process of nutrients uptake is affected by the growth of AMF in several soil conditions. A large number of AMF has the capability to reduce the manganese and aluminium toxicity for the host plants at low pH. Both the AMF and non-AMF plant roots secrets some substances that has the capability to decrease the activity of those substances that decline the growth of host plant (Mishra *et al.*, 2017). Arbuscular mycorrhizal fungi also help in the promotion of plants growth during stress conditions. During stressed environmental conditions, the volume of soil for the nutrient uptake increases due to the presence of AMF (Mitra *et al.*, 2021). The roots of AMF associated plants have more diameter, extra length and increased weight in comparison to non-mycorrhizal plants. Additionally, the less diameter of fungal hyphae can easily penetrate the plant roots and root hairs for the easily transport of nutrients.

Phosphorus

Plant's association with AM fungi has many beneficial effects, one of them is scavaging phosphorus and making it available to plant roots. It also aids in decreasing the effect of phosphorus deletion from rhizosphere, thereby uplifting phosphorus nutrition of plant. Nutrient exchange between plant and fungi is highly controlled process and is mediated by transport proteins like as phosphate transport and the P-type H^+-ATPase (Bucher *et al.*, 2007). AM fungi particularly assist those plants where P limits the plant growth. The AMF hyphae help in the exploration of the soil volume as compared to non-AMF plants. The effect of AMF for the root colonization decreases when the level of P increases in the soil (Püschel *et al.*, 2021). Generally, the fungal hyphae can transport P at greater distances from plant roots as compared to the non-AMF plant. It is reported that the phosphorous deletion zone for *Glomus mosseae* was >112mm and > 90mm for the *Acaulosporalaevis subterraneum* with hyphal compartment. Hyphae of *Glomus* species spread less as compared to *A. laevis*. Additionally, it has been reported that AMF hyphae could supply 80-90 per cent of phosphorus to plant to a distance of 100mm where even roots can't reach (Bever *et al.*, 2017). AMF provides much larger hyphal length to access more P and to facilitate the unsoluble P compounds to soluble form for more nutrient acquisition.

The AMF hyphae also secrete some volatile compounds, ethylene and some growth regulating compounds like cytokinins (Basu *et al.*, 2018). Moreover, the root exudates also contain organic substances, which help in enhancing the mineral acquisition and solubilization of the unsoluble compounds. The ectomycorrhizal fungi produced oxalic and citric acid that provides a direct evidence to solubilize sparingly soluble phosphorous sources. AM inoculated maize plants showed a comparable increased concentration of amino acids, reducing sugars and organic acids as compared with non-AM plants (Campos *et al.*, 2018). Furthermore, phosphorus solubalization has also been improved by coinoculation with phosphorus solubilizing microbes (Etesami *et al.*, 2021). The effect of *G. mosseae* was reported as much more effective than the *Glomus fasciculatum* when applied in soil with AMF+P-solubilizing bacteria. The increase in pH level in the rhizosphere was also found to increase the availability of Fe^{-} and Al phosphates but decreased the Ca-phosphates availability (Penn and Camberato, 2019). The more excretion of H^{+} is ubiquitous under many plant growths conditions that cause change in rhizospheric pH conditions during AMF colonization. However, the level of rhizospheric soil pH was found to decrease the uptake of phosphorus when *G. mosseae* was grown with ammonium sources.

In addition, the solubilization of sparingly soluble phosphorous compounds also increased by the activity of extracellular enzymes like phytase and phosphatase released from the AMF fungal hyphae (Zhang *et al.*, 2018). Phosphatase activity in rhizosphere soil is often greater than in bulk soil. In addition, external root surfaces have a high activity of alkaline/acid phosphatase under low or P deficient conditions, these enzymes have the potential to solubilize organic P (Meena *et al.*, 2017). The colonization of corn and soybean cultivars with *G. margarita* and *G. intraradices* showed greater root phosphatase activities than non-AMF plants and *G. margarita* roots had greater phosphatase activity than *G. intraradices* roots (Bhowmik and Das, 2018).

AM fungi supplies phosphorous to plant via direct and indirect pathway or the AM pathway, the former being usual pathway of taking up phosphorous from soil (Figure 2.2) (Smith *et al.*, 2011). Around 70 per cent of total phosphorous uptake is due to AM pathway (Yang *et al.*, 2012). In non-mycorrhizal plants, phosphate transporter genes in direct pathway are regulated by the phosphate starvation signalling pathway while in AM plants direct pathway can be controlled without influence of phosphate levels because of interaction with AM pathway. It has been confirmed that AM fungi colonisation leads to reduction in direct uptake of phosphate with no affect in distant non colonized roots (Watts-Williams *et al.*, 2015). Specific phosphate transporters play important role in AM pathway and are identified in *Medicago truncatula*, *Oryza sativa* and *Lotus japonicas* (Volpe *et al.*, 2016). PHT1 transporters reveal a definite expression pattern and are solely expressed for AMFs and are constrained to interface of symbionts in the peri-arbuscular membrane (Harrison *et al.*, 2002).

Absorption of phosphate begin with free phosphate uptake from soil via extra radical hyphae in mycorrhizal association (Bucher, 2007). Once the phosphate has been absorbed by the fungi, it is polymerised to polyphosphate chains and transferred to vacuolar compartment of intraradical hyphae. Polyphosphate so formed is hydrolysed to release phosphate in the interfacial apoplast from where PHT1 will lead phosphate to peri-arbuscular membrane. After reaching plant cytosol, phosphate could be transferred to all plant parts. Apart from Pi transporter, AM fungi induces ammonium transporter as well which are located in peri-arbuscular membrane and translocate ammonia into cortical cells (Breuillin-Sessoms *et al.*,

2015). Transporters supply phosphorous and ammonium to root cells and also functions for the sustenance of arbuscule (Berruti *et al.*, 2015; Breuillin-Sessoms *et al.*, 2015). Currently, sulphate transporters have also been identified.

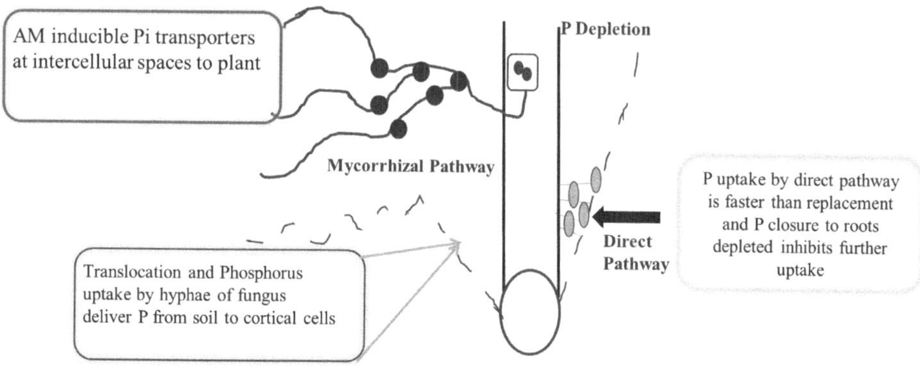

Figure 2.2: Mechanism of Phosphorus Translocation by Mycorrhizal Roots.

Carbon

The plant and the fungus regulate the flux of sucrose directed to the root. Carbon from plants is transferred to the fungus through the two membranes at the symbiotic interface. This carbon is released into the peri-arbuscular space (PAS), most likely as sucrose, before broken down to hexoses and absorbed through plasma membrane by AM fungi (Wang *et al.*, 2017). Hexoses are transformed inside the fungal cytoplasm into triacylglycerol lipid droplets (TAG) and glycogen granules suitable enough for transport to a long via hyphae (Jha and Sharma, 2019). Sucrose supplied to the AM root is acted upon by either sucrose synthase or invertase. A part of the novel clade of hexose transporters can also be expressed in the interface membrane of arbuscule as identified in the *Geosiphon pyriformis.* Mycoheterotrophic plants that interact with AM fungi, on the other hand, take carbon from AM fungi and the latter provide carbon through hyphal network as observed for carbon transfer by photosynthetic saprophytes of *Huperzia* species to mycoheterotrophic gametophytes. Nutrients uptake by fungi from soilare also transmitted to the plant cell, where these nutrients must pass through plasma membrane of fungus and travel a long way to the intraradical hyphae (IRH) including the arbuscules. From the arbuscles, nutrients then cross the fungal plasma membrane and the plant peri-arbuscular membrane to enter the plant cytoplasm.

Amelioration of Abiotic Stresses by Mycorrhizal Fungi

Application of agro-chemicals on field including pesticide and synthetic fertilizers, deforestation, inappropriate dumping of waste material and excessive irrigation has resulted in adverse effects on crop production (Plouznikof *et al.*, 2016; Sindhu and Sharma, 2019; Etasami and Glick, 2020; Sindhu *et al.*, 2020).

These human activities expose the plants to abiotic stresses. Plant genetic pathway perceives the stress signal by receptors of membrane, which as a result initiates either directly or indirectly the plant defence system (Rejeb *et al.*, 2014). Many studies have established the contribution of fungus for providing resistance/tolerance to the plant against abiotic stress including, salinity, drought and heavy metal (Figure 2.3).

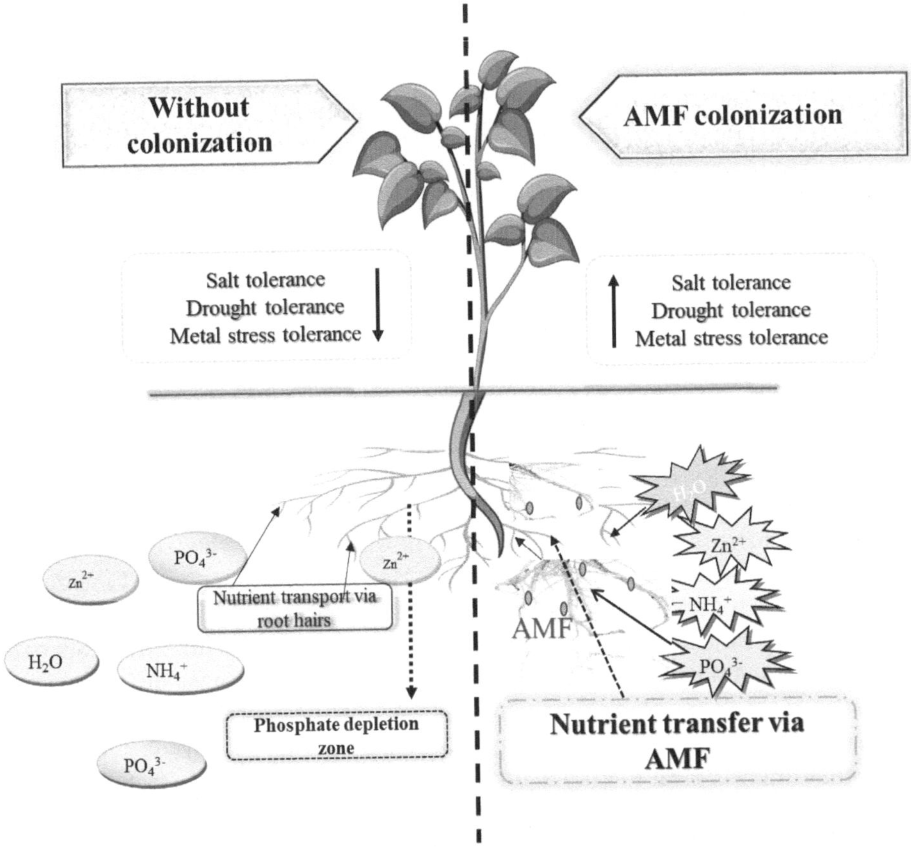

Figure 2.3: Absorption of Nutrient and Amelioration of Abiotic Stresses by Mycorrhizal Infected Roots.

Salinity Stress

Soil salinization is concerning situation of food security worldwide as higher soil salinity makes it tough for plant roots to absorb water from the soil, which can prove toxic to plant metabolism. Negative impacts of salinity on plants includes plasmolysis, ethylene production, imbalance in nutrient pool, hinderance to ROS (Reactive oxygen species) production and photosynthesis (Sairam and Tyagi, 2004; Munns and Tester, 2018). Thus, soil salinity adversely affects vegetative growth of plants along with suppressing net assimilation rate causing low crop yield (Hasanuzzaman *et al.*, 2013; Ahanger and Agarwal, 2017; Ahanger *et al.*, 2017a, b, 2018).

AMF is proved to be a good option for the mitigation of adverse effects induced upon soil salinity and for achieving better plant growth (Santander *et al.*, 2019). AMF reduces the salt stress by increasing the absorption water and minerals owing to high K/Na ratio, production of organic solutes and increased enzymatic activity of radicals like superoxide, peroxide and hydroxyl and singlet oxygen and alpha-oxygen for the quenching of ROS.Mycorrhiza infected roots balance the water potential in cell by hording the solutes like proline, glycine and glycine betaine without decreasing the turgor pressure (Munns and Gilliham, 2015). Also, salt tolerant AMF are well known and are continuously explored for the better potential of root colonization and symbiotic association with plants. Positive effects of AMF inoculation on plant prevents the lethal effects on photosynthesis by elevating stomatal conductance, rate of photosynthesis and leaf water relation under higher salinity conditions (Sheng *et al.*, 2011; Ait-El-Mokhtar *et al.*, 2019). In addition, growth parameters (fresh and dry weight and nitrogen concentration of shoot and roots) are improved by mycorrhizal inoculation (Wang *et al.*, 2018). Satander *et al.* (2019) reported the enhanced biomass production, proline concentration, nitrogen uptake, reduced Na^+ accumulation when lettuce and other plants were inoculated with mycorrhiza under salt stress. Besides this, increased nutrient uptake in cucumber was observed when it was inoculated with AMF(Sallaku *et al.*, 2019).

Heavy Metals

Enzymatic reactions, electron transport chain, nucleic acid metabolism and other metabolic reactions in a living organism requires certain elements like copper (Cu), iron (Fe), manganese (Mn), nickel (Ni), chromium (Cr), molybdenum (Mo), antimony (Sb) and zinc (Zn). However, their high concentration could be toxic as they can alter the structure of proteins by substituting with a vital element, causing symptoms like chlorosis, decreased root growth and seed germination, cell cycle arrest and interference in photosystems (Srinivas *et al.*, 2013). High concentration of heavy metals in ground water and soil surface is attributed to the anthropogenic activity. These heavy metals are ultimately taken up by the plant roots resulting in a stress condition (Mehmood *et al.*, 2019).

AMF strengthen the plant defence system under heavy metal contaminated soils and are recommended as potential organisms for improving plant growth and development. Food crops grown in heavy metal contaminated soil can cause lethal effects (Liu *et al.*, 2013; Yousaf *et al.*, 2016). Cadmium and zinc retard the growth of root and shoots and induces chlorosis, and may even cause death of contaminated plants (Moghadam, 2016). Some of the fungi have potential of heavy metal fixation by immobilising heavy metals in internal or external hyphae (Ouziad *et al.*, 2005), and stocking up them in the vacuole or may chelate with some other substances in the cytoplasm, thereby reducing the metal toxicity in the plants (Punamiya *et al.*,

2010). Beneficial mycorrhizal fungi directly or indirectly affect the morphological and physiological processes causing a rise in plant biomass and nutrient absorption which as a consequence, decreases the metal toxicity in host plants (Souza *et al.*, 2012; Kanwal *et al.*, 2015; Miransari, 2017). Additionally, fungi contribute towards metal dilution in plants by boosting the metal chelation in soil (Kapoor *et al.*, 2013; Audet, 2014). AMF complexes with Cd and Zn in mental hyphae (cell wall) and cortex, thereby confining heavy metal uptake by the plant, which eventually provides better growth and yield in plants (Andrade and Silveira, 2008; Garg and Chandel, 2012). Underlying mechanism behind the metal dilution by fungi is based upon boosting of plant growth and organic acid production that restricts the metal entry to outer regions of roots only, selective permeability of cell membrane, heavy metal chelation via metallothionine, confining heavy metals in hyphae or roots. Soil is also decontaminated by the secretion of a glycoprotein (Zhu *et al.*, 2017; Mishra *et al.*, 2017). Therefore, AM fungi are explored for the remediation of heavy metal contaminated sites, as plant-based strategy.

Rabie *et al.* (2005) conducted a green-house trialwith soil enriched with high levels of zinc, copper and lead. Results showed increased dry weight, chlorophyll and lipid content owing to mycorrhizal symbiosis between *Funneliformis mosseae* and wheat. Sonmez *et al.* (2009) studied the effect on wheat grown in high boron concentration inoculated with AMF and mycorrhizal fungi inoculation protected the plants by excessive uptake of metal ions. Shahabivand *et al.* (2012) inoculated *Funneliformis mosseae* on *Triticum aestivum* cv. Sardari39 plants, at four distinct concentrations of cadmium ranging 0, 0.3, 0.6 and 0.9mM and inoculation of fungus caused improved growth, performance index (PI) and chlorophyll content for all Cd concentrations as compared to the control. The colonization of *Vaccinium myrtillus*roots with metal tolerant Ericoid mycorrhizal fungus *Oidiodendron malus* reduced Cd content in Cd-exposed mycorrhizal roots as compared to non-mycorrhizal roots (Casarrubia *et al.*, 2020). Analysis of gene expression by using next generation sequencing technologies showed that Cd had a stronger impact on plant gene expression than symbiosis, whereas fungal gene expression was mainly regulated by symbiosis. The higher abundance of transcripts coding for stress-related proteins in non-mycorrhizal roots may be related to the higher Cd content.

Drought Stress

Drought is the most common abiotic stress that affects survival, growth and productivity of plants (Golldack *et al.*, 2014). Drought results in decreased stomatal conductance, chlorophyll content, water use efficiency and net photosynthesis rate (Abbaspour *et al.*, 2012). Besides this, drought induces production of reactive oxygen species (ROS) by impairing the electron transport system (Saraswathi and Paliwal, 2011). Drought also damages the stomata closing, which results in decreased uptake of CO_2 and an increase of reduction equivalents. With increasing accumulation of ROS, photosynthesis system gets damaged leading to production of secondary metabolites *i.e.*, isoprenoids, phenols, or alkaloids (Selmar and

Kleinwächter, 2013). Drought also affects enzymatic activity and reduces the nutrient uptake (Ahanger and Agarwal, 2017; Ahanger *et al.*, 2017a). However, different crops like wheat, barley, maize, soybean, strawberry and onion have adopted different drought stress mitigating strategies (Mena-Violante *et al.*, 2006; Ruiz-Lozano *et al.*, 2015; Yooyongwech *et al.*, 2016; Moradtalab *et al.*, 2019).

The presence of extra-radical hyphae fungi in the rhizosphere and association of AM fungi with plants, is one of the dominant mechanisms for mitigation of drought stress (Gianinazzi *et al.*, 2010; Gutjahr and Paszkowski, 2013; Rapparini and Penuelas, 2014; Calvo-Polanco *et al.*, 2016; Ruiz-Lozano *et al.*, 2016; Zhang *et al.*, 2016). Symbiosis between the plant and AMF regulates physio-biochemical processes including enhanced osmotic regulation (Kubikova *et al.*, 2001), Abscisic acid mediated regulation of stomata (Duan *et al.*, 1996), improved build-up of proline (Ruiz-Sánchez *et al.*, 2010; Yooyongwech *et al.*, 2013) or enhanced glutathione level (Rani, 2016). Fungal hyphae penetrate deep into the soil for the water for maintaining the plant water uptake. AMF produces trehalose sugar for preserving water loss and induces expression of various genes, 1-pyrroline-5-carboxylate synthetase (*p5cs*) gene translating rate limiting enzymes for proline (Farías-Rodríguez *et al.*, 1998; Jiménez Zacarías *et al.*, 2004) and the 9-cis-epoxycarotenoid dioxygenase (*nced*) genes encoding a key enzyme in the biosynthesis of abscisic acid (Li *et al.*, 2013). Production of abscisic acid and proline elevates the water flow through the cell, regulates the stomata closure by preventing water loss and results in antioxidant mechanism (Amanifar *et al.*, 2019; Jia-Dong *et al.*, 2019; Ouledali *et al.*, 2019). Moreover, AMF association with C3 and C4 plants *viz. Leymus chinensis* and *Hemarthria altissima*, respectively up regulates the antioxidant system providing enhanced plant growth and photosynthesis (Li *et al.*, 2019).

Effect of AM Fungi on Plant Defence and Resistance to Pathogens

AM fungi have also been found to suppress the diseases caused by phytopathogens and thus, they could be applied as biocontrol agents, an alternative to chemical pesticides (Bagy *et al.*, 2019; Nguvo and Gao, 2019). Plants in symbiosis with AM fungi produces isoflavonoid and phytoalexins, enzymes of the phenylpropanoid pathway, chitinases, β-1,3-glucanases, peroxidases, pathogenesis-related (PR) proteins, callose, hydroxyproline-rich glycoproteins (HRGP) and phenolics (Gianinazzi-Pearson *et al.*, 1994, Jung *et al.*, 2012), which inhibit the growth of pathogens (Morandi *et al.*, 1984). AM symbiosis lower down the yield loss due to diverse soil borne pathogems (De la Pena *et al.*, 2006; Sharma and Sharma, 2018). Pathogens as well as AM fungi are known to secrete effector molecules, which acts as defence inhibitors (Toro and Brachmann, 2016; Kamel *et al.*, 2017). Mycorrhiza enhances the plant resistance against pathogens by competing with pathogen for colonization sites on the plant roots. Mycorrhizal association in plants may have also enhanced resistance against the appearance of disease symptoms (Smith and Read, 1997a, b).

In addition, production of growth hormones like indole acetic acid, gibberellins and cytokinin is accelerated on symbiosis with AM fungi (Smith and Read, 1997a, b). Thus, mycorrhizae can be considered as an eco-friendly biological alternative of chemical compounds used for the growth, quality and productivity in crop production (Wood, 1992). AM fungal colonization of the roots increases the defence mechanism of plants against rhizosphere pathogens and thereby, AMF crops depicts enhanced disease resistance (Borowicz, 2001; Pozo and Azcon-Aguilar, 2007; Jung *et al.*, 2012; Cameron *et al.*, 2013). AMF establishment before any pathogen attack leads to systemic priming effect via reallocation of defence molecules (Jung *et al.*, 2012) in which plant defence response genes are expressed more rapidly in a high concentration than in non-inoculated plants (Song *et al.*, 2015). The protective action against pathogens by the plant occurs in a systemic action as observed by split root systems. This induction of systemic acquired resistance (SAR) usually results in improved plant health due to better nutrition (Jung *et al.*, 2012; Cameron *et al.*, 2013). Mycorrhizal plants induced for systemic acquired resistance or priming allows the mycorrhizal plants to respond quickly to pathogen attack (Conrath *et al.*, 2006).

Mycelia of AM fungi produces antimicrobial compounds when it comes in contact with the rhizospheric pathogens and resist the pathogen by competing for space and resources. Thus, inoculation with AM fungi results in plant protection. However, in few studies, it has been revealed that mycorrhizal crops may have no or negative effect like decreased growth and health of crop plants (Jacott *et al.*, 2017). AM colonization may also cause prominent alteration in morphology of root systems resulting in changed dynamics of microbial population, thus showing antagonistic activity towards certain root pathogens (Barea *et al.*, 2005). AM fungi as well as root pathogens compete for carbon source from root (Smith, 1987; Linderman, 1994); however, AM fungi has main access for the photosynthates and thus, the higher carbon demand may inhibit pathogen growth (Azcòn-Aguilar and Barea, 1996). Behn *et al.* (2008) observed a reduced disease incidence in spring wheat due to pathogen *Gaeumannomyces graminis* after inoculation with *Funneliformis mosseae,* and consequent increased the shoot biomass and seed production. The effect was increased after 4 weeks upon additional inoculation of *Pseudomonas fluorescens* strain RA56P along with the AM fungi. In another experiment, *Triticum aestivum* cv. Chinese spring inoculated with *F. mosseae* revealed a broad-spectrum defence (BSD) action against *Xanthomonas translucens*, and revealed a key role of genes and proteins in host immune system (Fiorilli *et al.*, 2018). Consortium of beneficial microorganisms showed a more pronounced effect than a single species of AM fungi in plant protection. Therefore, the concept of mycorrhizal inoculation with the plant associated beneficial microbes could be proved as a boom leading to improved productivity of plant communities (van der Heijden *et al.*, 1998; Maherali and Klironomos, 2007).

Inoculation Effects of Mycorrhizal Fungi and Plant Growth-promoting Bacteria

Soil harbours a variety of microbial diversity withwide range of interactions, which help in improving the growth and productivity of plants. Rhizosphere microbes assist plants either by directly involving in enhancing nutrient availability and phytohormone production, and indirectly via mitigation of biotic and abiotic stresses (Sindhu and Dadarwal, 1992; Enebe and Babalola, 2018; Bagyaraj and Jamaluddin, 2019; Sindhu and Sharma, 2019). The beneficial effects of rhizosphere microorganisms on crop plants due to microbial interactions can be clearly observed in different agro-ecosystems and natural ecosystems (Mendes *et al.*, 2013; Philippot *et al.*, 2013; Sindhu and Sharma, 2020; Sehrawat *et al.*, 2021a). Out of all rhizosphere colonizing microbes, arbuscular mycorrhizal fungi play dominant role in regulating plant growth as well as microbial population in soil (Table 2.1). Thus, mycorrhizal fungi and plant growth promoting bacteria play a key role in sustainable agricultural production and can enable growers to reduce or avoid the use of chemical fertilizers and pesticides (Raklami *et al.*, 2019; Phour *et al.*, 2020; Etesami *et al.*, 2021; Molina-Romero *et al.*, 2021).

Table 2.1: Inoculation Effect of Mycorrhizal Fungi to Plant Growth Promotion

Mycorrhizal Species	*Host Plant*	*Effect on Plant Growth*	*Reference*
Rhizophagus irregularis	*Cajanus cajan* L.	Enhanced nutrient level of P, N, Mg and Fe along with proline biosynthesis	Garg and Singh (2017)
Rhizophagus irregularis	*Solanum lycopersicum* L.	Increased fresh shoot and root weight, leaf area and number of leaves, growth hormones	Khalloufi *et al.* (2017)
Glomus mosseae	*Leymus chinensis*	Elevated rate of colonization, seedling weight, moisture content and N and P levels	Jixiang *et al.* (2017)
Funneliformis mosseae, Paraglomus occulatum	*Ponicirus trifoliata*	Enhanced water uptake, hyphal length and water potential of leaf	Zhang *et al.* (2018a)
Rhizophagus intaradices	*Triticum durum*	Increased grain biomass and copper, iron, manganes and zinc concentration in grains	Goicoechea *et al.* (2016, 2017)
Piriformaspora indica	*Ipomoea batatas* L.	Enhanced photosynthetic pigments, increased catalase activity in leaves and roots, improved biomass accumulation, increased resistance against herbivorous insects by inducing sporamin-dependent defense	Li *et al.* (2021)

Mycorrhizal Species	Host Plant	Effect on Plant Growth	Reference
Piriformospora indica	*Passiflora edulis*	Regulation of plant immunity and resistance in young plants, enhanced fruit size and quality, higher amounts of defense-related secondary metabolites,restriction of growth of herbivorous insect larvae	Yan *et al.* (2021)
Piriformospora indica	*Ponicirus trifoliata*	Better plant growth with increased plant height, leaf number and root biomass, higher N, P and Mg content in roots, increased soil phosphatase activity	Yang *et al.* (2021)

AM fungi possess efficient genes and metabolic pathways that help in increasing the cell number and activity of beneficial soil microbes, thus promoting the beneficial effects on plants (Gui *et al.*, 2017). Recently, plant growth promoting rhizobacteria (PGPR) has received attention because of the fact that their inoculation provides better plant growth by secreting plant growth-promoting substances and reducing the population and activity of the root pathogens (Choudhary and Sindhu, 2015; Goswami *et al.*, 2016; Olanrewaju *et al.*, 2017; Sehrawat *et al.*, 2021b). Another term "mycorrhiza helper bacteria" (MHB) is used for PGPR because of their ability to stimulate the growth of AMF hyphae leading toenhanced root penetration (Xie *et al.*, 2018). Some of the Gram-negative bacteria that depict interaction with the AMF include Proteobacteria (*Agrobacterium, Azotobacter, Azospirillum, Bradyrhizobium, Burkholderia, Enterobacter, Pseudomonas* and *Rhizobium*) (Ortao *et al.*, 2017). However, sufficient evidences are available in soil for the interaction of AMF with compatible Gram-positive bacteria. Signaling molecules like strigolactones for AMF and benzoxazinoids for PGPR are released by the plant to maintain a positive chemotaxis for the species-specific interactions (Akiyama *et al.*, 2004, Neal *et al.*, 2012). Synergistic relationship between AMF and PGPR supports plant growth more efficiently by increasing nutrients availability, permeability of hyphae leading to enhanced absorption of water and protection against biotic and abiotic stress as compared to individual inoculation (Divyananda *et al.*, 2006; Cely *et al.*, 2016) (Table 2.2).

Single Inoculation with Mycorrhizal Fungi

Mycorrhizal fungi boost the biomass of root and shoot, nutrient uptake and crop yield in different crop plants under different ecosystems (Table 2.1). *Piriformospora indica* fungus influence the growth of *Arabidopsis* and tobacco seedling by increasing the nitrate uptake via the enhanced gene expression of nitrate reductase, *Nia2* and the starch degrading enzyme glucan water dikinase, *SEX1* (Sherameti *et al.*, 2005). A very high activity of NADH-dependent nitrate reductase in the colonized roots resulted in a massive transfer of nitrogen into the aerial part of the seedlings (Johnson *et al.*, 2011). Thus, *Nia2, SEX1* and 2-nitropropane dioxygenase genes are

Table 2.2: Inoculation Effect of AMF on Host Plant Exposed to Various Biotic and Abiotic Stress

Host Plant	*Mycorrhizal Species*	*Mechanism*	*Effects on Plant Growth*	*References*
Solanum melongena	*Glomus intraradices*	Plant growth promotion	Enhanced crop growth, improved dry biomass, high water use efficiency and yield	Badr *et al.* (2020)
Lactuca sativa	*Claroideoglo musclaroideum*		Reduction of oxidative damage and enhanced growth development and yield	Santander *et al.* (2019)
Trigonella foenumgraecum L.	*Glomus monosporum, G. clarum, Gigaspora nigra* and *Acaulosporalaevis*	Heavy metal stress alleviation	Improved antioxidant enzymes activities and malondialdehyde content	Abdelhameed and Rabab (2019)
Cajanus cajan L.	*Rhizophagus irregularis*		Increased concentration of P, N, Mg, Fe, improved dry biomass and proline biosynthesis	Garg and Singh (2017)
Cucumis sativus L.	*Glomus etunicatum, G. intraradices* and *G. mosseae*	Salinity stress alleviation	Enhanced antioxidant enzymes activity, improved dry biomass and photosynthetic pigment	Hashem *et al.* (2018)
Lycopersicon esculentum L.	*Glomus fasciculatum*		Reduced pH and improved phosphatase enzyme activity	Sivakumar *et al.* (2019)
Solanum lycopersicum	*Funneliformis mosseae, F. geosporum, Claroideoglomus etunicatum, Glomus microaggregatum*	Drought stress alleviation	Improved water use efficiency, stomatal conductance and higher water potential of leaves	Bakr *et al.* (2018)
Solanum melongena	*Glomus intraradices*		Enhanced crop growth, dry biomass and yield	Badr *et al.* (2020)

upregulated in roots associated with *Piriformospora indica*. Also, stimulates gene expression of *uid A* encoding β-glucuronidase, which is regulated by *ArabidopsisNia2* promoter transgenic tobacco seedlings. Therefore, both nitrate and starch metabolic enzymes are regulated by *P. indica* along with inducing growth promoting effect (Sherameti *et al.*, 2005). *P. indica* is highly valuable in horticultural cultivation and interacts mutually with 30 families in plant kingdom. *P. indica* enhances plant growth, resistance to stress and phytochemical or secondary metabolite production (Verma *et al.*, 1998; Tsimilli-Michael and Strasser, 2013; Dolatabadi *et al.*, 2017).

Yao *et al.* (2005) studied the effect of two arbuscular mycorrhizal (AM) fungi, *Glomus intraradices* and *Gigaspora margarita* on endogenous indole acetic acid (IAA) and isopentenyl adenosine (iPAs) concentrations of inoculated plants. AM inoculation resulted in 13.5-30.1 per cent increased plant biomass along with significant increased, leaf number, leaflet number, total leaf area and first order lateral root number. Inoculation with AM fungi significantly increased P content and uptake of plants, and IAA concentration and iPAs concentration were also increased in mycorrhizal inoculated plant than non-mycorrhizal plants. In another study, Wahid *et al.* (2019) evaluated the effect of arbuscular mycorrhizal fungi on mung bean plant. For compost, cattle dung was taken and the cattle were fed with diet supplemented with powdered rock phosphate, thereafter phosphorus transfer and uptake were observed from alkaline soil. Results were positive including chlorophyll (a, b), shoot (3.04 g) and root (2.62 g) biomass, uptake of N (58.38 mg plant^{-1}) and P (4.61 mg plant^{-1}) and carotenoid contents. Further, 56 per cent root colonisation and nodulation, soil postharvest properties were improved with AM fungi inoculation in mung.

Similarly, for sweet sorghum inoculation with *Claroideoglomus etunicatum* BEG168 coupled with different concentration of MoS_2, resulted no toxic effect on plant biomass, root colonization and nitrogen and phosphorous uptake upon Mo accumulation (Shi *et al.*, 2020). AM fungi associated with sweet sorghum enhanced plant biomass,uptake of N, P, S and increased PSII efficiency and photosynthesis, and did not show any phytotoxic and fungitoxic effect. Adeyemi *et al.* (2021) carried out a pot experiment using three AM fungi *i.e.*, *Funneli formismosseae*, *Claroideoglomus etunicatum* and *Rhizophagus intraradices* along with different concentration of lead (0, 100 and 300 mg kg^{-1}) and soybean plant. Results depicted plant growth promotion and increased uptake of P and more accumulation of Pb rather than transfer in roots in comparison to non-inoculated plants. Treatment with *R. intraradices* was more tolerant to Pb toxicity than the other two AMF species and depicted better growth parameters.

Coinoculation Effects of Rhizospheric Microbes with Mycorrhizal Fungi

Synergistic interactions of beneficial microorganisms with mycorrhizal fungi were studied in tripartite symbiosis involving inoculation of crop plants with mycorrhizal fungi along with nitrogen-fixing or phosphorus-solubilizing bacteria

(Goel *et al.*, 2002; Etesami *et al.*, 2021). In majority of the conducted experiments, coinoculation of beneficial microorganisms improved the plant biomass and crop yield along with amelioration of the abiotic stresses (Sharma *et al.*, 2018; Sindhu *et al.*, 2020; Phour and Sindhu, 2020) (Table 2.3).

Toro *et al.* (1997) designed a microcosm experiment to evaluate nutrient cycling by inoculation of onion seedlings with AM fungus *Glomus intraradices* along with mycorrhiza helper bacteria *i.e. Enterobacter* sp. and *Bacillus subtilis.* Dual inoculation with *G. intraradices* and *B. subtilis* significantly increased biomass and N and P accumulation in plant tissues. In dually inoculated plants, at least 75 per cent of P was derived from the added rock phosphate, indicating that these mycorhizosphere interactions contributed to the biogeochemical P cycling and promoted a sustainable nutrient supply to plants. Coinoculation of AMF and PGPR *i.e.*, AMF + *Bacillus* sp. was demonstrated as an effective approach of promoting plant growth. Medina *et al.* (2003) carried out single or dual inoculation of three AMF with two of *Bacillus* strains strains (*Bacillus pumilus* and *B. licheniformis*) on *Medicago sativa*. Increased dry matter production was obtained from the dual inoculation of *Glomus deserticola* and *B. pumilus.* Tripartite symbiosis involving inoculation with AMF, *Frankia*, *Azospirillum* and *Phosphobacterium* in *Casuarina equisetifolia* resulted in increased height and biomass along with enhanced uptake of nutrients like N, P, K, Ca and Mg (Rajendran and Devaraj, 2004). Jaizme-Vega *et al.* (2006) studied the effect of combined inoculation of *Glomus mosseae* or *G. manihotis* (AMF) and a *Bacillus* sp. consortium against the nematode infestation damage in papaya. Significant drop in *Meloidogyne* infestation was observed that caused better plant growth. Yusran *et al.* (2009) evaluated the efficiency of *Pseudomonas* sp. and *Bacillus amyloliquefaciens* inoculation on *Fusarium oxysporum* f. sp. *radicis-lycopersici* infested tomato on mycorrhization, nutrient availability and plant growth.

P. indica depicts interactions with a diversity of microbes in soil, such as *Sebacina vermifera*, *Pseudomonas fluorescens*, *Chlamydomonas reinhardtii*, *Gaeumannomyces graminis*, *Aspergillus niger*, *Aspergillus sydowii* and *Rhizopus stolonifer* (Ansari *et al.*, 2014). In one of the coinoculation experiments in barley, no effect on plant growth was observed by using *P. indica* and *Azospirillum brasilense* Sp245, whereas coinoculation of *P. indica* with *Serratia liquefaciens* MG1, and *P. fluorescens* WS5 with *Burkholderia cepacia* LA3 exhibited inhibitory responses, but combined inoculation of *P. indica* with *Pseudomonas putida* IsoF showed increased stimulation of growth (Varma *et al.*, 2014).The inoculation of *P. indica* with several medicinal plants was reported to increase the production of secondary metabolites (Varma *et al.*, 2012b). Similarly, effect of *P. indica* inoculation on *Lantana camara* was evaluated on the induction of secondary metabolites like triterpenoids *i.e.*, oleanolic, ursolic, and betulinic acids. Results showed enhanced secondary metabolite production owing to plant fungus co-habitation. (Kumar *et al.*, 2016). Similarly, coinoculation of PGPR strains enhanced the dry matter and shoot nutrient. In another study, inoculation of consortium composed of PGPR rhizobia and AMF

Table 2.3: Coinoculation effect of AMF and PGPR in Protecting Plant against Biotic and Abiotic Atress

Mechanism	*Host Plant*	*AMF*	*PGPR Used for Coinoculation*	*References*
Plant growth promotion	Wheat and faba bean	*Glomus* sp., *Sclerocystis* sp. and *Acaulospora* sp.	*Ensifer meliloti* RhOF4 and *E. meliloti* RhOF155+*Acinetobacter* sp. BS17 and *Rahnella aquatilis* PGP27	Raklami *et al.* (2019)
	Lycopersicon esculentum	*Glomus mossea*	*Pseudomonas putida* and *Azotobacter chroococcum*	Ordookhani and Zare (2011)
Drought stress alleviating	*Dracocephalum moldavica*	*Claroideoglomus etunicatum*	*Micrococcus yunnanensis*	Ghanbarzadeh *et al.* (2020)
	Walnut seedling	*Glomus mosseae* and *G. etunicatum*	*Azotobacter chroococcum* and *Azospirillium lipoferum*	Behrooz *et al.* (2019)
Salinity stress alleviating	*Acacia gerrardii*	*Claroideoglomus etunicatum* + *Rhizophagus intraredices* + *Funneliformis mosseae*	*Bacillus subtilis*	Hashem *et al.* (2016)
	Casuarina obesa	*Rhizophagus fasciculatus* and *R. aggregatum*	*Pantoeaagglomerans* and *Bacillus* sp.	Diagne *et al.* (2020)
Heavy metal stress alleviating	*Trifolium repens*	AMF	*B. cereus* + *Candida parapsilosis*	Azcon *et al.* (2010)
	Pisum sativum L.	*Glomus* sp.	*Variovorax paradoxus* 5C-2 + *Rhizobium leguminosarum* bv.*viciae* RCAM1066	Belimov *et al.* (2020)
Pathogenicity against *Sclerotium rolfsii*	Onion	*Glomus mosseae* and *G. fasciculatum*	*Alcaligenes* sp., *Lichtheimia* sp. and *Brevibacterium* sp.	Nepomuceno *et al.* (2019)
Pathogenicity against root-knot nematode *Meloidogyne incognita*	Tomato	*Rhizophagus irregularis*	*Pseudomonas jessenii* R62 and *Pseudomonas synxantha* R81	Sharma and Sharma (2017)

enhanced shoot and root dry weight along with other growth parameters of faba bean and wheat (Raklami *et al.*, 2019). Synergistic relationship between phosphorus solubilizing bacteria (PSB) and AMF was reported in which PSB provided soluble phosphorus to AMF, which translocated it further into the host plant (Nacoon *et al.*, 2020). Etesami *et al.* (2021) reviewed the use of arbuscular mycorrhizal fungi (AMF), phosphate-solubilizing bacteria (PSB) and the addition of silicon (Si) as effective and economical ways to improve the availability and efficacy of P. Authors summarized the current knowledge concerning the interactions among AMF, PSB and Si, and concluded that the combined strategy of using Si along with AMF and PSB may be highly useful in improving the P availability and its uptake by plants compared to using either of them alone in P deficient soils.

Selection of Efficient AM Fungi

Selection of mycorrhizal fungi for field application must be based upon the fact that it must increase the nutrient uptake by plant along with stimulation of plant growth. *Glomus fasciculatum, G. mossae, G. etunicatum, G. tenue* and *Gigaspora margarita* are some of the spore types, which are usually applied in field. Fungi are screened for its infectiveness, effectiveness, colonization, competitiveness and persistence in rhizosphere (Sindhu and Dadarwal, 2000). Equally important characteristics for the selection of efficient AM fungi are its capability to stimulate plant growth, nutrient uptake and crop yield. Govinda Rao *et al.* (1983) provided a method for selection of fungi for symbiosis and recommended use of test host in pot culture followed by microplot and field trials.

For selection of efficient mycorrhizal fungi, seedlings are grown under unsterile soil in which a variety of AM fungi are added. Based on most probable number count, same number of infective propagules must be added. After the infection, seedlings were transferred to experimental field plots for plant growth observation. This method is adopted by farmers in India and other developing countries. This procedure is enormously good to select fungi for the pre-inoculation of transplanted crops grown under unsterile soil conditions. Therefore, selected fungi will effectively compete with indigenous AM fungi. This procedure has allowed the selection of fungi for economically important forest trees like *Leucaena, Tamarindus indica, Casuarina equisetifolia*, root stocks of citrus and mango. AM fungi inoculation in citrus root led to 4 to 6 months early budding, which has great impact in citriculture.

Management and Application of Mycorrhizae: Problems and Potential

Improved plant growth obtained in pot studies with mycorrhizal inoculation in less fertile soil, in the 1980s has increased the attention of scientists and farmers for using this organism as biofertilizers. Mycorrhiza fungi can either induce positive or negative effects on crop production and factors so involved includes host dependency on mycorrhizal fungi, soil nutrient composition, the fungal inoculum efficacy. Agriculture management practices like crop rotation and tillage and fallowing has a negative impact on mycorrhizal fungi in the field. With the current state of technology, inoculation is most feasible for transplanted crops and in areas where soil disturbance has greatly reduced the native inoculum potential. The first step in any inoculation program is to obtain an isolate that is both infective and able to penetrate and spread in the root along with its potential to stimulate the plant growth, apart from providing tolerance to stress to host. Different isolates will have varying characteristics thus screening is essential to get the best isolate. Screening under actual cropping conditions is best because indigenous mycorrhizal fungi, pathogens and soil chemical and physical properties will influence the results of symbiotic parameters. Alimi *et al.* (2021) reported that the fungal diversity and spatial structure of AM fungal communities were significantly influenced by soil properties. They identified twenty AM fungal species belonging to ten genera from the rhizosphere of eleven indigenous legume species of South Africa. *Glomus* and *Acaulospora* were the dominant genera, while *Acaulospora colombiana, A. mellea,* and *Claroideoglomus etunicatum* were found as the prevalent species. The AM fungal spore density, diversity and community structure differed markedly among the legume species.

The process of isolation and inoculum preparation of ectomycorrhizal fungi and arbuscular mycorrhizal fungi is linked with different problems. As ectomycorrhizal fungi can be cultured using artificial media, therefore, isolation can be done by employing sporocarps or mycorrhizal short roots after disinfection on an agar growth medium. The produced biomass can be used as inoculum. But inoculum often consists of the fungal material mixed with a carrier or bulking material such as peat. However, in AM fungi is in obligatory association with host and thus can not be cultured in laboratory. Spores can be retrieved from soil by sieving and latter applied for pot cultures on host plant either in sterilised soil or on plant growth medium. Alternatively, aeroponic or hydroponic system are likely as they don't require any supporting substratum thus permitting breaking of colonised roots into a high propagule number inoculum. The goal of inoculation is to introduce propagules of selected mycorrhizal fungi into the rhizosphere of target plant. The most common method is to place inoculum below the seed or seedling prior to planting. Alternative strategies include coating of seed or encasing somatic embryos with inoculum. Such technologies should allow for more widespread use of mycorrhizal fungi as inoculum in the future

Conclusion

Nitrogen, phosphorus and potassium are vital nutrients in crop nutrition, which are applied to crops in the form of chemical fertilizers. However, adverse environmental effects of chemical-based fertilizers have necessitated in finding a sustainable approach for efficient nutrient availability in agriculture to meet the ever-increasing global demand of food. Mycorrhizae develop symbiotic associations between root of plant and fungi involving two-way exchange of nutrients, thereby establishing a critical linkage between the plant root and soil. This mycorrhizal association confers plants better stress tolerance and competitive advantage over non-mycorrhizal plants. These mycorrhizal association differes in the morphology, form and functions. Ectomycorrhizal fungi generally include basidiomycetes and grow in between root cortical cells to form hartig net. Arbuscular mycorrhiza fungi are mostly from order Glomales and differentiates to branched structures called arbuscules in cells of root cortex. Nutrient uptake by mycorrhizal fungi leads to enhanced plant growth and improves crop yield dramatically. However, the positive or negative effects of mycorrhizal fungi on crop production depend on various factors including host dependency on mycorrhizal fungi, soil nutrient composition and the fungal inoculum efficacy. Agriculture management practices like crop rotation, tillage and fallowing may result in potential reduction of native mycorrhizal fungi inoculum in the field. However, the combination inoculation of AMFand PGPR may be utilized as a strategy for improving the availability of nutrients leading to enhanced plant growth in P-deficient soils and minimizing chemical fertilization to exercise sustainable agriculture. Thus, inoculation with mycorrhizal fungi in the field could be a beneficial step for enhancing crop productivity and has emerged as a promising technique that may assist in solving problems encountered with the crop production economy and food shortages.

REFERENCES

Abbaspour, H., Saeidi-Sar, S., Afshari, H. and Abdel-Wahhab, M.A. 2012. Tolerance of mycorrhiza infected pistachio *Pistacia vera* L. seedling to drought stress under glasshouse conditions. *J. Plant Physiol.* 169: 704–709.

Abdelhameed, R.E. and Rabab, A.M. 2019. Alleviation of cadmium stress by arbuscular mycorrhizal symbiosis. *Int. J. Phytoremed.* doi: 10.1080/15226514.2018.1556584

Adeyemi, N.O., Atayese, M.O., Sakariyawo, O.S., Azuz, J.O. and Ridwan, M. 2021. Arbuscular mycorrhizal fungi species differentially regulate plant growth, phosphorus uptake and stress tolerance of soybean in lead contaminated soil. *J. Plant Nut.*44 (11): 1633–1648.

Agerer, R. 1991. Characterization of ectomycorrhiza. *Methods Microbiol.* 23: 25–73.

Ahanger, M.A. and Agarwal, R.M. 2017. Potassium up-regulates antioxidant metabolism and alleviates growth inhibition under water and osmotic stress in wheat *Triticum aestivum* L. *Protoplasma* 254(4): 1471–1486.

Ahanger, M.A., Alyemeni, M.N., Wijaya, L., Alamri, S.A., Alam, P. and Ashraf, M. 2018. Potential of exogenously sourced kinetin in protecting *Solanum lycopersicum* from NaCl-induced oxidative stress through up-regulation of the antioxidant system, ascorbate–glutathione cycle and glyoxalase system. *PLoS ONE* 13(9): e0202–e0175.

Ahanger, M.A., Tittal, M., Mir, R.A. and Agarwal, R.M. 2017a. Alleviation of water and osmotic stress-induced changes in nitrogen metabolizing enzymes in *Triticum aestivum* L. cultivars by potassium. *Protoplasma* 254(5): 1953–1963.

Ahanger, M.A., Tomar, N.S., Tittal, M., Argal, S. and Agarwal, R.M. 2017b. Plant growth under water/salt stress: ROS production; antioxidants and significance of added potassium under such conditions. *Physiol. Mol. Biol. Plants* 23(4): 731–744.

Ait-El-Mokhtar, M., Laouane, R.B., Anli, M., Boutasknit, A., Wahbi, S. and Meddich, A. 2019. Use of mycorrhizal fungi in improving tolerance of the date palm *Phoenix dactylifera* L. seedlings to salt stress. *Sci. Hori.* 253: 429–438.

Akiyama, K., Matsuzaki, K. and Hayashi, H. 2005. Plant sesquiterpenes induce hyphal branching in arbuscular mycorrhizal fungi. *Nature* 435(7043): 824–827.

Allen, E.B., Allen, M.F., Helm, D.J., Trappe, J.M., Molina, R. and Rincon, E. 1995. Patterns and regulation of mycorrhizal plant and fungal diversity. *Plant Soil*170(1): 47–62.

Allen, M.F. 1999. The Ecology of Mycorrhizae; Cambridge Univ. Press: Cambridge, UK, pp. 189.

Alimi, A.L., Adeleke, R. and Moteetee, A. 2021. Soil environmental factors shape the rhizosphere arbuscular mycorrhizal fungal communities in South African indigenous legumes (*Fabaceae*). *Biodiversitas* 22(5): 2466-2476. doi: 10.13057/biodiv/d220503

Amanifar, S., Khodabandeloo, M., Fard, E.M., Askari, M.S. and Ashrafi, M. 2019. Alleviation of salt stress and changes in glycyrrhizin accumulation by arbuscular mycorrhiza in liquorice *Glycyrrhiza glabra* grown under salinity stress. *Environ. Exp. Bot.* 160: 25–34.

Andrade, S.A.L. and Silveira, A.P.D. 2008. Mycorrhiza influence on maize development under Cd stress and P supply. *Braz. J. Plant Physiol.* 20(1): 39–50.

Ansari, M.W., Gill, S.S. and Tuteja, N. 2014. *Piriformospora indica* a powerful tool for crop improvement. *Proc. Indian Natl. Sci. Acad.* 80: 317–324.

Audet, P. (2014). "Arbuscular mycorrhizal fungi and metal phytoremediation: eco-physiological complementarity in relation to environmental stress," In: Emerging technologies and management of crop stress tolerance. Ahmad, P. and Rasool, S. (eds). (San Diego: Academic Press), pp. 133–160. doi: 10.1016/B978-0-12-800875-1.00006-5

Azcon, R., Perálvarez, M.C., Roldán, A. and Barea, J. 2010. Arbuscular mycorrhizal fungi, *Bacillus cereus*, and *Candida parapsilosis* from a multi contaminated soil alleviate metal toxicity in plants. *Microbiol. Ecol.* 59: 668–77.

Azcon-Aguilar, C. and Barea, J.M. 1996. Arbuscular mycorrhizas and biological control of soil-borne plant pathogens: An overview of the mechanisms involved. *Mycorrhiza* 6: 457–464.

Badr, M.A., El-Tohamy, W.A., Abou-Hussein, S.D. and Gruda, N.S. 2020. Deficit irrigation and arbuscular mycorrhiza as a water-saving strategy for eggplant production. *Horticulturae* 6: 45.

Bagy, H.M.M.K., Hassan, E.A., Nafady, N.A. and Dawood, M.F.A. 2019. Efficacy of arbuscular mycorrhizal fungi and endophytic strain *Epicoccum nigrum* ASU11 as biocontrol agents against blackleg disease of potato caused by bacterial strain *Pectobacterium carotovora* subsp. *atrosepticum* PHY7. *Biol. Control* 134: 103–113.

Bagyaraj, D.J. 1992. Methods in Microbiology, Academic Press, London.

Bagyaraj, D.J. and Jamaluddin. 2019. Microbes for plant stress management. NIPA Publishers Co-Published with USA, FL: CRC Press: NIPA Publishers.

Bakr, J., Peék, Z., Helyes, L. and Posta, K. 2018. Mycorrhizal inoculation alleviates water deficit impact on field-grown processing tomato. *Polish J. Environ. Stud.* 27: 1949–1958.

Balestrini, R. and Bonfante, P. 2014. Cell wall remodeling in mycorrhizal symbiosis: a way towards biotrophism. *Front. Plant Sci.* 5: 237.

Bardgett, R.D. and van der Putten, W.H. 2014. Belowground biodiversity and ecosystem functioning. *Nature* 515: 505–511.

Barea, J.M., Pozo, M.J., Azcon, R. and Azcon-Aguilar, C. 2005. Microbial co-operation in the rhizosphere. *J. Exp. Bot.* 56: 1761–1778.

Basu, S., Rabara, R.C. and Negi, S. 2018. AMF: The future prospect for sustainable agriculture. *Physiol. Mol. Plant Physiol.* 102: 36–45.

Becard, G., Taylor, L.P., Douds, D.D., Pfeffer, P.E. and Doner, L.W. 1995. Flavonoids are necessary plant signal compounds in arbuscular mycorrhizal symbioses. *Mol. Plant-Microbe Interact.* 8: 252–258.

Behn, O. 2008. Influence of *Pseudomonas fluorescens* and arbuscular mycorrhiza on the growth, yield, quality and resistance of wheat infected with *Gaeumannomycesgraminis*. *J. Plant Dis. Prot.* 115: 4–8.

Behrooz, A., Vahdati, K., Rejali, F., Lotfi, M., Sarikhani, S. and Leslie, C. 2019. Arbuscular mycorrhiza and plant growth-promoting bacteria alleviate drought stress in walnut. *Hort. Sci.* 54: 1087–1092.

Belimov, A.A., Shaposhnikov, A.I., Azarova, T.S., Makarova, N.M., Safronova, V.I. and Litvinskiy, V.A. 2020. Microbial consortium of PGPR, rhizobia and arbuscular mycorrhizal fungus makes pea mutant SGECdt comparable with Indian mustard in cadmium tolerance and accumulation. *Plants* 9: 1–21.

Bender, S.F. and van der Heijden, M.G.A. 2015. Soil biota enhance agricultural sustainability by improving crop yield, nutrient uptake and reducing nitrogen leaching losses.*J. Appl. Ecol.* 52: 228–239.

Berruti, A., Lumini, E., Balestrini, R. and Bianciotto, V. 2015. Arbuscular mycorrhizal fungi as natural biofertilizers: let's benefit from past successes. *Front Microbiol.* 6: 155.

Besserer, A., Becard, G., Jauneae, A., Roux, C. and Sejalon-Delmas, N. 2008. GR24, a synthetic analogue of strigolactones, stimulates the mitosis and growth of the arbuscular mycorrhizal fungus *Gigasporaroseae* by boosting its energy metabolism. *Plant Physiol.* 148: 402–413.

Bever, J.D., Bauer, J.T., House, G.L., Cheeke, T., Koziol, L., Tipton, A., Schultz, P.A., Copprick, P.R., Duell, E.B., Zaiger, K.L. and Wilson, G.W. 2017. Soil microbial communities: Critical roles in control of non-native invasive species and restoration of ecosystem functions. Indiana University-Bloomington Bloomington United States.

Bhowmik, S.N. and Das, A. 2018. Biofertilizers: a sustainable approach for pulse production. In: Legumes for Soil Health and Sustainable Management. Springer, Singapore, pp. 445–485.

Bonfante, P. and Genre, A. 2010. Mechanisms underlying beneficial plant—Fungus interactions in mycorrhizal symbiosis. *Nat. Commun.* 1: 48.

Borowicz, V.A. 2001. Do arbuscular mycorrhizal fungi alter plant-pathogen relations. *Ecology*82: 3057–3068.

Breuillin-Sessoms, F., Floss, D.S., Gomez, S.K., Pumplin, N., Ding, Y. and LevesqueTremblay, V. 2015. Suppression of arbuscule degeneration in *Medicago truncatula* phosphate transporter4 mutants is dependent on the ammonium transporter 2 family protein AMT2; 3. *Plant Cell* 274: 1352–66.

Brooker, R.W., Bennett, A.E., Cong, W.F., Daniell, T.J., George, T.S., Hallett, P.D., Hawes, C., Iannetta, P.P., Jones, H.G., Karley, A.J. and Li, L. 2015. Improving intercropping: a synthesis of research in agronomy, plant physiology and ecology. *New Phytol.* 206(1): 107–117.

Brundrett, M.C. 2009. Mycorrhizal associations and other means of nutrition of vascular plants: understanding the global diversity of host plants by resolving conflicting information and developing reliable means of diagnosis. *Plant Soil* 320(1): 37–77.

Brundrett, M.C. and Tedersoo, L. 2018. Evolutionary history of mycorrhizal symbioses and global host plant diversity. *New Phytol.* 220: 1108–1115.

Bucher, M. 2007. Functional biology of plant phosphate uptake at root and mycorrhiza interfaces, *New Phytol.* 173(1): 11–26.

Calvo-Polanco, M., Sánchez-Castro, I., Cantos, M., García, J.L., Azcón, R., Ruiz-Lozano, J.M., Beuzón, C.R. and Aroca, R. 2016. Effects of different arbuscular mycorrhizal fungal backgrounds and soils on olive plants growth and water relation properties under well-watered and drought conditions. *Plant Cell Environ.* 39: 2498–2514.

Cameron, D.D., Neal, A.L., Van Wees, S.A. and Ton, J. 2013. Mycorrhiza-induced resistance: More than the sum of its parts? *Trends Plant. Sci.* 18: 539–545.

Campos, P., Borie, F., Cornejo, P., López-Ráez, J.A., López-García, Á. and Seguel, A. 2018. Phosphorus acquisition efficiency related to root traits: is mycorrhizal symbiosis a key factor to wheat and barley cropping? *Front. Plant Sci.* 9: 752.

Casarruba, S., Martino, E., Daghino, S., Kohler, A., Morin, E., Khouja, H-R., Murat, C., Barry, K.W., Lindquist, E.A., Martin, F.M. and Perotto, S. 2020. Modulation of plant and fungal gene expression upon Cd exposure and symbiosis in ericoid mycorrhizal *Vaccinium myrtillus. Front. Microbiol.* 11: 341.

Cavagnaro, T.R., Bender, S.F., Asghari, H.R. and van der Heijden, M.G.A. 2015. The role of arbuscular mycorrhizas in reducing soil nutrient loss. *Trends Plant Sci.* 20: 283–290.

Cavagnaro, T.R., Jackson, L.E., Six, J., Ferris, H., Goyal, S. and Asami, D. 2006. Arbuscular mycorrhizas, microbial communities, nutrient availability, and soil aggregates in organic tomato production. *Plant Soil* 282: 209–225.

Cely, M.V.T., Siviero, M.A., Emiliano, J., Spago, F.R., Freitas, V.F., Barazetti, A.R., Goya, E.T., de Souza Lamberti, G., dos Santos, I.M.O., de Oliveira, A.G. and Andrade, G. 2016. Inoculation of *Schizolobium parahyba* with mycorrhizal fungi and plant growth-promoting rhizobacteria increases wood yield under field conditions. *Front. Plant Sci.* 7: 1708.

Choudhary, S.R. and Sindhu, S.S.2015. Suppression of *Rhizoctonia solani* root rot disease of cluster bean (*Cyamposis tetragonoloba*) and plant growth promotion by rhizosphere bacteria. *Plant Pathol. J.* **14:** 48-57.

Christie, P., Li, X. and Chen, B. 2004. Arbuscular mycorrhiza can depress translocation of zinc to shoots of host plants in soils moderately polluted with zinc. *Plant Soil* 261: 209–217.

Conrath, U., Beckers, G.J.M., Flors, V., Garcia-Agustin, P., Jakab, G. and Mauch, F. 2006. Priming: getting ready for battle. *Mol. Plant-Microbe Interact.* 19: 1062–1071.

Danesh, Y.R. 2015. The *Piriformospora indica*, magic fungus and its role on sustainable agriculture. *Intern. J. Agric. Technol.* 11(2): 517–523.

Das, A., Prasad, R., Srivastava, R.B., Deshmukh, S., Rai, M.K. and Varma, A. 2013. Cocultivation of *Piriformospora indica* with medicinal plants: case studies. In *Piriformospora indica,* Springer, Berlin, Heidelberg, pp. 149–171.

Das, A., Sherameti, I. and Varma, A. 2012. Contaminated soil: physical, chemical and biological components. In Bio-geo interactions in metal-contaminated soils, Springer, Berlin, Heidelberg, pp. 1–15.

De la Pena, E., Echeverria, S.R., van der Putten, W.H., Freitas, H. and Moens, M. 2006. Mechanism of control of root-feeding nematodes by mycorrhizal fungi in the dune grass *Ammophila arenaria. New Phytol.* 169: 829–840.

Dearnaley, J.D., Martos, F. and Selosse, M.A. 2012. 12 orchid mycorrhizas: molecular ecology, physiology, evolution and conservation aspects. In *Fungal associations,* Springer, Berlin, Heidelberg, pp. 207–230.

Den Camp, R.O., Streng, A., De Mita, S., Cao, Q., Polone, E., Liu, W., Ammiraju, J.S.,Kudrna, D., Wing, R., Untergasser, A. and Bisseling, T. 2011. Lys M-type mycorrhizal receptor recruited for rhizobium symbiosis in non-legume *Parasponia*. *Science*331(6019): 909–912.

Denton, B. 2007. Advances in phytoremediation of heavy metals using plant growth promoting bacteria and fungi MMG 445. *Basic Biotechnol.* 3: 1–5.

Diagne, N., Ndour, M., Djighaly, P.I., Ngom, D., Ngom, M.C.N., Ndong, G., Svistoonoff, S. and Cherif-Silini, H. 2020. Effect of plant growth promoting rhizobacteria (PGPR) and arbuscular mycorrhizal fungi (AMF) on salt stress tolerance of *Casuarina obesa* (Miq.) *Front. Sustain.Food Syst.* 4: 601004.

Dickson, S., Smith, F.A. and Smith. S.E. 2007. Structural differences in arbuscular mycorrhizal symbioses: more than 100 years after Gallaud, where next? *Mycorrhiza*17(5): 375–393.

Divyananda, M.C., Harinikumar, K.M. and Bagyaraj, D.J. 2006. Influence of *Glomus leptotichum* and plant growth promoting rhizo microorganisms PGPRs on growth and nutrition of teak. *J. Soil Biol. Ecol.* 26: 66–74.

Dolatabad, H.K., Goltapeh, E.M., Safari, M. and Golafaie, T.P. 2017. Potential effect of *Piriformospora indica* on plant growth and essential oil yield in *Mentha piperita*. *Plant Pathol.* 7: 96–104.

Dong, Y., Zhu, Y.G., Smith, F.A., Wang, Y. and Chen, B. 2008. Arbuscular mycorrhiza enhanced arsenic resistance of both white clover *Trifolium repens* L. and ryegrass *Lolium perenne* L. plants in an arsenic-contaminated soil. *Environ. Pollut.* 155: 174–181.

Duan, X., Neuman, D.S., Reiber, J.M., Green, C.D., Arnold, M. and Saxton, A.M. 1996. Mycorrhizal influence on hydraulic and hormonal factors implicated in the control of stomatal conductance during drought. *J. Exp. Bot.* 47(303): 1541–1550.

Duhamel, M. and Vandenkoornhuyse, P. 2013. Sustainable agriculture: possible trajectories from mutualistic symbiosis and plant neodomestication. *Trends Plant Sci.* 18: 597–600.

Edwards, C.A. 2020. Sustainable agricultural systems. CRC Press.

Egger, K.N. and Fortin, J.A. 1988. Ectendomycorrhzae: Diversity and classification. In: Lalonde, M. and Oiche, Y. (Eds.), Canadian workshop on Mycorrhzae in forestry. C.R.B.F. Universite Laval; Ste-Foy, pp. 113–114.

Enebe, M.C. and Babalola, O.O. 2018. The infuence of plant growth-promoting rhizobacteria in plant tolerance to abiotic stress: a survival strategy. *Appl. Microbiol. Biotechnol.* 102: 7821–35.

Etasami, H. and Glick, B.R. 2020. Halotolerant plant growth-promoting bacteria: Prospects for alleviating salinity stress in plants. *Environ. Exp. Botany* 178: 104–124.

Etesami, H, Jeong, B.R. and Glick, B.R. 2021. Contribution of arbuscular mycorrhizal fungi, phosphate–solubilizing bacteria, and silicon to P uptake by plant. *Front. Plant Sci.* 12: Article 699618. doi: 10.3389/fpls.2021.699618doi: 10.3389/fpls.2021.699618

Evelin, H., Kapoor, R. and Giri, B. 2009. Arbuscular mycorrhizal fungi in alleviation of salt stress: a review. *Ann. Bot.* 104: 1263–1280.

Fakhro, A., Andrade-Linares, D.R., von Bargen, S., Bandte, M., Büttner, C., Grosch, R., Schwarz, D. and Franken, P. 2010. Impact of *Piriformospora indica* on tomato growth and on interaction with fungal and viral pathogens. *Mycorrhiza* 20(3): 191–200.

Farias-Rodriguez, R., Mellor, R.B., Arias, C. and Pena Cabriales, J.J. 1998. The accumulation of trehalose in nodules of several cultivars of common bean *Phaseolus vulgaris* and its correlation with resistance to drought stress. *Physiol. Plant.* 102: 353–359.

Fiorilli, V., Vannini, C., Ortolani, F., Garcia-Seco, D., Chiapello, M., Novero, M., Domingo, G., Terzi, V., Morcia, C. and Bagnaresi, P. 2018. Omics approaches revealed how arbuscular mycorrhizal symbiosis enhances yield and resistance to leaf pathogen in wheat. *Sci. Rep.* 8: 9625.

Floss,D.S.,Levy,J.G.,L´evesque-Tremblay,V.,Pumplin,N. andHarrison,M.J.2013.DELLA protein regulate arbuscule formation in arbuscular mycorrhizal symbiosis. *Proc. Natl. Acad. Sci. U.S.A.* 110(51): 5025–5034.

Garcia, S.N., Osburn, B.I. and Jay-Russell, M.T. 2020. One health for food safety, food security, and sustainable food production.*Front. Sustain. Food Syst.* 4: 1.

Garcia-Garrido, J.M. and Ocampo, J.A. 2002. Regulation of the plant defence response in arbuscular mycorrhizal symbiosis. *J. Exp. Bot.* 53(373): 1377–1386.

Garg, N. and Chandel, S. 2012. Role of arbuscular mycorrhizal AM fungi on growth, cadmium uptake, osmolyte, and phytochelatin synthesis in *Cajanus cajan* L. Millsp. under NaCl and Cd stresses. *J. Plant Growth Regul.* 31(3): 292–308.

Garg, N. and Singh, S. 2017. Arbuscular mycorrhiza *Rhizophagus irregularis*, and silicon modulate growth, proline biosynthesis and yield in *Cajanus cajan* L. Millsp. (pigeon pea) genotypes under cadmium and zinc stress. *J. Plant Growth Regul.* 37: 1–18.

Genre, A., Chabaud, M., Timmers, T., Bonfante, P. and Barker, D.G. 2005. Arbuscular mycorrhizal fungi elicit a novel intracellular apparatus in *Medicago truncatula* root epidermal cells before infection. *Plant Cell* 17(12): 3489–3499.

Gerdemann, J.W. 1975. The Development and Function of Roots, Academic Press, New York, USA, pp. 575–591.

Ghanbarzadeh, Z., Mohsenzadeh, S., Rowshan, V. and Zarei, M. 2020. Mitigation of water deficit stress in *Dracocephalum moldavica* by symbiotic association with soil microorganisms. *Sci. Hortic.* 272: 109549.

Gianinazzi, S., Golotte, A., Binet, M.N., van Tuinen, D., Redecker, D. and Wipf, D. 2010. Agroecology: the key role of arbuscular mycorrhizas in ecosystem services. *Mycorrhiza* 20: 519–530.

Gianinazzi-Pearson, V., Gollotte, A., Dumas-Gaudot, E., Franken, P. and Gianinazzi, S. 1994. Gene expression and molecular modifications associated with plant responses to infection by arbuscular mycorrhizal fungi. In Advances in Molecular Genetics of Plant-Microbe Interactions; Daniels, M., Downic, J.A. and Osbourn, A.E.(eds.), Kluwer Academic Publishers: Dordrecht, The Netherlands, pp. 179–186.

Giovannetti, M., Sbrana, C. and Logi, C. 1994. Early processes involved in host recognitionby arbuscular mycorrhizal fungi. *New Phytol.* 127(4): 703–709.

Godfray, H.C.J., Beddington, J.R., Crute, I.R., Haddad, L., Lawrence, D., Muir, J.F., Pretty, J., Robinson, S., Thomas, S.M. and Toulmin, C. 2010. Food security: the challenge of feeding 9 billion people. *Science*327(5967): 812–818.

Goel A.K., Sindhu S.S. and Dadarwal K.R. 2002. Stimulation of nodulation and plant growth of chickpea (*Cicer arietinum*) by *Pseudomonas* spp. antagonistic to fungal pathogens. *Biol. Fertil Soils***36:** 391-396.

Goh, H.H., Sloan, J., Malinowski, R. and Fleming, A. 2014. Variable expansin expression in *Arabidopsis* leads to different growth responses. *J. Plant Physiol.* 171(3-4): 329–339.

Goicoechea, N. and Antol, M.C. 2017. Increased nutritional value in food crops. *Microbiol. Biotechnol.* 10: 1004–1007.

Goicoechea, N., Bettoni, M., Fuertes-Mendiza´bal, T., Gonzalez-Murua, C. and Aranjuelo, I. 2016. Durum wheat quality traits affected by mycorrhizal inoculation, water availability and atmospheric CO_2 concentration. *Crop Past. Sci.* 67: 147–155.

Golldack, D., Li, C., Mohan, H. and Probst, N. 2014. Tolerance to drought and salt stress in plants: Unraveling the signaling networks. *Front. Plant Sci.* 5: 151.

Goswami, D., Thakker, J.N. and Dhandhukia, P.C. 2016. Portraying mechanics of plant growth promoting rhizobacteria PGPR: a review. *Cogent. Food Agric.* 2: 1127500.

Govinda Rao, Y.S., Bagyraj, D.J. and Rai, P.V. 1983. Selection of an efficient VA mycorrhizal fungus for finger millet: II. Screening under field conditions. *Zentralblatt fur Mikrobiol.* 138(6): 415–419.

Gui, H., Hyde, K., Xu, J. and Mortimer, P. 2017. Arbuscular mycorrhiza enhances the rate of litter decomposition while inhibiting soil microbial community development. *Sci. Rep.* 7: 42184.

Gutjahr, C. and Paszkowski, U. 2013. Multiple control levels of root system remodeling in arbuscular mycorrhizal symbiosis. *Front. Plant Sci.* 4: 204.

Gutjahr, C., Gobbato, E., Choi, J., Riemann, M., Johnston, M.G., Summers, W.,Carbonnel, S., Mansfield, C., Yang, S.Y., Nadal, M. and Acosta, I. 2015. Rice perception of symbiotic arbuscular mycorrhizal fungi requires the karrikin receptor complex. *Science* 350(6267): 1521–1524.

Hakeem, K.R., Akhtar, M.S. and Abdullah, S.N.A. 2016. Plant, soil and microbes -Vol 1, Implications in crop science. Springer International Publishing AG, Gewerbestrasse 11, 6330 Cham, Switzerland, pp. 366.

Harrison, M.J. and van Buuren, M.L. 1995. A phosphate transporter from the mycorrhizal fungus *Glomus versiforme. Nature* 3786557: 626–9.

Harrison, M.J., Dewbre, G.R. and Liu, J. 2002. A phosphate transporter from *Medicago truncatula* involved in the acquisition of phosphate released by arbuscular mycorrhizal fungi. *Plant Cell* 1410: 2413–2.

Hart, M.M., Aleklett, K., Chagnon, P.L., Egan, C., Ghignone, S., Helgason, T., Lekberg, Y., Öpik, M., Pickles, B.J. and Waller, L. 2015. Navigating the labyrinth: a guide to sequence based, community ecology of arbuscular mycorrhizal fungi. *New Phytol.* 207(1): 235–247.

Hasanuzzaman, M., Gill, S.S. and Fujita, M. 2013. "Physiological role of nitric oxide in plants grown under adverse environmental conditions," in Plant acclimation to environmental stress. Tuteja, N.and Gill, S.S(eds.), NY: Springer Science+Business Media, pp. 269–322.

Hashem, A., Adb-Allah, E.F., Alqarawi, A.A., Al-Huqail, A.A. and Shah, M.A. 2016. Induction of osmoregulation and modulation of salt stress in *Acacia gerrardii*Benth by arbuscular mycorrhizal fungi and *Bacillus subtilis* (BERA 71). *Biomed. Res. Int.*https: //doi.org/10.1155/2016/6294098.

Hashem, A., Alqarawi, A.A., Radhakrishnan, R., Al-Arjani, A.F., Aldehaish, H.A. and Egamberdieva, D. 2018. Arbuscular mycorrhizal fungi regulate the oxidative system, hormones and ionic equilibrium to trigger salt stress tolerance in *Cucumis sativus* L. *Saudi J. Biol. Sci.* 25(6): 1102–1114.

Högberg, P. 1982. Mycorrhizal associations in some woodland and forest trees and shrubs in Tanzania. *New Phytol.* 92: 407–415.

Hua, T., Zhang, R., Sun, H. and Liu, C. 2020. Alleviation of boron toxicity in plants: Mechanisms and approaches. *Crit. Rev. Environ. Sci. Technol.*1–41.

Hughes, J.K., Hodge, A., Fitter, A.H. and Atkin, O. 2008. Mycorrhizal respiration: implications for global scaling relationships. *Trends Plant Sci.* 13(11): 583–588.

Jacott, C.N., Murray, J.D. and Ridout, C.J. 2017. Trade-offs in arbuscular mycorrhizal symbiosis: disease resistance, growth responses and perspectives for crop breeding. *Agronomy* 7: 75.

Jaizme-Vega, M.C., Rodriguez-Romero, A.S. and Barroso-Nunez, L.A. 2006. Effect of the combined inoculation of arbuscular mycorrhizal fungi and plant growth promoting rhizobacteria on papaya *Carica papaya* L. infected with the root-knot nematode *Meloidogyne incognita*. *Fruits*61: 151–62.

Javaid, A. 2009. Arbuscular mycorrhizal mediated nutrition in plants. *J. Plant Nutr.* 32: 1595–1618.

Jha, A.B. and Sharma, P. 2019. Regulation of osmolytes syntheses and improvement of abiotic stress tolerance in plants. In: Approaches for enhancing abiotic stress tolerance in plants. CRC Press.pp. 311–338.

Jia-Dong, H., Tao, D., Hui-Hui, W., Zou, Y.N., Wu, Q.S. and Kamil, K. 2019. Mycorrhizas induce diverse responses of root TIP aquaporin gene expression to drought stress in trifoliate orange. *Sci. Hortic.* 243: 64–69.

Jimenez-Zacarias, J.J., Altamirano-Hernández, J. and Pena Cabriales, J.J. 2004. Nitrogenase activity and trehalose content of nodules of drought-stressed common beans infected with effective Fix+ and ineffective Fix⁻ rhizobia. *Soil Biol. Biochem.* 36: 1975–1981.

Jin, Y., Liu, H., Luo, D., Yu, N., Dong, W., Wang, C., Zhang, X., Dai, H., Yang, J. and Wang, E. 2016. DELLA proteins are common components of symbiotic rhizobial and mycorrhizal signalling pathways. *Nat. Commun.* *7*(1): 1–14.

Jixiang, L., Yingnan, W., Shengnan, S., Chunsheng, M. and Xiufeng, Y. 2017. Effects of arbuscular mycorrhizal fungi on the growth, photosynthesis and photosynthetic pigments of *Leymus chinensis* seedlings under salt-alkali stress and nitrogen deposition. *Sci. Total Environ.* 576: 234–241.

Johns, C.D. 2014. Agricultural application of mycorrhizal fungi to increase crop yields, promote soil health and combat climate change; Future Directions international: Dalkeith, Australia.

Johnson, J.M, Sherameti, I., Ludwig, A., Nongbri, P.L., Sun, C., Lou, B., Varma, A. and Oelmu¨ller, R. 2011. Protocols for *Arabidopsis thaliana* and *Piriformospora indica* co-cultivation—a model system to study plant beneficial traits. *J. Endocytobiosis Cell Res.* 101–113.

Jung, S.C., Martinez-Medina, A., Lopez-Raez, J.A. and Pozo, M.J. 2012. Mycorrhiza-induced resistance and priming of plant defenses. *J. Chem. Ecol.* 38: 651–654.

Kamel, L., Tang, N.W., Malbreil, M., San Clemente, H., Le Marquer, M. and Roux, C. 2017. The comparison of expressed candidate secreted proteins from two arbuscular mycorrhizal fungi unravels common and specific molecular tools to invade different host plants. *Front. Plant Sci.* 8: 124.

Kanamori, N., Madsen, L.H., Radutoiu, S., Frantescu, M., Quistgaard, E.M.H. Miwa, H., Downie, J.A., James, E.K., Felle, H.H., Haaning, L.L., Jensen, T.H., Sato, S., Nakamura, Y., Tabata, S., Sandal, N. and Stougaard, J. 2006. A nucleoporin is required for induction of Ca2þ spiking in legume nodule development and essential for rhizobial and fungal symbiosis. *Proc. Natl. Acad. Sci.* 103(2): 359–364.

Kanwal, S., Bano, A. and Malik, R.N. 2015. Effects of arbuscular mycorrhizal fungi on metals uptake, physiological and biochemical response of *Medicago sativa* L. with increasing Zn and Cd concentrations in soil. *Am. J. Plant Sci.* 6: 2906–2923.

Kape,R.,Wex,K.,Parniske,M.,Go¨rge,E.,Wetzel,A. andWerner,D.1993.Legumeroot metabolites and VA-mycorrhiza development. *J. Plant Physiol.* 1: 54–60.

Kapoor, R., Evelin, H., Mathur, P. and Giri, B. 2013. Arbuscular mycorrhiza: approaches for abiotic stress tolerance in crop plants for sustainable agriculture, In: Plant acclimation to environmental stress. Tuteja, N. and Gill, S.S. (eds.), LLC: Springer Science+Business Media, 359–401.

Khalloufi, M., Martínez-Andújar, C., Lachaâl, M., Karray-Bouraoui, N., PérezAlfocea, F. and Albacete, A. 2017. The interaction between foliar GA3 application and arbuscular mycorrhizal fungi inoculation improves growth in salinized tomato *Solanum lycopersicum* L. plants by modifying the hormonal balance. *J. Plant Physiol.* 214: 134–144.

Kistner, C., Winzer, T., Pitzschke, A., Mulder, L., Sato, S., Kaneko, T., Tabata, S., Sandal, N., Stougaard, J., Webb, K.J., Szczyglowski, K. and Parniske, M. 2005. Seven *Lotusjaponicus* genes required for transcriptional reprogramming of the root during fungal and bacterial symbiosis. *Plant Cell* 17(8): 2217–2229.

Kretzschmar, T., Kohlen, W., Sasse, J., Borghi, L., Schlegel, M., Bachelier, J.B.,Reinhardt, D., Bours, R., Bouwmeester, H.J. and Martinoia, E. 2012. A petunia ABCprotein controls strigolactone- dependent symbiotic signaling and branching. *Nature* 483: 341–344.

Krishnamurthy, K.V. and Senthilkumar, S. 2005. Mycorrhiza: Role and applications, Allied Publishers Pvt Ltd, New Delhi, pp. 66–90.

Krüger, M., Krüger, C., Walker, C., Stockinger, H. and Schüßler, A. 2012. Phylogenetic reference data for systematics and phylotaxonomy of arbuscular mycorrhizal fungi from phylum to species level. *New Phytol.* 193(4): 970–984.

Kubikova, E., Moore, J.L., Ownlew, B.H., Mullen, M.D. and Augé, R.M. 2001. Mycorrhizal impact on osmotic adjustment in *Ocimumbasilicum* during a lethal drying episode. *J. Plant Physiol.* 158: 1227–1230.

Kumar, P., Chaturvedi, R., Sundar, D. and Bisaria, V.S. 2016. *Piriformospora indica* enhances the production of pentacyclic triterpenoids in *Lantana camara* L. suspension cultures. *Plant Cell, Tissue and Organ Culture* (PCTOC). 125(1): 23–29.

Kuo, C.H. 2015. Scrambled and not-so-tiny genomes of fungal endosymbionts. *Proc. Natl. Acad. Sci. U.S.A.* 112(25): 7622–7623.

Lal, R. 2020. Managing soils for resolving the conflict between agriculture and nature: The hard talk. *Eur. J. Soil Sci.* 71: 1–9.

Li, H., Chen, X.W. and Wong, M.H. 2015. Arbuscular mycorrhizal fungi reduced the ratios of inorganic/organic arsenic in rice grains. *Chemosphere* 145: 224–230.

Li, J., Meng, B., Chai, H., Yang, X., Song, W., Li, S., Lu, A., Zhang, T. and Sun, W. 2019. Arbuscular mycorrhizal fungi alleviate drought stress in C3 *Leymus chinensis* and C4 *Hemarthriaaltissima* grasses via altering antioxidant enzyme activities and photosynthesis. *Front. Plant. Sci.* 10: 10.

Li, Q., Kuo, Y-W., Lin, K-H., Huang, W., Deng, C., Yeh, K-W. and Chen, S-P. 2021. *Piriformospora indica* colonization increases the growth, development, and herbivory resistance of sweet potato (*Ipomoea batatas* L.). *Plant Cell Rep.* 40: 339–350.

Li, T., Hu, J.L., Hao, Z.P., Li, H., Wang, Y.S. and Chen, B.D. 2013. First cloning and characterization of two functional aquaporin genes from an arbuscular mycorrhizal fungus *Glomus intraradices. New Phytol.* 197: 617–630.

Linderman, R.G. 1994. Role of VAM fungi in biocontrol. In: Mycorrhizae and plant health; APS: St Paul, Brazil, pp. 1–26.

Liu, R.J. and Chen, Y.L. 2007. Mycorrhizology. China Science Press, Beijing, pp. 447.

Liu, X., Song, Q., Tang, Y., Li, W., Xu, J. and Wu, J. 2013. Human health risk assessment of heavy metals in soil–vegetable system: a multi-medium analysis. *Sci. Total Environ.* 463: 530–540.

Lohse, S., Schliemann, W. and Ammer, C. 2005. Organization and metabolism of plastidsand mitochondria in arbuscular mycorrhizal roots of *Medicago truncatula*. PlantPhysiol. 139(1): 329–340.

Lopez-Raez, J.A., Shirasu, K. and Foo, E. 2017. Strigolactones in plant interactions with beneficial and detrimental organism: the Yin and Yang. *Trends Plant Sci.* 22: 527–537.

Louis, I. 1988. Ecto- and ectendomycorrhizae in the tropical dipterocarp *Shoreaparvifolia*. Mycolog.

Maherali, H. and Klironomos, J. 2007. Influence of phylogeny on fungal community assembly and ecosystem functioning. *Science* 316: 1746–1748.

Maillet,F.,Poinsot,V.,Andr´e,O.,Puech-Pag‘es,V.,Haouy,A.,Gueunier,M.,Crome r,L.,Giraudet, D., Formey, D., Niebel, A. and Martinez, E.A. 2011. Fungal lipochitooligosaccharide symbiotic signals in arbuscular mycorrhiza. *Nature* 469(7328): 58–63.

Manchanda, G. and Garg, N. 2007. Endomycorrhizal and rhizobial symbiosis: How much do they share? *J. Plant Interact.* 2(2): 79–88.

Medina, A., Probanza, A., Gutierrez-Mañero, F.J. and Azcon, R. 2003. Interactions of arbuscular-mycorrhizal fungi and *Bacillus* strains and their effects on plant growth, microbial rhizosphere activity thymidine and leucine incorporation and fungal biomass ergosterol and chitin. *Appl. Soil Ecol.* 22: 15–28.

Meding, S.M. and Zasoski. R.J. 2008. Hyphal-mediated transfer of nitrate, arsenic, cesium, rubidium, and strontium between arbuscular mycorrhizal forbs and grasses from California oak woodland. *Soil Biol. Biochem.* 40: 126–134.

Meena, V.S., Meena, S.K., Verma, J.P., Kumar, A., Aeron, A., Mishra, P.K., Bisht, J.K., Pattanayak, A., Naveed, M. and Dotaniya, M.L. 2017. Plant beneficial rhizospheric microorganism PBRM strategies to improve nutrients use efficiency: a review. *Ecol. Eng.* 107: 8–32.

Mehmood, A., Mirza, M.A., Choudhary, M.A., Kim, K.H., Raza, W., Raza, N., Lee, S.S., Zhang, M., Lee, J.H. and Sarfraz, M. 2019. Spatial distribution of heavy metals in crops in a wastewater irrigated zone and health risk assessment. *Environ. Res.* 168: 382–388.

Mena-Violante, H.G., Ocampo-Jimenez, O., Dendooven, L., Martinez-Soto, G., Gonzalez-Castafeda, J. and Davies, F.T. 2006. Arbuscular mycorrhizal fungi enhance fruit growth and quality of chile ancho *Capsicum annuum* L. cv. San Luis plants exposed to drought. *Mycorrhiza* 16: 261–267.

Mendes, R., Garbeva, P. and Raaijmakers, J.M. 2013. The rhizosphere microbiome: significance of plant benefcial, plant pathogenic, and human pathogenic microorganisms. *FEMS Microbiol. Rev.* 37: 634–63.

Miransari, M. 2010. Contribution of arbuscular mycorrhizal symbiosis to plant growth under different types of soil stress. *Plant Biol.* 12: 563–569.

Miransari, M. 2017. Arbuscular mycorrhizal fungi and heavy metal tolerance in plants, In: Arbuscular mycorrhizas and stress tolerance of plants. Wu Q.S. (ed.), Singapore: Springer Nature, pp. 174–161.

Mishra, J., Singh, R. and Arora, N.K. 2017. Alleviation of heavy metal stress in plants and remediation of soil by rhizosphere microorganisms, *Front. Microbiol.* 8: 1706.

Mishra, J., Singh, R. and Arora, N.K. 2017. Plant growth-promoting microbes: diverse roles in agriculture and environmental sustainability. In: Probiotics and plant health. Springer, Singapore, pp. 71–111.

Mitra, D., Djebaili, R., Pellegrini, M., Mahakur, B., Sarker, A., Chaudhary, P., Khoshru, B., Gallo, M.D., Kitouni, M., Barik, D.P. and Panneerselvam, P. 2021. Arbuscular mycorrhizal symbiosis: plant growth improvement and induction of resistance under stressful conditions. *J. Plant Nutr.*1–37.

Moghadam, H.R.T. 2016. Application of super absorbent polymer and ascorbic acid to mitigate deleterious effects of cadmium in wheat. *Pesqui. Agropecu. Trop.* 1: 9–18.

Mohanta, T.K. and Bae, H. 2015. Functional genomics and signaling events in mycorrhizal symbiosis. *J. Plant Interact.* 10(1): 21–40.

Molina, R., Massicote, H.B. and Trappe, J.M. 1992. Specificity phenomena in mycorrhizal symbioses: community-ecological consequences and practical implications. In: Mycorrhizal functioning – an integrative plant-fungal process,Allen, M.F. (ed). Chapman and Hall, New York, pp. 357–423.

Molina-Romero, D., Juárez-Sánchez, S., Venegas, B., Ortíz-González, C.S., Baez, A. and Morales-García, Y.E. 2021. A bacterial consortium interacts with different varieties of maize, promotes the plant growth, and reduces the application of chemical fertilizer under field conditions. *Front. Sustain. Food Syst.* 4: 293.

Moradtalab, N., Roghieh, H., Nasser, A., Tobias, E.H. and Günter, N. 2019. Silicon and the association with an arbuscular-mycorrhizal fungus *Rhizophagusclarus* mitigate the adverse effects of drought stress on strawberry. *Agronomy*. 9: 41.

Morandi, D., Baily, J.A. and Gianinazzi-Pearson, V. 1984. Isoflavonoid accumulation in soybean roots infected with vesicular-arbuscular mycorrhizal fungi. *Physiol. Plant Pathol.* 24: 357–364.

Mori, N., Nishiuma, K., Sugiyama, T., Hayashi, H. and Akiyama, K. 2016. Carlactone-typestrigolactones and their synthetic analogues as inducers of hyphal branching inarbuscular mycorrhizal fungi. *Phytochemistry*130: 90–98.

Morton, J.B. and Benny, G.L. 1990. Revised classification of arbuscular mycorrhizal fungi (Zygomycetes): a new order, Glomales, two new suborders, Glominaeae and Gigasporineae and two new families, Acaulosporaceae and Gigasporaceae, with an emendation of Glomaceae. *Mycotaxon* 37: 471–491.

Mukherjee,A. andAn´e,J.M.2011.Germinating spore exudates from arbuscular mycorrhizal fungi: molecular and developmental responses in plants and theirregulation by ethylene. *Mol. Plant Microbe. Interact.* 24(2): 260–270.

Munns, R. and Gilliham, M. 2015. Salinity tolerance of crops - What is the cost? *New Phytol.* 208: 668–673.

Munns, R. and Tester, M. 2018. Mechanisms of salinity tolerance. *Annu. Rev. Plant Biol.* 59: 651–681.

Nacoon, S., Jogloy, S., Riddech, N., Mongkolthanaruk, W., Kuyper, T.W. and Boonlue, S. 2020. Interaction between phosphate solubilizing bacteria and arbuscular mycorrhizal fungi on growth promotion and tuber inulin content of *Helianthus tuberosus* L. *Sci. Rep.* 10: 1–10.

Nair, M.G., Safir, G.R. and Siqueira, J.O. 1991. Isolation and identification of vesicular-arbuscular mycorrhiza-stimulatory compounds from clover (*Trifolium repens*) roots. *Appl. Environ. Microbiol.* 57(2): 434–439.

Naito, M., Morton, J.B. and Pawlowska, T.E. 2015. Minimal genomes of mycoplasma-related endobacteria are plastic and contain host-derived genes for sustained life within Glomeromycota. *Proc. Natl. Acad. Sci. U.S.A.* 112(25): 7791–7796.

Najafi, A., Ardakani, M.R., Rejali, F. and Sajedi, N. 2012. Response of winter barley to co-inoculation with *Azotobacter* and Mycorrhiza fungi influenced by plant growth promoting rhizobacteria. *Ann. Biol. Res.* 3: 4002–4006.

Neal, A.L., Ahmad, S., Gordon-Weeks, R. and Ton, J. 2012. Benzoxazinoids in root exudates of maize attract *Pseudomonas putida* to the rhizosphere. *PLoS ONE* 7: e35498.

Nepomuceno, R.A., Brown, C.M.B., Mojica, P.N. and Brown, M.B. 2019. Biological control potential of vesicular arbuscular mycorrhizal root inoculant (VAMRI) and associated phosphate solubilizing bacteria, *Pseudochrobactruma saccharolyticum* against soilborne phytopathogens of Onion (*Allium cepa* L. var. Red Creole). *Arch. Phytopathol. Plant Prot.* 52: 714–732.

Nguvo, K.J. and Gao, X. 2019. Weapons hidden underneath: Bio-control agents and their potentials to activate plant induced systemic resistance in controlling crop Fusarium diseases. *J. Plant Dis. Prot.* 126: 177–190.

Nunes, J.L.D., de Souza, P.V.D., Marodin, G.A.B. and Fachinello, J.C. 2010. Effect of arbuscular mycorrhizal fungi and indole butyric acid interaction on vegetative growth of 'Aldrighi' peach rootstock seedlings. *Cienc. Agrotecnol.* 34: 80–86.

Oberwinkler, F., Riess, K., Bauer, R., Selosse, M.A., Weiß, M., Garnica, S. and Zuccaro, A. 2013. Enigmatic sebacinales. *Mycol. Prog.* 12(1): 1–27.

Oelmuller, R., Sherameti, I., Tripathi, S. and Varma, A. 2009. *Piriformospora indica*, a cultivable root endophyte with multiple biotechnological application. *Symbiosis* 49: 1–17.

Okayama, M., Yamato, M., Yagame, T. and Iwase, K. 2012. Mycorrhizal diversity and specificity in *Lecanorchis* (Orchidaceae). *Mycorrhiza* 22(7): 545–553.

Olanrewaju, O.S., Glick, B.R. and Babalola, O.O. 2017. Mechanism of action of plant growth promoting bacteria. *World J. Microbiol. Biotechnol.* 33: 197

Oldroyd, G.E. 2013. Speak, friend, and enter: signalling systems that promote beneficial symbiotic associations in plants. *Nat. Rev. Microbiol.* 11(4): 252–263.

Ordookhani, K. and M. Zare. 2011. Effect of *Pseudomonas, Azotobacter* and arbuscular mycorrhiza fungi on lycopene, antioxidant activity and total soluble solid in tomato (*Lycopersicon esculentum* F1 Hybrid, Delba). *Adv. Environ. Biol.* 5: 1290–1294.

Ordookhani, K., Khavazi, K., Moezzi, A. and Rejali, F. 2010. Influence of PGPR and AMF on antioxidant activity, lycopene and potassium contents in tomato. *Afr. J. Agric. Res.* 5: 1108–1116.

Ortas, I., Rafique, M. and Ahmed, I.A.M. 2017. Application of arbuscular mycorrhizal fungi into agriculture. *Arbuscular Mycorrhizas and Stress Tolerance of Plants.* pp. 305–327.

Oueslati, O. 2003. Allelopathy in two durum wheat *Triticum durum* L. varieties. *Agric. Ecosyst. Environ.* 96: 161–163.

Ouledali, S., Ennajeha, M., Ferrandino, A., Khemiraac, H., Schubert, A. and Secchi, F. 2019. Influence of arbuscular mycorrhizal fungi inoculation on the control of stomata functioning by abscisic acid ABA in drought-stressed olive plants. *South Afr. J. Bot.* 121: 152–158.

Ouziad, F., Hildebrandt, U., Schmelzer, E. and Bothe, H. 2005. Differential gene expressions in arbuscular mycorrhizal-colonized tomato grown under heavy metal stress. *J. Plant Physiol.* 162: 634–649.

Palm, C., Blanco-Canqui, H., Declerck, F., Gatere, L. and Grace, P. 2014. Conservation agriculture and ecosystem services: an overview. *Agric. Ecosyst. Environ.* 187: 87–105.

Parniske, M. 2004. Molecular genetics of the arbuscular mycorrhizal symbiosis. *Curr. Opin. Plant Biol.* 7(4): 414–421.

Paszkowski, U. 2006. A journey through signaling in arbuscular mycorrhizal symbioses. *New Phytol.* 172(1): 35–46.

Paul, B.K., Vanlauwe, B., Ayuke, F., Gassner, A., Hoogmoed, M., Hurisso, T.T., Koala, S., Lelei, D., Ndabamenye, T., Six, J. and Pulleman, M.M. 2013. Medium-term impact of tillage and residue management on soil aggregate stability, soil carbon and crop productivity. *Agric. Ecosyst. Environ.* 164: 14–22.

Penn, C.J. and Camberato, J.J. 2019. A critical review on soil chemical processes that control how soil pH affects phosphorus availability to plants. *Agriculture*96: 120.

Pepe, A., Giovannetti, M. and Sbrana, C. 2018. Lifespan and functionality of mycorrhizal fungal mycelium are uncoupled from host plant lifespan. *Sci. Rep.* 8: 10235.

Peskan-Berghofer, T., Shahollari, B., Going, P.H., Hehl, S., Markert, C., Blanke, V., Kost, G., Varma, A. and Oelmuller, R. 2004. Association of *Piriformospora indica* with *Arabidopsis thaliana* roots represents a novel system to study beneficial plant-microbe interactions and involves early plant protein modifications in the endoplasmic reticulum and at the plasma membrane. *Physiol. Plant.* 122: 456-477.

Peterson, R.L., Massicotte, H.B. and Melville, L.H. 2004. Anatomy and cell biology. *Mycologist*19: 133.

Philippot, L., Raaijmakers, J.M., Lemanceau, P. and van Der Putten, W.H. 2013. Going back to the roots: the microbial ecology of the rhizosphere. *Nature Rev. Microbiol.* 11: 789.

Phillips, D.A. and Tsai, S.M. 1992. Flavonoids as plant signals to the rhizosphere microbes. *Mycorrhiza* 1: 55–58.

Phour M. and Sindhu, S.S. 2020.Exploiting plant growth promoting bacteria for salt tolerance and crop productivity in mustard. *Applied Soil Ecology* 149: (in May issue).

Phour, M., Sehrawat, A., Sindhu, S.S. and Glick, B.R. 2020. Interkingdom signaling in plant-rhizomicrobiome interactions for sustainable agriculture. *Microbiol. Res.* 241: 126589.

Plouznikoff, K., Declerck, S. and Calonne-Salmon, M. 2016. Mitigating abiotic stresses in crop plants by arbuscular mycorrhizal fungi; Springer Nature: Basel, Switzerland, pp. 341–400.

Powell, C.L. and Bagyaraj, D.J. 1984. Vasicular Mycorrhiza. CRC Press: Boca Raton, FL, USA.

Pozo, M.J. and Azcon-Aguilar, C. 2007. Unraveling mycorrhiza-induced resistance. *Curr. Opin. Plant Biol.* 10: 393–398.

Pumplin, N. and Harrison, M.J. 2009. Live-cell imaging reveals periarbuscular membrane domains and organelle location in *Medicago truncatula* roots during arbuscular mycorrhizal symbiosis. *Plant Physiol.* 151(2): 809–819.

Punamiya, P., Datta, R., Sarkar, D., Barber, S., Patel, M. and Da, P. 2010. Symbiotic role of *Glomus mosseae* in phytoextraction of lead in vetiver grass *Chrysopogonzizanioides* L. *J. Hazard. Mater.* 177: 465–474.

Püschel, D., Bitterlich, M., Rydlová, J. and Jansa, J. 2021. Drought accentuates the role of mycorrhiza in phosphorus uptake. *Soil Biol. Biochem.* 157: 108243.

Rabie, G.H. 2005. Contribution of arbuscular mycorrhizal fungus to red kidney and wheat plants tolerance grown in heavy in metal polluted soil. *Afr. J. Biotechnol.* 4: 332–345.

Rajendran, K. and Devaraj, P. 2004. Biomass and nutrient distribution and their return of *Casuarina equisetifolia* inoculated with biofertilizers in farm land. *Biomass Bioenergy* 26: 235–249.

Raklami, A., Bechtaoui, N., Tahiri, A.I., Anli, M., Meddich, A. and Oufdou, K. 2019. Use of rhizobacteria and mycorrhizae consortium in the open field as a strategy for improving crop nutrition, productivity and soil fertility. *Front. Microbiol.* 10: 1106.

Rani, B. 2016. Effect of arbuscular mycorrhiza fungi on biochemical parameters in wheat *Triticum aestivum* L. under drought conditions. Doctoral dissertation, CCSHAU, Hisar.

Rapparini, F. and Penuelas, J. 2014. "Mycorrhizal fungi to alleviate drought stress on plant growth," In: Use of Microbes for the Alleviation of Soil Stresses, Miransari,M. (ed.). New York, NY: Springer.

Rasmussen, S.R., Fuchtbauer, W., Novero, M., Volpe, V., Malkov, N., Genre, A., Bonfante, P., Stougaard, J. and Radutoiu, S. 2016. Intraradical colonization by arbuscular mycorrhizal fungi triggers induction of a lipochitooligosaccharide receptor, *Sci. Rep.* 6: 29733.

Read, D.J., Lewis, D.H., Fitter, A.H. and Alexander, I.J. 1992. Mycorrhizas in Ecosystems; CAB International: Wallingford, Oxon, UK.

Redecker, D. and Schüßler, A. 2014. Glomeromycota. In Systematics and evolution, Springer, Berlin, Heidelberg, pp. 251–269.

Redecker, D., Schüßler, A., Stockinger, H., Stürmer, S.L., Morton, J.B. and Walker, C. 2013. An evidence-based consensus for the classification of arbuscular mycorrhizal fungi (Glomeromycota. *Mycorrhiza* 23(7): 515–531.

Rejeb, I., Pastor, V. and Mauch-Mani, B. 2014. Plant responses to simultaneous biotic and abiotic stress: Molecular mechanisms. *Plants* 3: 458–475.

Remy, W., Taylor, T.N. Hass, H. and Kerp, H. 1994. Four hundred-million-year-old vesicular arbuscular mycorrhizae. *Proc. Nat. Acad. Sci. U.S.A.* 91(25): 11841–11843.

Rillig, M.C. and Mummey, D.L. 2006. Mycorrhizas and soil structure. *New Phytol.* 171: 41–53.

Rozp[1]dek,P.,Domka,A.M.,Nosek,M.,Wazÿny, R., Jêdrzejczyk,R.J.,Wiciarz,M. and Turnau, K. 2018. The role of strigolactone in the cross-talk between *Arabidopsis thaliana* and the endophytic fungus *Mucor* sp. *Front. Microbiol.* 9: 441.

Ruiz-Lozano, J.M., Aroca, R., Zamarreno, A.M., Molina, S., Andreo-Jimenez, B. and Porcel, R. 2015. Arbuscular mycorrhizal symbiosis induces strigolactonebiosynthesis under drought and improves drought tolerance in lettuce and tomato. *Plant Cell Environ*39(2): 441–452.

Ruiz-Sanchez, M., Aroca, R., Munoz, Y., Polon, R. and Ruiz-Lozano, J.M. 2010. The arbuscular mycorrhizal symbiosis enhances the photosynthetic efficiency and the antioxidative response of rice plants subjected to drought stress. *J. Plant Physiol.* 167: 862–869.

Sairam, R.K. and Tyagi, A. 2004. Physiology and molecular biology of salinity stress tolerance in plants. *Curr. Sci.* 86: 407–421.

Sallaku, G., Sanden, H., Babaj, I., Kaciu, S., Balliu, A. and Rewald, B. 2019. Specific nutrient absorption rates of transplanted cucumber seedlings are highly related to RGR and influenced by grafting method, AMF inoculation and salinity. *Sci. Hortic.* 243: 177–188.

Santander, C., Ruiz, A., Garcia, S., Aroca, R., Cumming, J. and Cornejo, P. 2019. Efficiency of two arbuscular mycorrhizal fungi inocula to improve saline stress tolerance in lettuce plants by changes of antioxidant defense mechanisms. *J. Sci. Food Agric.* 100: 1577–1587.

Santander, C., Sanhueza, M., Olave, J., Borie, F., Valentine, C. and Cornejo, P. 2019. Arbuscular mycorrhizal colonization promotes the tolerance to salt stress in lettuce plants through an efficient modification of ionic balance. *J. Soil Sci. Plant Nutr.* 19(2): 321–331.

Saraswathi, S.G. and Paliwal, K. 2011. Drought induced changes in growth, leaf gas exchange and biomass production in *Albizia lebbeck* and *Cassia siamea* seedlings. *J. Environ. Biol.* 32: 173–178.

Schüßler, A., Schwarzott, D. and Walker, C. 2001. A new fungal phylum, the Glomeromycota: phylogeny and evolution. *Mycol. Res.* 105(12): 1413–1421.

Sehrawat, A., Glick, B.R. and Sindhu, S.S. 2021b. Hydrogen cyanide production by soil bacteria: Biological control of pests and plant growth promotion. *Pedosphere* (in press).

Sehrawat, A., Phour, M. and Sindhu, S.S. 2021a. Rhizosphere microbiome's contribution towards soil fertility and nutrient acquisition by crop plants. In: Microbial Biotechnology for sustainable agriculture, Arora, N.K. and Bouizgarne, B.(eds.), Springer Nature Singapore Pte Ltd. (in press).

Sehrawat, A., Phour, M., Kumar, R. and Sindhu, S.S. 2020. Bioremediation of pesticides by rhizosphere microorganisms to improve food safety. In: Microbial rejuvenation of polluted environment. Panpatte, D.G. (eds.). Springer Nature Singapore Pte Ltd. pp. 23-84. https: //doi.org/10.1007/978-981-15-7447-4_2

Selmar, D. and Kleinwachter, M. 2013. Influencing the product quality by deliberately applying drought stress during the cultivation of medicinal plants. *Ind. Crops Prod.* 42: 558–566.

Serfling, A., Wirsel, S.G., Lind, V. and Deising, H.B. 2007. Performance of the biocontrol fungus *Piriformospora indica* on wheat under greenhouse and field conditions. *Phytopathology*97(4): 523–531.

Shahabivand, S., Maivan, H.Z., Goltapeh, E.M., Sharifi, M. and Aliloo, A.A. 2012. The effects of root endophyte and arbuscular mycorrhizal fungi on growth and cadmium accumulation in wheat under cadmium toxicity. *Plant Physiol. Biochem.* 60: 53–58.

Sharda, J.N. and Koide, R.T. 2010. Exploring the role of root anatomy in P-mediated control of colonization by arbuscular mycorrhizal fungi. *Botany* 88: 165–173.

Sharma, I.P. and Sharma, A.K. 2017. Physiological and biochemical changes in tomato cultivar PT-3 with dual inoculation of mycorrhiza and PGPR against root-knot nematode. *Symbiosis* 71: 175–183.

Sharma, I.P. and Sharma, K. 2018. Mycorrhizal colonization and phosphorus uptake in presence of PGPRs along with nematode infection. *Symbiosis*77: 185–187.

Sharma, R. andSindhu, S.S. 2020. Plant growth promoting rhizobacteria (PGPR): A sustainable approach for managing soil fertility and crop productivity. In: Microbes for humankind and application, Malik, D.K., Rathi, M., Kumar, R. and Bhatia, D. (eds). Daya Publishing House, New Delhi. pp. 97–130.

Sharma, R., Sindhu, S. and Sindhu, S.S. 2018. Bioinoculation of mustard (*Brassica juncea* L.) with beneficial rhizobacteria: A sustainable alternative to improve crop growth. *Intern. J. Curr. Microbiol. Appl. Sci.*7(5): 1375-1386.

Shaul-Keinan, O., Gadkar, V., Ginzberg, I., Grünzweig, J.M., Chet, I., Elad, Y.,Wininger, S., Belausov, E., Eshed, Y., Atzmon, N. and Ben-Tal, Y. 2002. Hormoneconcentrations in tobacco roots change during arbuscular mycorrhizal colonizationwith *Glomus intraradices. New Phytol.* 154(2): 501–507.

Sheng, M., Tang, M., Zhang, F. and Huang, Y. 2011. Influence of arbuscular mycorrhiza on organic solutes in maize leaves under salt stress. *Mycorrhiza* 21: 423–430.

Sherameti, I., Shahollari, B., Venus, Y., Altschmied, L., Varma, A. and Oelmuller, R. 2005. The endophytic fungus *Piriformospora indica* stimulates the expression of nitrate reductase and the starch-degrading enzyme glucan-water dikinase in tobacco and *Arabidopsis* roots through a homeodomain transcription factor that binds to a conserved motif in their promoters. *J. Biol. Chem.* 280(28): 26241–26247.

Shi, Z., Zhang, J., Lu, S, Li, Y. and Wang, F. 2020. Arbuscular mycorrhizal fungi improve the performance of sweet sorghum grown in a Mo-contaminated soil. *J. Fungi* 6(2): 44.

Shokri, S. and Maadi, B. 2009. Effects of arbuscular mycorrhizal fungus on the mineral nutrition and yield of *Trifolium alexandrinum* plants under salinity stress. *J. Agron.* 8: 79–83.

Simon, L., Bousquet, J., Levesque, C. and Lalonde, M. 1993. Origin and diversification of endomycorrizal fungi and coincidence with vascular land plants. *Nature* 363: 67–69.

Simpson, R.T., Frey, S.D., Six, J. and Thiet, R.K. 2004. Preferential accumulation of microbial carbon in aggregate structures of no-tillage soils. SSSA. 68: 1249–1255.

Sindhu, S., Dahiya, A., Gera, R. and Sindhu, S.S. 2020. Mitigation of abiotic stress in legume-nodulating rhizobia for sustainable crop production. *Agric. Res.* 9(4): 444-459.

Sindhu S.S. and Dadarwal K.R. 1992.*Rhizobium* strain and host plant cultivar interaction affecting relative efficiency and effectivity in chickpea (*Cicer arietinum* L.). Bisen, P.S. (ed). In: *Frontier in Microbial Technology,* CBS Publishers, New Delhi. pp. 105-109.

Sindhu, S.S. and Dadarwal, K.R. 2000. Competition for nodulation among rhizobia in legume-*Rhizobium* symbiosis. *Indian J. Microbiol.* 40(4): 211–246.

Sindhu, S.S. and Sharma, R. 2019. Amelioration of biotic stress by application of rhizobacteria for agriculture sustainability. Sayyed, R.Z. and Tabassum, B. (eds). In: *Plant Growth Promoting Rhizobacteria for Sustainable Stress Management, Microorganisms for Sustainability.*Springer Nature Singapore Pte Ltd. Chapter 5, pp. 111 – 168.

Siqueira, J.O., Nair, M.G., Hammerschmidt, R., Safir, G.R. and Putnam, A.R. 1991.Si gnificanceofphenoliccompoundsinplant-soil-microbialsystems.*Crit.Rev.Plant Sci.* 10(1): 63–121.

Sivakumar, K., Kumaresan, G. and Sugapriya, N. 2019. Arbuscular mycorrhizal fungi (AMfungi) and phosphate solubilizing bacteria (PSB) on tolerance of tomato under salt stress. *J. Pharmacogn. Phytochem.* 8: 4717–4721.

Smith, G.S. 1987. Interactions of nematodes and mycorrhizal fungi. Vistas on Nematology, eds., Veech, J.A. and Dickson, D.W. Soc. Nemat., Hyattsville, MD, pp. 292–300.

Smith, S.E. and Read, D.J. 1997a. Mycorrhizal symbiosis, 2nd edn. Academic Press, London, pp. 605.

Smith, S.E. and Read, D.J. 1997b. Vesicular-arbuscular mycorrhizas. In: Mycorrhizal symbiosis, 2nd edn. Academic Press, London, pp. 9–160.

Smith, S.E. and Read, D.J. 2008. Microbial Symbiosis 3rd Edition, Academic Press, London.

Smith, S.E., Barker, S.J. and Zhu, Y-G. 2006. Fast moves in arbuscular mycorrhizal symbiotic signalling. *Trends Plant Sci.* 11(8): 369–371.

Smith, S.E., Jakobsen, I., Gronlund, M. Smith, F.A. 2011. Roles of arbuscular mycorrhizas in plant phosphorus nutrition: interactions between pathways of phosphorus uptake in arbuscular mycorrhizal roots have important implications for understanding and manipulating plant phosphorus acquisition, *Plant Physiol.* 156(3): 1050–1057.

Song, Y., Chen, D., Lu, K., Sun, Z. and Zeng, R. 2015. Enhanced tomato disease resistance primed by arbuscular mycorrhizal fungus. *Front. Plant. Sci.* 6: 6.

Sonmez, O., Aydemir, S. and Kaya, C. 2009. Mitigation effects of mycorrhiza on boron toxicity in wheat *Triticum durum* plants. *New Zeeland. J. Crop Hortic. Sci.* 37: 99–104.

Souza, L.A., Andrade, S.A.L., Souza, S.C.R. and Schiavinato, M.A. 2012. Evaluation of mycorrhizal influence on the development and phytoremediation potential of *Canavalia gladiata* in Pb contaminated soils. *Int. J. Phytorem.* 15: 465–476.

Srinivas, J., Purushotham, A.V. and Murali Krishna, K.V.S.G. 2013. The effects of heavy metals on seed germination and plant growth on Coccinia, Mentha and Trigonella plant seeds in Timmapuram. *Int. Res. J. Environ. Sci.* 2: 20–24.

Stirling, G.R., Hayden, H.L., Pattison, A.B. and Stirling, A.M. 2016. Soil Health, Soil Biology, Soilborne Diseases and Sustainable Agriculture: A Guide. CSIRO Publishing, Melbourne, VIC, Australia.

Sun, J., Miller, B., Granqvist, E., Wiley-Kalil, A., Gobbato, E., Maillet, F., Cottaz, S., Samain, E., Venkateshwaran, M., Fort, S., Morris, R.J., Ane, J-mM., Denarie, J. and Oldroyd, G.E.D. 2015. Activation of symbiosis signaling by arbuscular mycorrhizal fungi in legumes and rice. *Plant Cell.* 27(3): 823–838.

Takacs, T. and Voros, I. 2003. Effect of metal non-adapted arbuscular mycorrhizal fungi on Cd, Ni and Zn uptake by ryegrass. *Acta Agron. Hung.* 51: 347–354.

Taylor, D.L., Bruns, T. D. and Hodges, S.A. 2004. Evidence for mycorrhizal races in a cheating orchid. *Proc. R. Soc. Lond. Series B: Biological Science* 271(1534): 35–43.

Tedersoo, L. and Smith, M.E. 2013. Lineages of ectomycorrhizal fungi revisited: foraging strategies and novel lineages revealed by sequences from belowground. *Fungal Biol. Rev.* 27(3-4): 83–99.

Toro, K.S. and Brachmann, A. 2016. The effector candidate repertoire of the arbuscular mycorrhizal fungus *Rhizophagusclarus. BMC Genomics.* 17: 101.

Toro, M., Azcon, R. and Barea, J-M. 1997. Improvement of arbuscular mycorrhiza development by inoculation of soil with phosphate solubilizing bacteria to improve rock phosphate bioavailability (P32) and nutrient cycling. *Appl. Environ. Microbiol.* 63(11): 4408–4412.

Torres-Cortés, G., Ghignone, S., Bonfante, P. and Schüßler, A. 2015. Mosaic genome of endobacteria in arbuscular mycorrhizal fungi): Transkingdom gene transfer in an ancient mycoplasma-fungus association. *Proc. Natl. Acad. Sci. U.S.A.* 112(25): 7785–7790.

Tsimilli-Michael, M. and Strasser, R.J. 2013. Biophysical phenomics: evaluation of the impact of mycorrhization with *Piriformospora indica*. In: *Piriformospora indica,* Springer, Berlin, Heidelberg, pp. 173–190.

van der Heijden, M.G., Martin, F.M., Selosse, M.A. and Sanders, I.R. 2015. Mycorrhizal ecology and evolution: the past, the present and the future. *New Phytol.* 205(4): 1406–1423.

van der Heijden, M.G.A., Klironomos, J.N., Ursic, M., Moutoglis, P., Streitwolf-Engel, R., Boller, T., Wiemken, A. and Sanders, I.R. 1998. Mycorrhizal fungal diversity determines plant biodiversity, ecosystem variability and productivity. *Nature* 396: 69–72.

Vangelisti, A., Natali, L. and Bernardi, R. 2018. Transcriptome changes induced byarbuscular mycorrhizal fungi in sunflower (*Helianthus annuus* L.) roots. *Sci. Report.* 8(1): 4.

Varma, A., Bakshi, M., Lou, B., Hartmann, A. and Oelmueller, R. 2012. *Piriformospora indica*: a novel plant growth-promoting mycorrhizal fungus. *Agric. Res.* 1(2): 117–131.

Varma, A., Chordia, P., Bakshi, M. and Oelmüller, R. 2013. Introduction to sebacinales. In *Piriformospora indica,* Springer, Berlin, Heidelberg, pp. 3–24.

Varma, A., Chordia, P., Bakshi, M. and Oelmuller, R. 2013b. Introduction to Sebacinales, In: Varma, A., Kost, G. and Oelmuller, R. (eds.), *Piriformospora indica:* Sebacinales and their biotechnological applications. Springer, Berlin, Heidelberg, 3: 24.

Varma, A., Sherameti, I., Tripathi, S., Prasad, R., Das, A., Sharma, M., Bakshi, M., Johnson, J.M., Bhardwaj, S., Arora, M., Rastogi, K., Agrawal, A., Kharkwal, A.C., Talukdar, S., Bagde, U.S., Bisaria, V.S., Upadhyaya, C.P., Won, P.S., Chen, Y., Ma, J.,Lou, B., Adya, A., Zhang, L., Meghvansi, M.K.,Sree, K.S.,Gosal, S.K., Srivastava, R.B., Johri, A.K., Kumar, M., Dua, M., Cruz, C. and Oelmüller, R. 2012b. The symbiotic fungus *Piriformospora indica*: review, In: Hock,B. (ed.): Fungal associations. Springer, Berlin, Heidelberg, pp.231–254.

Varma, A., Singh, A., Sahay, N.S., Sharma, J., Roy, A., Kumari, M., Rana, D., Thakran, S., Deka, D., Bharti, K. and Hurek, T. 2001. *Piriformospora indica*: an axenically culturable mycorrhiza-like endosymbiotic fungus. In Fungal Associations, Springer, Berlin, Heidelberg, pp. 125–150.

Varma, A., Sree, K.S., Arora, M., Bajaj, R., Prasad, R. and Kharkwal, A.C. 2014. Functions of novel symbiotic fungus-*Piriformospora indica. In Proc. Indian Natl. Sci. Acad.* 80(2): 429–441.

Varma, A., Verma, S., Sudha, Sahay, N., Butehorn, B. and Franken, P. 1999. *Piriformospora indica*, a cultivable plant-growth-promoting root endophyte. *Appl. Environ. Microbiol.* 65(6): 2741–2744.

Verma, S., Varma, A., Rexer, K.H., Hassel, A., Kost, G., Sarbhoy, A., Bisen, P., Bütehorn, B. and Franken, P. 1998. *Piriformospora indica*, gen. et sp. nov., a new root-colonizing fungus. *Mycologia* 90(5): 896–903.

Volpe, V., Giovannetti, M., Sun, X-G., Fiorilli, V. and Bonfante, P. 2016. The phosphate transporters LjPT4 and MtPT4 mediate early root responses to phosphate status in non-mycorrhizal roots. *Plant Cell Environ.* 393: 660–71.

Wahid, F., Sharif, M., Fahad, S., Adnan, M., Khan, I.A., Aksoy, E., Ali, A., Sultan, T., Alam, M., Saeed, M., Ullah, h., Basir, A., Noor, M. and Khan, N.A. 2019. Arbuscular mycorrhizal fungi improve the growth and phosphorus uptake of mung bean plants fertilized with composted rock phosphate fed dung in alkaline soil environment. J. *Plant Nut.* 42: 15.

Wang, W., Shi, J., Xie, Q., Jiang, Y., Yu, N. and Wang, E. 2017. Nutrient exchange and regulation in arbuscular mycorrhizal symbiosis. *Mol. Plant.* 109: 1147–1158.

Wang, Y., Wang, M., Li, Y., Wu, A. and Huang, J. 2018. Effects of arbuscular mycorrhizal fungi on growth and nitrogen uptake of *Chrysanthemum morifolium* under salt stress. *PLoS ONE*13(4): e0196408.

Watts-Williams, S.J., Jakobsen, I., Cavagnaro, T.R. and Grønlund, M. 2015. Local and distal effects of arbuscular mycorrhizal colonization on direct pathway Pi uptake and root growth in *Medicago truncatula*. *J. Exp. Bot.* 6613: 4061–73.

Willis, A., Rodriguesb. B.F. and Harrisa, P.J.C. 2013. The ecology of arbuscular mycorrhizal fungi. *Crit. Rev. Plant Sci.* 32: 1–20.

Wipf, D., Krajinski, F., van Tuinen, D., Recorbet, G. and Courty, P. 2019. Trading on the arbuscular mycorrhiza market: From arbuscules to common mycorrhizal networks. *New Phytol.* 223.

Wood, T. 1992. VA mycorrhizal fungi: challenges for commercialization. In: Arora, D.K., Elander, P. and Mukerji, K.G. (eds.). Handbook of applied mycology, fungal biotechnology IV. Marcel Dekker Inc., New York, pp. 823–848.

Wu, Q., Zou, Y. and He, X. 2011b. Differences of hyphal and soil phosphatase activities in drought stressed mycorrhizal trifoliate orange *Poncirus trifoliata* seedlings. *Sci. Hortic.* 129: 294–298.

Wu, Q.S., Li, G.H. and Zou, Y.N. 2011a. Roles of arbuscular mycorrhizal fungi on growth and nutrient acquisition of peach *Prunus persica* L. Batsch seedlings. *J. Anim. Plant Sci.* 21: 746–750.

Wuepper, D., Borrelli, P. and Finger, R. 2020. Countries and the global rate of soil erosion. *Nat. Sustain.* 3: 51–55.

Xie, L., Lehvavirta, S., Timonen, S., Kasurinen, J., Niemikapee, J. and Valkonen, J.P.T. 2018. Species-specific synergistic effects of two plant growth—promoting microbes on green roof plant biomass and photosynthetic efficiency. *PLoS ONE*13(12): e0209432.

Xing, X., Gai, X., Liu, Q., Hart, M.M. and Guo, S. 2015. Mycorrhizal fungal diversity and community composition in a lithophytic and epiphytic orchid. *Mycorrhiza* 25(4): 289–296.

Yadav, V., Kumar, M., Deep, D.K., Kumar, H., Sharma, R., Tripathi, T., Tuteja, N., Saxena, A.K. and Johri, A.K. 2010. A phosphate transporter from the root endophytic fungus *Piriformospora indica* plays a role in phosphate transport to the host plant. *J. Biol. Chem.* 285: 26532–26544.

Yan, C., Rizwan, H.M., Liang, D., Reichelt, M., Mithofer, A., Scholz, S.S., Oelmuller, R. and Chen, F. 2021. The effect of the root-colonizing *Piriformospora indica* on passion fruit (*Passiflora edulis*) development: Initial defense shifts to fitness benefits and higher fruit quality. *Food Chem.* 15: 129671.

Yang, L., Zou, Y-N., Tian, Z-H., Wu, Q-S. and Kuca, K. 2021. Effects of beneficial endophytic fungal inoculants on plant growth and nutrient absorption of trifoliate orange seedlings. *Sci. Hortic.*277: 109815.

Yang, S-Y., Gronlund, M., Jakobsen, I., Grotemeyer, M.S., Rentsch, D. and Miyao, A. 2012. Nonredundant regulation of rice arbuscular mycorrhizal symbiosis by two members of the phosphate transporter1 gene family. *Plant Cell.* 2410: 4236–51.

Yao, Q., Zhu, H. and Chen, J. 2005. Growth responses and endogenous IAA and iPAs changes of litchi (*Litchi chinensis*Sonn.) seedlings induced by arbuscular mycorrhizal fungal inoculation. *Sci. Hortic.* 105: 145–151.

Yaseen, T., Burni, T. and Hussain, F. 2012. Effect of arbuscular mycorrhizal inoculation on nutrient uptake, growth and productivity of chickpea *Cicer arietinum* varieties. *Intern. J. Agron. Plant Prod.* 3: 334–345.

Yooyongwech, S., Phaukinsang, N., Cha-Um, S. and Supaibulwatana, K. 2013. Arbuscular mycorrhiza improved growth performance in *Macadamia tetraphylla* L. grown under water deficit stress involves soluble sugar and proline accumulation. *Plant Growth Regul.* 69: 285–293.

Yooyongwech, S., Samphumphuang, T., Tisarum, R., Theerawitaya, C. and Chaum, S. 2016. Arbuscular mycorrhizal fungi AMF improved water deficit tolerance in two different sweet potato genotypes involves osmotic adjustments via soluble sugar and free proline. *Sci. Hort.* 198: 107–117.

Yousaf, B., Liu, G., Wang, R., Imtiaz, M., Zia-ur-Rehman, M. and Munir, M.A.M. 2016. Bioavailability evaluation, uptake of heavy metals and potential health risks via dietary exposure in urban-industrial areas. *Environ. Sci. Pollut. Res.* 23: 22443–22453.

Yusran, Y., Roemheld, V. and Mueller, T. 2009. Effects of *Pseudomonas* sp. "Proradix" and *Bacillus amyloliquefaciens* FZB42 on the establishment of AMF infection, nutrient acquisition and growth of tomato affected by *Fusarium oxysporum*Schlechtf.sp. *radicis-lycopersici* Jarvis and Shoemaker. UC Davis: Department of Plant Sciences, UC Davis. https: //escholarship.org/uc/item/8g70p0zt.

Zak, B. 1974. Ectendomycorrhiza of Pacific madrone (*Arbutus menziesii*). *Trans. Br. Mycol. Soc.* 62: 202–205.

Zak, B. 1976. Pure culture synthesis of Pacific madrone ectendomycorrhizae. *Mycologia*68: 362.

Zhang, F., Jia-Dong, H.E., Qiu-Dan, N.I., Qiang-Sheng, W.U. and Zou, Y.N. 2018a. Enhancement of drought tolerance in trifoliate orange by mycorrhiza: changes in root sucrose and proline metabolisms. *Not. Bot. Horti. Agrobot.* 46: 270.

Zhang, L., Feng, G. and Declerck, S. 2018. Signal beyond nutrient, fructose, exuded by an arbuscular mycorrhizal fungus triggers phytate mineralization by a phosphate solubilizing bacterium. *The ISME Journal.* 1210: 2339–2351.

Zhang, Q., Blaylock, L.A. and Harrison, M. 2010. Two *Medicago truncatula* half-ABC transporters are essential for arbuscular development in arbuscular mycorrhizal symbiosis. *Plant Cell*25(5): 1483–97.

Zhang, X., Li, W., Fang, M., Jixian, Y. and Meng, S. 2016. Effects of arbuscular mycorrhizal fungi inoculation on carbon and nitrogen distribution and grain yield and nutritional quality in rice *Oryza sativa* L. *J. Sci. Food Agric.* 97: 2919–2925.

Zhou, Y., Ge, S., Jin, L., Yao, K., Wang, Y., Wu, X., Zhou, J., Xia, X., Shi, K., Foyer, C.H. andYu, J. 2019. A novel CO_2-responsive systemic signaling pathway controlling plantmycorrhizal symbiosis. *New Phytol.* 224(1): 106–116.

Zhu, X., Song, F. and Liu, F. 2017. Arbuscular mycorrhizal fungi and tolerance of temperature stress in plants, In: Arbuscular Mycorrhizas and Stress Tolerance of Plants, Wu, Q. (ed.). Springer, Singapore, pp. 163–194.

Zuccaro, A., Lahrmann, U., Guldener, U., Langen, G., Pfiffi, S., Biedenkopf, D., Wong, P., Samans, B., Grimm, C., Basiewicz, M. and Mu-rat, C. 2011. Endophytic life strategies decoded by genome and transcriptome analyses of the mutualistic root symbiont *Piriformospora indica. PLoSPathol.* 7(10): e1002290.

Chapter 3

Fermented Plant Foods: Cereals and Vegetables

*Nikhil Mehta, Priyanka Rao and Raman Saini**

Department of Biotechnology,
Kurukshetra University, Kurukshetra – 136 119, Haryana
**e-mail: ramanksaini@rediffmail.com*

ABSTRACT

The history of fermented food is with the origin of the microorganisms. The human race utilizes fermented plant products as an important source of food. Plants and plant products are subjected to microorganisms as soon as grown. If they are harvested and consumed immediately (e.g., fresh fruits and berries), there has been little, if any, or no fermentation. However, when stored for some period of time, the micro flora in the plant products starts the fermentation process. Fermentation of the plant products is acceptable as long as the products are attractive in flavour and aroma and do not contain any toxic products. Uncontrolled fermentation may result an unacceptable aroma and taste and lead to fermented food as spoiled. So, the fermentation of the plant's products should be controlled by their natural flora. Fermentation of the plant products not only protects them from getting spoiled but also adds nutritive value to the products prepared from them. In this regard the current chapter highlights the microbiology of various plant-based fermented foods that are consumed in various parts of the world along with highlighting a few fermented foods consumed in different parts of the world. A brief insight to the biochemistry behind the action of various microorganisms has also been discussed to learn more about the action of the fermenting microbes.

Introduction

Campbell Platt defined fermented foods as '*those foods that have been subjected to the action of microorganisms or enzymes so that desirable biochemical changes cause significant modification in the food*' (Campbell Platt, 1987). The process of fermentation is immemorial and was carried out unknowing. Mostly the fermentation is used as a cheap method of preserving the food and protecting them from spoilage before the advancement in the freezing, canning and refrigeration. The simple result of the fermentation was the lowering of the pH of the product by using the lactic acid bacteria and also the production of the alcohol by the yeasts (Bamforth and Ward, 2014). Biochemically that was the conversion of the carbohydrates into the alcohols, acids and carbon dioxide. In this way the population of the spoilage microorganisms was reduced without affecting the nutritional value of the product. The advancement of the technology created tools like refrigeration and also increased the knowledge about the microbiology behind such processes, which in turn has increased the consistency in the processing of food. More advancement removed the dependency of the process on the indigenous bacteria and replaced it with the spawned starter cultures which help in controlling the desired end products. Fermentation in the strictest sense of the word is anaerobic, but most people extend the use of the term to embrace aerobic processes and indeed related non microbial processes, such as those completed by isolated enzymes (Caballero *et al.,* 2003). The present chapter deals with an insight to the microbiology and biochemistry behind the fermentation of plant and plant-derived food products along with discussing few examples of fermented food that is prepared and consumed in different parts of the world.

Microbiology of Fermentation

Bacteria

A diverse range of bacteria are used in fermentation technology, but most of them are classified into one of the three phyla, the *Firmicutes*, the *Proteobacteria*, and the *Actinobacteria*. The most important of these are lactic acid bacteria that belong to the *Firmicutes* and are Gram positive bacteria that are the main organisms used in the manufacture of fermented foods. Other bacteria that are the part of same phylum are the genera, *Bacillus* and *Staphylococcus* used for manufacturing of some fermented foods. The *Proteobacteria* contains Gram negative bacteria that are involved in mainly fermentation to produce vinegar and other alcoholic products. Finally, the *Actinobacteria* contains several genera relevant to fermented foods, including *Propionibacterium*, *Kocuria*, *Micrococcus*, and *Bifidobacterium* (Gibbons and Rinker, 2015).

The Lactic Acid Bacteria

The term, lactic acid bacteria, used to describe a group of functionally and genetically related bacteria. This term carries significant meaning to the food fermentation workers, therefore used freely in the text. Lactic acid bacteria contain lactic acid producing, low per cent G+C, non spore forming, Gram positive rods and cocci sharing many biochemical, physiological, and genetic properties. Most lactic acid bacteria are catalase negative, acid tolerant, aero tolerant, facultative anaerobes. They are classified as heterotrophic chemo-organotrophs, based on their carbon and energy needs as they require pre formed organic carbon both as a source of carbon and energy. The lactic acid bacteria are fastidious and grow only in nutrient rich, well fortified media (or food) under optimum conditions. These organisms can be isolated not only from plant material, milk, and meat, but also salt brines, low pH foods, and ethanolic environments (Holzapfel and Wood, 2014).

Different nutrient metabolism pathways have made the lactic acid bacteria to be used in fermented foods as they can metabolize sugars and make lactic acid and other metabolic end products. Two fermentative pathways exist in lactic acid bacteria that are, homofermentative pathway where more than 90 per cent of the sugar substrate is converted exclusively to lactic acid and the heterofermentative pathway with four end products–lactic acid, acetic acid, ethanol, and carbon dioxide. Lactic acid bacteria may possess either one or the other of these two pathways (*i.e.*, they are obligate homofermentative or obligate heterofermentative) or may show both pathways (referred to as being facultative heterofermentative).

The lactic acid bacteria group consists of 40 genera, of which 12 are used in fermented foods. Seven genera of lactic acid bacteria, *Lactobacillus*, *Lactococcus*, *Leuconostoc*, *Oenococcus*, *Pediococcus*, *Streptococcus*, and *Tetragenococcus*, are used directly in food fermentations. Whereas, genera like *Enterococcus*, *Carnobacterium*, *Aerococcus*, *Vagococcus*, and *Weissella*, are not widely found in foods due to their less significance.

Lactococcus

The genus *Lactococcus* consists of seven phylogenetically distinct species, *Lactococcus lactis*, *Lactococcus garviae*, *Lactococcus piscium*, *Lactococcus plantarum*, *Lactococcus fugiensis*, *Lactococcus chungangensis*, and *Lactococcus raffinolactis*. They all are non motile, ovoid cocci, obligately homofermentative, and facultative anaerobes, with an optimum growth temperature near 30°C. They have a distinctive microscopic morphology, usually appearing as ovoid cocci in pairs or short chains. One particular species, *L. lactis*, is among the most important of all lactic acid bacteria having foursubspecies: *L. lactis* subsp. *lactis*, *L. lactis* subsp. *cremoris*, *L. lactis* subsp. *hordinae*, and *L. lactis* subsp. *tructae*. However, only *L. lactis* subsp. *lactis* and *L. lactis* subsp. *cremoris* are used as starter cultures. *L. lactis* subsp. *lactis*, is an important commercially used variant due to its ability to metabolize citrate.

Plant material has long been considered to be the original habitat of *L. lactis* but now milk has become their new habitat. Due to this, the genomes of *L. lactis* reflect gene decay, implying that genes useful in their original habitat (*i.e.*, plant) are no longer essential in the new environment (*i.e.*, milk). Also, plasmid borne genes in lactococci encode for proteins involved in lactose transport and metabolism and casein hydrolysis and utilization. Although, *L. lactis* subsp. *lactis* and *L. lactis* subsp. *cremoris* differ in just a few seemingly minor physiological respects, such as the highest temperature at which most strains of *L. lactis* subsp. *lactis* are able to grow is at 40°C, whereas most *L. lactis* subsp. *cremoris* strains do not grow above 38°C. Also, *L. lactis* subsp. *lactis* has greater tolerance to salt (up to 4 per cent) than does *L. lactis* subsp. *cremoris*. It has recently been noted that there are strains of *L. lactis* subsp. *lactis* that have a *lactis* genotype, but a *cremoris* like phenotype. Likewise, there are strains of *L. lactis* subsp. *cremoris* that have a *cremoris* genotype, but a *lactis* like phenotype.

Streptococcus

The genus, *Streptococcus*, consists of many species that are facultative anaerobe, non-motile and with homofermentative metabolism. Only one species of this genus is used in fermentation, which is *Streptococcus thermophilus*. The bacterium is very much similar to the *L. lactis* and is also found in milk converting lactose into lactic acid. But the pathway of the *S. thermophilus* is different from the lactococci and also it works on higher temperature than its counterpart. Nutritional requirements of *S. thermophilus* are also more than the lactococci and it requires amino acids in the medium for its optimum growth. It is also characterized by bile sensitivity, salt tolerance, and a limited metabolic diversity.

Leuconostoc (Lc)

Leuconostoc genus contains mesophilic, facultative aerobic bacteria that are heterofermentative and metabolize sugars using phosphoketolase pathway. Plasmids are commonly found in this genus and reported to code for citrate and lactose metabolism genes along with bacteriocin coding genes. The genus contains 12 species whereas many other species were moved to other genera like *Oenococcus* and *Weissella* after taxonomical revisions. Of the 12 species, those involved in plant-based fermentation are *Leuconostoc mesenteroides* subsp. *mesenteroides, Leuconostoc kimchii*, and *Leuconostoc fallax* which are capable of fermenting plant-based sugars. Due to the heterofermentative nature, the bacteria produce various end products like CO_2, ethanol, lactic acid, and acetic acid. The genus is particularly important for fermentation due to its ability to impart diverse flavour and aroma to the fermented foods.

Pediococcus

Pediococcus genus is similar to the cocci-shaped lactic acid bacteria and phylogenetically located in the *Lactobacillaceae* family. The difference is in their ability to tolerate high acid and salt conditions. Of the 12 species of *Pediococcus*, only 2 species (*P. acidilactici* and *P. pentosaceus*) are found naturally in plant products and play important role in fermentation. The genes for the key metabolic pathways of fermentation are located on the plasmids present in the bacteria. Also, plasmids encode the genes for bacteriocin production that additionally increases the shelf life of the fermented products.

Lactobacillus (Lb)

Lactobacillus is heterogeneous group of bacteria with similarity being rod-shaped bacteria. Many species of this genus are found commonly in plant and vegetable materials and also in juices and fermented foods. Although they are very important part of fermented foods, they are also responsible for the spoilage of fermented and non-fermented foods. Currently there are more than 250 species of *Lactobacillus* and they may be either obligate homofermentative, or facultative heterofermentative or obligate heterofermentative in their mode of fermenting sugars. Being natural inhabitant of plant products, they can ferment vegetables without the need of any starter cultures. However, now a days, starter cultures are preferred that contain *L. plantarum* and *L. brevis* for obtaining a controlled fermented pickle. The species are mostly acid and salt tolerance, hence can ferment the products to their end stage without the requirement of any other bacteria. The genes required for sugar metabolisms are encoded by the plasmids, which also codes for the bacteriocins and antibiotic resistance.

Other Bacteria Important in Food Fermentations

Apart from the lactic acid bacteria, some other bacteria are also involved in the fermentation process as culture adjuncts or as primary culture but they are used for single purpose for most cases. These non-lactic acid bacteria may be Gram negative or Gram positive and are involved mainly for a single function to perform.

Bacillus

Most of the species of the genus *Bacillus* are involved in food spoilage and only one species that is *B. subtilis* has been reported to be involved in fermentation process. The species is mainly involved in the formation of Asian fermented food natto that is made from the soybeans.

Bifidobacterium

The use of *Bifidobacterium* species in the fermentation is not clear as such and also these are not commonly found in the raw food materials. But some of the species have been found fermentative in nature that may produce acetic acid and lactic acid. Also, these bacterial species have probiotic property, so their addition to the fermentation adds to the food nutritive values. So, due to these properties, these bacteria are now commonly used as starter culture on many commercially important fermented products. There are about 50 species of the genus *Bifidobacterium*, which were earlier classified in the genus *Lactobacillus*, but later these were placed as a separate genus based on differences with the *Lactobacillus*. *Bifidobacterium* are anaerobic and growth optimum is near the neutral pH. These bacteria rarely contain any plasmid and sugar fermentation is by a unique pathway that results in the production of both lactic acid and acetic acid.

Yeasts and Moulds Used in the Manufacture of Fermented Foods

There are more than a million fungi known to exist, but only a few are used in the fermentation technology that is classified as Ascomycota and Zygomycota. The fungi commonly used are yeasts that are unicellular and non-filamentous and moulds that are multi-cellular and filamentous. Important genera of these two groups that are involved in fermentation are *Saccharomyces*, *Kluyveromyces*, and *Zygosaccharomyces* belong to yeast and *Aspergillus*, *Penicillium*, and *Rhizopus* belong to moulds group (Kirk *et al.*, 2008; Kurtzman *et al.*, 2011).

Saccharomyces

The most important microorganism that is used in fermentation is the yeast of genus *Saccharomyces*. These are used in fermentation even more than the bacteria and the fermented products are popular as well as impact the economy of a country. *Saccharomyces cerevisiae* is the most commonly used species of all and its whole genome has been already studied. The speciation has been made by studying the carbon assimilation and fermentation pattern. Some strains are halotolerant and osmophilic and used in fermentation where there is high amount of sugar and salts in the substrate or the end product. In some of the plant products, the natural presence of the yeast is enough to carry out the fermentation while in most cases; a starter culture of the yeast is required for initiating the fermentation process (Naumov *et al.*, 2000; Hittinger, 2013).

Fermentation Basics

Biochemically, fermentation is energy yielding reaction where electron accepter is an organic molecule. So, in this reaction either pyruvic acid formed in glycolytic pathway act as electron accepter to produce lactic acid or acetaldehyde formed by pyruvate decarboxylation act as electron acceptor to produce ethanol. So, the above definition of fermentation is true regarding these reactions (Bosma *et al.*, 2017). However, there are many fermented foods that are made by pathways that do not fit in the above definition. In these fermentations, instead of glycolytic products, proteins and polysaccharides are metabolized by the fungi. So, in this chapter, term fermentation is used for all the metabolic and enzymatic processes that are involved in the fermentation of food. The process of fermentation by the microbes is for the energy production in the form of ATP that is to be utilized by the microbe itself. However, the end products of the metabolic reactions impart flavour and aroma to the food but sometimes metabolic end-products may spoil the food and thus, requires controlled fermentation process. Therefore, there is a need of understanding the substrates and the pathways associated with them for consistency in the fermented foods with desired chemical, physical, biological and nutritional characteristics (Kowalczyk *et al.*, 2016).

Homofermentation

Homofermentation involves the metabolism of the sugar by using the glycolytic Embden-Meyerhof-Parnas (EMP) pathway. The pathway yields two moles of ATP along with two moles of pyruvate. Lactate dehydrogenase enzyme then hydrolyses the pyruvate to yield lactate which is then converted into lactic acid. The end product of the homofermentative pathway is always lactic acid and the pH of the product is very low due to high amount of acid production. Examples of homofermentative microbes are *L. lactis, S. thermophilus, Lb.helveticus, Lb. delbrueckii, P. acidilactici* and *Lb.* sakei (Gänzle, 2015).

Heterofermentation

Heteroferementation involves the metabolism of the sugar by phosphoketolase pathway which utilizes phosphoketolase enzyme instead of aldolase. The end products of the reaction are acetate and lactate in the equimolar concentration along with evolution of CO_2. These end products depend upon the type of the substrate used and the culture conditions provided. Most of the lactic acid bacteria that are used in fermentation are heterofermentative in nature and used predominantly as starter cultures. Under conditions like low sugar or aerobiosis, some homofermentative bacteria also divert some lactate towards the end-products such as acetaldehyde, formate or acetate. Such pathways are very important for the fermentation process as these end products imparts flavours and texture to the fermented foods. Examples of heterofermentative bacteria are *Lc. mesenteroides,Lc. lactis, Lc. mesenteroides, Lc. kimchii, Oenococcus oeni*, and *Lb. sanfranciscensis* (Zaunmüller *et al.*, 2006).

Sugar Metabolism by *Saccharomyces cerevisiae*

Saccharomyces cerevisiae utilizes the sugars in the similar fashion as the homofermentative lactic acid bacteria but they can utilize wide range of sugars. But the difference in the pathway is that the pyruvate is not reduced to lactate, instead the end product is ethanol and CO_2. Also, the yeast has functional electron chain and citric acid pathway, making them survive in aerobic conditions. The yeast has 18 different transport proteins that enable them to metabolize in high and low concentration of sugars. Due to this there is complete metabolism of the sugars ensuring high quality end products (Conrad *et al.*, 2014).

Other Metabolic Systems

The lactic acid and ethanol are the products of the sugar metabolism that forms the basis of most foods and beverages. But there are other metabolites like diacetyl which imparts flavour to the fermented foods. These metabolites are synthesised by using the components other than sugars. In this regard, fats and proteins are converted to the products that impart strong flavour and aroma to the foods.

Protein Metabolism by Lactic Acid Bacteria

Lactic acid bacteria are not able to utilize the inorganic nitrogen, so they require organic source of nitrogen like free amino acids. Hence, for the low concentration of the amino acids in the substrate, lactic acid bacteria have developed proteolytic systems for degradation of the peptides into free amino acids and transport system for up taking them. More than 20 different peptidases have been identified in lactococci and lactobacilli that generated a pool of amino acids from proteins. The protein metabolism systems in vegetables and bread are less important than the casein metabolism in milk. The casein is utilized in three steps that are casein hydrolyzation by extracellular proteinases to form peptides, transportation of the peptides into cells via peptide transport systems and hydrolyzation of peptides by intracellular peptidases to form free amino acids. This metabolism system is an important system for flavour and aroma enhancement (Liu *et al.,* 2010).

The Citrate Fermentation

Citrate is a common constituent of the plants that is fermented by the lactic acid bacteria to produce diacetyl, a flavouring and aromatic compound. The pathway of citrate to diacetyl is of no significance to the microbe as no energy is produced in this pathway. However, the protons formed during this pathway may initiate the proton motive force which when utilized properly may result in overall increase in the energy of the cell. Thus, citrate fermentation results in an increase in the metabolic energy available to the cell along with adding flavour and aroma to the fermented food (Pedersen *et al.,* 2012).

Fermented Plant Products

The process of fermentation of vegetables is almost same for different vegetables grown around the world; however, grains are fermented in different ways to produce different kind of final fermented foods. The diversity in the food product is contributed by the production process and microbiological variables. Vegetable's fermentation is an old process that was a means of food preservation involving salting and packing in jars for storage at ambient temperature. So, in this regard different plant-based products are fermented to prepare foods that are traditional to different parts of the world (Table 3.1). Some of them have been described below.

Boza

Boza is a traditional fermented beverage that is produced from millet, maize, wheat, rye, or rice and other cereals (Zorba *et al.,* 2003; Kohajdová, 2010; LeBlanc and Todorov, 2010; Nyanzi and Jooste, 2012; Kancabas and Karakaya, 2013). It is a viscous pale yellow colour liquid with a sweet or sour taste (Altay *et al.,* 2013) that is consumed in countries of the Balkan region including Bulgaria, Romania, Albania and Turkey (Cosansu, 2009; Kivanc *et al.,* 2011; Nyanzi and Jooste, 2012). The preparation of Boza involves milling of cereals and cooking in water in open boiler. After that, bran and hull are removed and sugar is added to the cooled mixture. Fermentation is carried out at 30°C for 24 hr (Todorov, 2010; Altay *et al.,* 2013) by back-slopping or adding starter cultures as yoghurt or sourdough (Nyanzi and Jooste, 2012). After fermentation, the boza is cooled and stored in plastic bottles and has a shelf life of 15 days (Todorov, 2010; Nyanzi and Jooste, 2012). Fermentation is carried out predominantly by various *Lactobacillus* and *Leuconostoc* species along with some yeasts belonging to the genus *Saccharomyces* and *Candida* (Östürk *et al.,* 2013, Vashuda and Mishra, 2013). The bacteria used in fermentation also produce bacteriocins that increase the shelf life of the final product (Kivanc *et al.,* 2011).

Table 3.1: Fermented Plant Foods of different Regions of the World

Fermented Food	*Substrate*	*Microorganism*	*Country*	*Food Form*
Boza	Wheat, rye, millet, maize	*Lactobacullus spp., Leuconostoc, Sachcaromyces cerevisiae*	Bulgaria, Romania, Turkey, Albania	Beverage
Burong mustala	Mustard	*Lactobacillus brevis, Pediococcus cerevisiae*	Philippines	Salad, side dish
Bushera	Sorghum, millet	*Lactobacullus spp., Leuconostoc, Sachcaromyces cerevisiae*	Uganda	Beverage
Caper berries	Caper berries	*Lactobacillus plantarum*	Greece, Italy	Brine of capers
Cucumbers	Cucumbers	*Lactobacillus plantarum, Pediococcus pentosaceus*	Asia, USA	Pickle
Dakguadong	Mustard leaf	*Lactobacillus plantarum*	Thailand	
Dhamuoi	Cabbage, vegetables	*Leuconostoc mesenteroides, Lactobacillus plantarum*	Vietnam	Salad, side dish
Dhokla	Rice, Bengal gram	*Lactic acid bacteria, yeasts*	India, Sri Lanka	Soft spongy food
Dosa	Rice ad blackgram	*Lactobacillus fermentum, Leuconostoc spp., Sacharomyces* spp.	India	Crisp pancake
Gundruk	Cabbage, radish, leafy vegetables	*Lb. plantarum, Lb. casei*subsp. *casei, Lc. pseudoplantarum, Lb. fermentum, P. pentosaceus*	Nepal	Pickle
Hardaliye	Vegetables	*Lb. paracasei*subsp. *paracasei, Lb. casei Lb. pontis, Lb. brevis, Lb. acetotolerans, Lb. sanfranciscoensis*	Turkey	Beverage

Fermented Food	***Substrate***	***Microorganism***	***Country***	***Food Form***
Idli	Rice and black gram	*Leuconostoc mesenteroides, Lactobacillus delbruckei, Lactobacillus fermentum, Lactobacillus lactis, Sacharomyces cerevisiae*	India	Soft spongy food
Injera	Tef	Yeasts, *Lactobacillus* sp.	Ethiopia, Sudan	Bread
Jiang-gua	Cucumber	*Enterococcus casseliflavus, Leuconostoc lactis,Lc. mesenteroides, Lb. pentosus, Lb. plantarum, Lb. paraplantarum, Lactococcus lactis* subsp. *lactis, Weissella hellenica, Weissella cibaria*	China	Pickle
Kanji	Carrots	*Lactobacillus plantarum, Lactobacillus brevis*	India, Pakistan	Beverage
Kenkey	Maize, sorghum	*Lactobacillus* sp., yeasts	Ghana, Botswana	Gruel or dumpling
Khalpi	Cucumbers	*Lactobacillus plantarum, Lb. brevis, Leuconostoc fallax, Pediococcus pentosaceus*	Nepal	Pickle
Kimchi	Vegetables	*Leuconostoc mesenteroides, Leuconostoc kimchii, Leuconostoc citreum, Leuconostoc gasicomitatum, Lc. pseudomesenteroides, Lactobacillus plantarum, L. brevis, Lactobacillus curvatus, Lactobacillus sakei, Lactobacillus maltaromicus, Lactobacillus bavaricus, P. pentosaceus, Weissella confusa, Weissella kimchii, Weissella koreensis*	Korea	Salad, side dish
Kirsa	Sorghum	*Lactobacillus bulgaricus*, yeasts	Sudan, Ethiopia	Bread
Kishk	Wheat	*Lactobacillus plantarum, L. casei, L. brevis, Bacillus subtilis, Saccharomyces cerevisiae*	Egypt, Syria	Dry balls
Kocho	False banana	*Lactobacillus* sp.	Ethiopia	Bread
Koko	Maize, sorghum	*Sacharomyces* spp., *Lactobacillus* sp.	Ghana	Dumpling
Kvass	Rye, Barley	*Lactobacillus casei, Leuconostoc mesenteroides, Saccharomyces cerevisiae*	Europe	Non-alcoholic beverage

Fermented Food	**Substrate**	**Microorganism**	**Country**	**Food Form**
Mahewu	Maize	*Lactobacillus delbruckii, Lactobacillus bulgaricus, Streptococcus lactis*	East African countries	Beverage
Mawe	Maize, sorghum, millet	*Lactobacillus fermentum, Candida krusei, Saccharomyces cerevisiae*	South Africa, Benin	Dough
Ogi	Maize, sorghum, millet	*Lactobacillus*spp., *Aerobacter, Corynebacterium*, yeast, moulds	Nigeria, Benin	Gruel
Olives	Olives	*Lactobacillus plantarum, Lb. paracasei, Lb. pentosus, Lb. casei, Lb. vaccinostercus, Lb. suebicus, Lb. paracollinoides, Lactobacillus brevis, Pediococcus cerevisiae, Leuconostoc mesenteroides, Lc. Lactis*	Spain, Italy	Brine of olives
Onion and garlic	Onion and garlic	*Lactobacillus plantarum*		Sour onion Sour garlic
Pak-sian-dong	Leaves of Pak-sian	*Lactobacillus brevis, Pediococcus cerevisiae, Lactobacillus plantarum*	Thailand	-
Poi	Taro	*Lactobacillus* sp., *Geotrichum* sp., *Saccharomyces cerevisiae*	Polynesia	Viscous liquid
Pozol	Maize	*Lactobacillus pantarum, Lactococcus lactis, Lactobacillus casei, Lactobacillus delbruckii, Lactobacillus fermentum*	Mexico	Beverage
Proviva	Oat	*Lactobacillus plantarum*	Sweden	Beverage
Puto	Rice	*Leuconostoc mesenteroides, Lc. Pseudomesenteroides, Lc. Citreus, Lc. fallax*	Philippines	Cake
Salgam	Black/Violet carrots, turnip, bulgur flour, sour dough, salt and water	*Lb. plantarum, Lb. paracasei subsp. paracasei, Lb. fermentum, Lb. brevis*	Turkey	Beverage
Sauerkraut	Cabbage	*Leuconostoc mesenteroides, Lactobacillus plantarum, Lactobacillus brevis*	China, Korea	Dried salted vegetable
Sinki	Radish	*Lactobacillus plantarum, Lb. brevis, Lb. fermentum, Leuconostoc fallax, Pediococcus pentosaceus*	India, Nepal, Bhutan	Pickle

Fermented Food	*Substrate*	*Microorganism*	*Country*	*Food Form*
Suan-tsai	Mustard leaves	*Pediococcus pentosaceus, Tetragenococcus halophilus*	Taiwan	Pickle
Sunki	Leaves of otaki-turnip	*Lactobacillus plantarum, Lactobacillus brevis, Pediococcus pentosaceus, Bacillus coagulans*	Japan	Salad, side dish
Sweet cherry	Sweet cherry	*Lactobacillus plantarum, Leuconostoc mesenteroides, Pediococcus acidilactici*	Asia	Brine of sweet cherry
Sweet potato	Sweet potato	*Lactobacillus plantarum*		Pickle
Tarhana	Wheat, milk	*Lactobacillus* sp. *Sachcaromyces cerevisiae*	Greece, Turkey	Dough
Tempeh	Soy beans	*Rhizopus oligisporus, Mucor* sp.	Indonesia, USA, Netherlands	Tempeh cake
Togwa	Maize	*Lactobacillus pantarum, Lactobacillus brevis, Lactobacillus fermentum, Lactobacillus cellobiosus, Sacharomyces cerevisiae, Candida tropicalis*	East Africa	Beverage
Tursu	Cucumbers, cabbage, green tomatoes, green peppers and other vegetables	*Lactobacillus plantarum, Lb. brevis, Leuconostoc mesenteroides, Pediococcus pentosaceus, Enterococcus fecalis*	Turkey	Pickle
Uji	Sorghum, maize, cassava	*Leuconostoc mesenteroides, Lactobacillus plantarum*	Kenya, Uganda, Tanzania	Non-alcoholic beverage
Yosa	Oat bran	*Lactobacillus acidophilus, Bifidobacterium*	Finland	Beverage

Bushera

Bushera is a traditional fermented beverage made from millet or sorghum flour and widely consumed in Uganda (Muyanja *et al.*, 2003). The production process involves mixing the germinated grains in boiled water and then cooling it. Then the grains are powdered and fermented up to 6 days to get the beverage with desired taste (Vashuda and Mishra, 2013). The bacteria that are involved in fermentation are *Lb. plantarum, Lb. paracasei, Lb. fermentum, Lb. brevis, Lb. delbruckii,* and *Streptococcus thermophillus* (Nyanzi and Jooste, 2012).

Cucumbers

Cucumbers are fermented to form dark green pickle that is consumed widely in Asia, Africa and Latin America. The production process involves washing of the cucumbers and then addition of the brine to it. Starter culture of *Lb. plantarum* or *P. pentosaceus* is added to the brine when the pH is around 4.7 (Steinkraus, 2002). The fermentation is carried out for two to four weeks until a final pH of 3.1-3.4 is reached and then the fermentation is stopped by adding salt to it (Tamang *et al.*, 2005).

Dhokla

Dhokla is a traditional fermented food made from rice and chickpea and is widely consumed in India and Sri Lanka (Das *et al.*, 2013). The production involves mixing the flours of the rice and chickpea and then keeping it for fermentation at room temperature for few hours to overnight for fermentation. The preparation is similar to the idli where fermented batter is poured into the greased pan and steamed to get soft and spongy dhokla (Kohajdová, 2010; Das *et al.*, 2013). The microorganisms involved in fermentation are lactic acid bacteria and yeast that bring about the changes making the food rich in flavour and texture.

Dosa

Dosa is a traditional fermented dish prepared from the rice and blackgram and consumed widely in Indian sub-continent (Das *et al.*, 2013). The production involves preparing the suspension of the ingredients separately with water and mixing them. Then the mixed suspension is fermented for 8-20 hours at room temperature for auto-fermentation. The dosa is prepared by spreading the thin layer of batter on flat heated plate with little oil that transforms the sol into a gel within a few minutes to form a crisp pancake like product (Kohajdová, 2010; Das *et al.*, 2013). The microbes involved in the fermentation are bacteria like *Lb. fermentum*, *Lc.Mesenteroides* and *Streptococcus faecalis* along with yeast like *Debareomyces hansenii*, *Trichosporon beigelii* and *Saccharomyces cerevisiae* (Kohajdová, 2010).

Gundruk

Gundruk is a traditional fermented vegetable such as radish and cabbage, consumed as pickle in Nepal (Tamang *et al.*, 2005). The dish is served as appetizer with the main food to add on minerals to the diet which lacks them. The production involves fermenting the shredded vegetables in earthen pots with warm water for 5-7 days at 18°C. The microorganisms involved in fermentation are *Lactobacillus* and *Pediococcus* species that develop mild acidic taste (Steinkraus, 2002; Tamang *et al.*, 2005).

Idli

Idli is a traditional fermented soft and spongy snack food consumed widely in Indian sub-continent and is made from rice and black gram (Das *et al.*, 2013;

Manickavasagan *et al.*, 2013). Black gram can be substituted with other legumes like chickpea, green gram and soybean whereas small proportion of rice with millet (Maheswari and Shetty, 2013). The idli production involves grinding of the soaked ingredients separately and then mixing the batter together. Fermentation could be carried out naturally or sometimes sour buttermilk or yeasts can be added to it for enhancing fermentation (Sekar and Mariappan, 2007). The batter is fermented overnight at room temperature and then the batter is steamed in greased pans for few minutes to get soft, fluffy idlis with good taste and texture (Aachary *et al.*, 2011; Manickavasagan *et al.*, 2013). Fermentation process is a mixed auto-fermentation involving bacteria like *Lc. mesenteroides, Lb. delbrueckii, Lb. fermenti, Lb. lactis* and yeast like *Saccharomyces cerevisiae* (Nout, 2009; Kohajdová, 2010; Das *et al.*, 2013). Lactic acid bacteria reduce the pH of the batter from 6 to 4.2 which then favour the growth of the yeast for further fermentation of the batter for enhanced flavour and leavening of the batter, also increasing the vitamin and amino acid content of the food (Nout, 2009).

Injera

Injera is traditional fermented bread made from teff and sorghum and is widely consumed in Ethiopia (Anyango *et al.*, 2011). The production process involves mixing the flour with water and to prepare the dough by adding starter culture in the form of previously fermented dough (Abiyu *et al.*, 2013). The dough is fermented for 2-3 days after which made the bread from it. The fermented bread is further stored at room temperature for 3 days for enhancing the flavour of the product (Kohajdová, 2010). Fermentation is carried out by fungi like *Candida* sp., *Hormodendrum* sp., *Rhodotorula* sp.,*Penicillium* sp., and *Aspergillus* sp. along with some unidentified bacteria, which make the injera highly nutritional by increasing the amount of calcium and iron in it and also giving it slightly sour taste (Champagne, 2009; Kohajdová, 2010).

Kanji

Kanji is a traditional ready-to-serve beverage of India and Pakistan which is deep purple in colour. The production process involves mixing of finely grated carrot with water, salt, mustard seed and chilli powder followed by placing the mixture in sealed earthenware vessel. Fermentation is carried out for 7 to 10 days by *Lactobacillus plantarum* and *Lactobacillus brevis*. The kanji is thought to have high nutritional value with soothing effect but it is to be consumed within 3-4 days after which it goes bad (Ray and Panda, 2007).

Kenkey

Kenkey is a fermented food made from maize and is widely consumed in Ghana (Nout, 2009). The production involves mixing of raw fermented and partially cooked dough, and making balls of it which is wrapped in leaves and boiled for few hours. After that final product, dumpling is made which is eaten with stews and soups (Annan *et al.*, 2003). The fermentation involves mostly lactic acid bacteria like

Lb. reuteri and *Lb. fermentum* along with yeast *Candida krusei* (Kohajdová, 2010; Nyanzi and Jooste, 2012). Fermentation results in increase in the lysine content along with flavouring agents (Nyanzi and Jooste, 2012).

Kimchi

Kimchi is a traditional fermented dish prepared from the vegetables and served with almost every dish in Korea, its estimated production in Korea is around 1.5 million tonnes (Cho *et al.,* 2009). The production process involves cutting and brining (salting) of the vegetables followed by rinsing and draining of the water. After that some minor ingredients like spices are added to it and fermented in jars. Fermentation has been reported to increase the nutritional value of the food by increasing nutritional components along with phytochemicals, thus making it one of the healthiest food of the world (Lee *et al.,* 2011). Fermentation is done by lactic acid bacteria that vary with the pH during production process. Early fermentation is carried out by the *Lc. Mesenteroides* while the final product is formed by *Lb. sakei* (Cho *et al.,* 2009). Other bacterial species such as *Achromobacter, Flavobacterium* and *Pseudomonas* have also been reported from the kimchi fermentation (Kim and Chun, 2005).

Kirsa

Kirsa is traditional bread made from sorghum or millet and is widely consumed in Sudan (Asmahan and Muna, 2009). The production process involves mixing of the millet or sorghum flour with water in 1:2 and then addition of the starter culture in the form of fermented dough from the previous batch (Rahman *et al.,* 2010). The fermentation is carried out for 12-19 hours until pH-4.0 is obtained after which the dough is baked into breads (Asmahan and Muna, 2009). Fermentation is carried out mainly by *Lb. fermentum* and *Lb. amylovorus* with other bacteria and yeast like *Lb. brevis, P. pentosaceas, Acetobacters* sp. and *Saccharomyces cerevisiae* (Asmahan and Muna, 2009; Rahman *et al.,* 2010).

Kishk

Kishk is traditional food made from wheat and is consumed widely in countries like Jordan, Syria, Lebanon and Egypt (Kohajdová, 2010; Mahasneh and Abbas, 2010; El-Nawawy *et al.,* 2012). The production process involves boiling, drying, milling and sieving of the wheat to remove bran. Then started culture is added to the wheat paste in the form of yogurt to carry out fermentation. After fermentation, the product formed is dried, powdered and stored in the shape of balls (Kohajdová, 2010). Fermentation is carried out by *Lb. plantarum, Lb. casei Lb. brevis, Bacillus subtilis* and yeasts which not only enhances the mineral in the fermented food but also increases the bioavailability of the wheat nutrients (Blandino *et al.,* 2003).

Kocho

Kocho is a traditional food made from Ensete false banana and is widely consumed in Ethiopia (Steinkraus, 2002). The production process involves fermenting the peeled banana (pulp) in the pits for three to six weeks. Fermentation results in increased CO_2 which creates an anaerobic atmosphere and an increase in temperature. The resulted fermented product has a low pH and a strong odour with paste like consistency (Ray and Panda, 2007).

Kvass

Kvass is traditional beverage made from barley and rye and is widely consumed in European countries (Nyanzi and Jooste, 2012). The beverage is similar to bosa in terms of usage and composition and is non-alcoholic in nature (Kohajdová, 2010). The production process involves the dilution of the flour batter in hot water and its clarification. After that sugar is added to it and fermentation is started by addition of the yeast to it. The fermentation is stopped by cooling the kvass to 4°C to obtain a nutritionally rich beverage (Nyanzi and Jooste, 2012). Fermentation is carried out predominantly by *Lb. casei, Lc. mesenteroides* and *Saccharomyces cerevisiae* which are not killed after fermentation thus providing a probiotic drink for the consumers (Nyanzi and Jooste, 2012).

Mahewu

Mahewu is a fermented beverage that is made from the maize and is widely consumed in Africa and Gulf countries (Chelule *et al.*, 2010; Kohajdová, 2010; Nyanzi and Jooste, 2012). The production involves mixing the maize porridge with water and then adding the flour of wheat, millet or sorghum to it, and allowed to ferment (Kohajdová, 2010; Vashuda and Mishra, 2013). The fermentation is carried out by natural flora consisting mainly of bacterium *Lactococcus lactis* (Champagne, 2009) whereas commercial production of mahewu employs *Lb. delbruckii* (Nyanzi and Jooste, 2012).

Mawe

Mawe is fermented dough made form maize and consumed widely in Togo and Benin (Greppi *et al.*, 2013). Production of mawe involves washing of maize, extracting endosperm and dough kneading. The dough is fermented at 30°C for about 3 days to reach the desired pH of 3.5-4.0 (Nout, 2009). Mawe so prepared is further used for preparation of other foods like albo, agidi, akassa and koko (Greppi *et al.*, 2013). Fermentation involves hetero-fermentative bacteria like *Lb. buchneri, Weissella confuse, Lb. curvatus, Lb. brevis, Lb. cellobiosus, Lb. fermentum* and yeast species like *Saccharomyces cerevisiae, Candida glabrata, C. kefyr, C. krusei* (Nyanzi and Jooste, 2012; Greppi *et al.*, 2013).

Ogi

Ogi is traditional fermented gruel made from maize and is widely consumed in Nigeria (Greppi *et al.*, 2013). The production process involves washing, steeping, milling and sieving of the maize. After that fermentation is carried out for 1-3 days at room temperature to obtain the final product (Evans *et al.*, 2013). Fermentation is spontaneous and is carried out by bacteria like *Lb. plantarum, Lb. brevis, Lb.*

fermentum and yeasts like *Saccharomyces cerevisiae, Candida krusei and C. tropicalis* which decreases the pH to 3.6 and also produce lactic acid and acetic acid during fermentation (Kohajdová, 2010).

Onion and Garlic

Onions are fermented to form the sour onions using yellow or white storage onions. The production process involves the slicing of onions followed by salt addition and anaerobic fermentation at 18°C. Fermentation is carried out by *Lb. plantarum* added as started culture in the form of brine of sauerkraut. The fermented onions have a pH of 3.25-3.35 with lactic acid content of 1.2-1.5 g/100ml. The sour onions produced have similar tartaric acid taste like that of sauerkraut with no pungency left of the raw onions (Roberts and Kidd, 2005). Garlic fermentation is carried out for a week by using garlic that are blanched in hot water (90°C) and addition of starter culture of *Lb. plantarum.* The fermented blanched garlic can be stored at 30°C in 3 per cent NaCl solution without addition of any preservative (De Castro *et al.,* 1998).

Pozol

Pozol is a traditional beverage made from maize and is widely consumed in Mexico (Diaz-Ruíz *et al.,* 2003). The production process involves cooking of maize grains in 1 per cent lime water for 90 minutes followed by grounding the dehusked grains to make dough (Nyanzi and Jooste, 2012). The dough is then made into balls and wrapped in banana leaf for fermentation. The fermentation is done for up to 4 days to attain a desired pH of 3.7-4.7 (Prado *et al.,* 2008; Vashuda and Mishra, 2013). Fermentation is carried out by a variety of lactic acid bacteria like *L. lactis, S. suis, Lb. plantarum, Lb. casei, Lb. alimentarium* and *Lb. delbruekii* however amylolytic enterococci and streptococci importance has been reported in recent studies (Díaz-Ruíz *et al.,* 2003; Champagne, 2009; Kohajdová, 2010).

Proviva

Proviva is a probiotic beverage made from oats and is widely consumes in Sweden (Prado *et al.,* 2008). The production involves addition of malted barley to the oats and fermenting it with the bacterium *Lb. plantarum* 299v (Nyanzi and Jooste, 2012). The final product has a bacterial count of 5 x 10^{10} cfu and is a mixture of 5 per cent oat meal and rest is fruit juice (Nyanzi and Jooste, 2012).

Puto

Puto is a fermented food made from rice and is widely consumed in Philippines as snack or in breakfast. The production involves washing, soaking of rice, grounding and then addition of coconut milk and sugar. The batter so formed is then fermented for few hours till it is acidified. Then the batter is steamed and served (Kohajdová, 2010). Fermentation involves lactic acid bacteria like *Lc. fallax, Lc. citreus, Lc. pseudomesenteroides,* and *Lc. mesenteroides* (Kelly *et al.,* 1995).

Sauerkraut

Sauerkraut is a form of fermented cabbage that is consumed in China and Korea (Liu *et al.*, 2011). The fermentation of the cabbage results not only increase digestibility of the cabbage but also used for long time as a preservation technique to protect cabbage from spoilage (Steinkraus, 2002; Viander *et al.*, 2003). Currently the sauerkraut production is a big part of food industry in Vietnam and Korea. The fermentation process involves the addition of the salt (2.25-2.5 per cent) to the shredded cabbage followed by anaerobic fermentation in a tank which is carried out at 18°C for three weeks (Viander *et al.*, 2003). The fermentation is carried out by *Lc. mesenteroides*, which produces lactic acid, acetic acids, and CO_2 which maintains anaerobic conditions and well as decrease the pH quickly. At lower pH, *Lactobacillus* and *Pediococcus* species continue to decrease the pH until it reaches 3.5-3.8 (Yoon *et al.*, 2006; Xiong *et al.*, 2012).

Sinki

Sinki is a traditional pickle prepared from radish and consumed mainly in India, Bhutan and Nepal (Dahal *et al.*, 2005). The fermentation procedure involves shredding of sundried radish and fermenting it for 12 days at 3°C in sealed containers. Fermentation is done by *Lb. fermentum* and *Lb. brevis*, followed by *Lb. plantarum* until the pH reaches 3.3.

Sunki

Sunki is traditional fermented vegetable prepared for the otaki-turnip leaves and is consumed with rice in Japan. The fermentation procedure involves mixing of boiled otaki turnip with zumi and addition of previous year sunki as starter culture to ferment it for two months. Fermentation is carried out at low temperatures during winter and involves *Lb. plantarum, Lb. brevis, Bacillus coagulans* and *P. pentosaceus* (Battcock and Azam-Ali, 2001).

Sweet Potato

Sweet potatoes are fermented to form pickles. The production procedure involves the blanching of the sweet potato followed by brining with NaCl solution (2-10 per cent) and inoculation with *Lb. plantarum* at 28± 2°C for 28 days. The fermentation results depend upon the choice of sweet potatoes used. The anthocyanin rich sweet potatoes fermentation results in lower pH (2.5-2.8), lactic acid (1.0–1.3 g/kg), starch (56–58 g/kg), and titratable acidity (1.5–1.7 g/kg) as compared with the β-carotene-rich sweet potatoes fermentation leads pH (2.9–3.0), lactic acid (2.6-3.2 g/kg), starch (58–68 g/kg), and titratable acidity (2.9–3.7 g/kg) (Panda *et al.*, 2007). β-carotene rich fully boiled potatoes when fermented under same conditions for 2 days results in formation of lacto-juice that is biochemically similar to the sweet potato pickle formed by fermentation (Panda and Ray, 2007).

Uji

Uji is a popular non-alcoholic fermented beverage that is made from sorghum, maize and millet. It is widely consumed in breakfast in East Africa in fermented as well as unfermented form. The production process of fermented uji involves grinding of soaked grains followed by filtration to remove the coarse particles and bran. The fermentation is carried out for few days in submerged culture state (Kohajdová, 2010). Starter culture of *Lactobacillus* species is added to it in the form of small amount of previously fermented uji. The dominant *Lactobacillus* species is *Lb.plantarum* while *Lb. cellobiosus* and *Lb. buchneri* are also the part of fermentation process (Kohajdová, 2010; Nyanzi and Jooste, 2012). Some part of the supernatant is boiled and mixed with the remaining part to obtain thin cooked beverage which is then consumed by adding some salt or sugar to it for taste (Nout, 2009).

Yosa

Yosa is a fermented drink made from oat bran using probiotic bacteria (Rivera-Espinosa and Gallardo-Navarro, 2010; Zubaidah *et al.,* 2012). The flavour of the drink is similar to yoghurt and is consumed widely in Finland and other Scandinavian countries (Rivera-Espinoza and GallardoNavarro, 2010). The production process involves the cooking of the rice bran in water and then fermenting it with bacteria *Lb. acidophilus* and *B. lactis* (Nyanzi and Jooste, 2012; Zubaidah *et al.,* 2012). Nutritionally yosa is a rich source of fibre and β-glucan along with potential probiotic beverage (Zubaidah *et al.,* 2012).

Conclusion

It is concluded that the fermentation of the plant products is very significant as this not only helps in preservation of the food, but also adds to the nutritive value of the product formed. Fermentation of vegetables and grains increases their shelf life along with increased digestibility and mineral content. Different metabolic pathways involved in fermentation process decreases the harmful microbe content from the food and also increases the probiotic potential by incorporating the good bacteria to the final product. Thus, fermentation is a cheap means of preservation and increasing the quality of the plant-based food products along with increasing the palpability of the food.

Authors' Contributions

NM- Initial writeup of the manuscript; NM, PR and RS- collected the literature; RS- conceptualize, editing and final writing of the manuscript.

REFERENCES

Aachary, A.A., Gobinath, D., Prapulla, S.G. 2011. Short chain xylooligosaccharides: a potential prebiotic used to improve batter fermentation and its effect on the quality attributes of idli, a cereal-legume-based Indian traditional food. Int. J. Food Sci. Tech. 46: 1346–1355.

Abiyu, H.T., Woldegiorgus, A.Z., Haki, G.D. 2013. Preparation of injera pre-fermented flour: nutritional and sensory quality. Int. J. Sci. Inno. Discover. 3: 165–175.

Annan, N.T., Poll, L., Plahar, W.A., Jakobsen, M. 2003. Aroma characteristics of spontaneously fermented Ghanian maize dough kenkey. Eur. Food Res. Technol. 217: 53–60.

Anyango, J.O., De Kock, H.L., Taylor, R.N. 2011. Evaluation of the functional quality of cowpea-fortifi ed traditional African sorghum foods using instrumental and descriptive sensory analysis. LWT - Food Sci. Technol. 44: 2126–2133.

Asmahan, A.A., Muna, M.M. 2009. Use of starter cultures of lactic acid bacteria and yeasts in the preparation of kisra, a Sudanese fermented food. Pak. J. Nutr. 8: 1349–1359.

Bamforth, C.W., Ward, R.E. (2014). Oxford Handbook of Food Fermentations. New York: Oxford University Press.

Battcock, M., Azam-Ali, S. 2001. Fermented Fruits and Vegetables: A Global Perspective. FAO Agricultural Services Bulletin. No. 134. 96 pp. Rome.

Blandino, A., Al-Aseeri, M.E., Pandiella, S.S., Cantero, D., Webb., C. 2003. Cereal-based fermented foods and beverages. Food Res. Int. 36: 527–543.

Bosma, E.F., Forster, J., Nielsen, A.T. 2017. Lactobacilli and pediococci as versatile cell factories – Evaluation of strain properties and genetic tools. Biotechnol. Adv. 35: 419–442.

Caballero, B., Trugo, L.C., Finglas, P.M. (2003). Encyclopaedia of Food Sciences and Nutrition. Oxford: Academic Press.

Campbell Platt, G. (1987). Fermented Foods of the World: A Dictionary and Guide. London: Butterworths.

Champagne, C.P. 2009. Some technological challenge sin the addition of probiotic bacteria to foods. pp. 761–804. *In*: D. Charalampopoulos and R.A. Rastall (eds.). Prebiotics and Probiotics Science and Technology. Springer, New York.

Chelule, P.K., Mdongwa, H.P., Carries, S., Gqaleni N. 2010. Lactic acid fermentation improves the quality of amahewu, traditional South African maize-based porridge. Food Chem. 122: 656–661.

Cho, K.M., Math, R.K., Islam, S.M.A., Lim, W.J., Hong, S.Y., Kim, J.M., Yun, M.G., Cho, J.J., Yun, H.D. 2009. Novel multiplex PCR for the detection of lactic acid bacteria during kimchi fermentation. Mol. Cell. Probes 23: 90–94.

Conrad, M., Schothorst, J., Kankipati, H.N., Zeebroeck, G.V., Rubio Texeira, M., Thevelein, J.M. 2014. Nutrient sensing and signaling in the yeast *Saccharomyces cerevisiae.* FEMS Microbiol. Rev. 38: 254–299.

Cosansu, S. 2009. Determination of biogenic amines in a fermented beverage, boza. J. Food Agric. Environ. 7: 54–58.

Dahal, N.R., Karki, T.B., Swamylingppa, B., Li, Q., Gu, G. 2005. Traditional foods and beverages of Nepal—a review. Food Rev. Int. 21(1): 1–25.

Das, A., Raychaudhuri, U., Chakraborty, R. 2013. Cereal based functional food of Indian subcontinent: a review. J. Food Sci. Tech. 49(6): 665-672.

De Castro, A., Montano, A., Sanchez, A.H., Rejano, L. 1998. Lactic acid fermentation and storage of blanched garlic. Int. J. Food Microbiol. 39: 205–211.

Diaz-Ruíz, G., Guyot, J.P., Ruíz-Terán, F., Morlon-Guyot, J., Wacher, C. 2003. Amylolytic lactic acid bacteria from pozol: a natural potential to produce complementary foods? 2nd International Workshop. Food-based Approaches for a Healthy Nutrition, Ouagadougou, Bukina Faso, pp. 411–417.

El-Nawawy, M.A., Ibrahim, R., Al-Bonayan, A.M., El-Beialy, A.R. 2012. Development of functional food products. Int. J. Dairy Sci. Res. 1: 12–17.

Evans, E., Musa, A., Abubakar, Y., Mainuna, B. 2013. Nigerian indigenous fermented foods: Processes and prospects. pp. 153–179. *In*: H.A. Makun (ed.). Food Safety in Developing Countries. InTech, Rijeka.

Gänzle, M.G. 2015. Lactic metabolism revisited: metabolism of lactic acid bacteria in food fermentations and food spoilage. Curr. Opin. Food Sci. 2: 106–117. Holzapfel, W.H. and B.J.B. Wood, eds. 2014. Lactic Acid Bacteria: Biodiversity and Taxonomy, 1 st edn. John Wiley, Oxford.

Gibbons, J.G., Rinker, D.C. 2015. The genomics of microbial domestication in the fermented food environment. Curr. Opin. Genet. Dev., 35: 1 8.

Greppi, A., Rantisou, K., Padonou, W., Hounhouigan, J., Jespersen, L. 2013. Yeast Dynamics during spontaneous fermentation of mawe and tchoukoutou, two traditional products from Benin. Int. J. Food Microbiol. 165: 200–207.

Hittinger, C.T. 2013. *Saccharomyces* diversity and evolution: a budding model genus. Trend Genet. 29: 309–317.

Holzapfel, W.H., Wood, B.J.B. 2014. Lactic Acid Bacteria: Biodiversity and Taxonomy, 1st edn. John Wiley and Sons, Oxford

Kancabas, A., Karakaya, S. 2013. Antiotensin-converting enzyme (ACE)-inhibitory activity of Boza, a traditional fermented beverage. J. Sci. Food Agric. 93: 641–645.

Kelly, W.J., Asmundson, R.V., Harrison, G.L., Huang, C.M. 1995. Differentiation of dextranproducing *Leuconostoc* strains from fermented rice cake (puto) using pulsed-field gel electrophoresis. Int. J. Food Microbiol. 26: 345–352.

Kim, M., Chun, J. 2005. Bacterial community structure in kimchi, a Korean fermented vegetable food, as revealed by 16S rRNA gene analysis. Int. J. Food Microbiol. 103: 91–96.

Kirk, P.M., Cannon, P.F., Minter, D.M., Stalpers, J.A. 2008. Dictionary of Fungi, 10th edn. CAB International, Oxford.

Kivanc, M., Yilmaz, M., Caktr, E. 2011. Isolation and identifi cation of lactic acid bacteria from boza, and their microbial activity against several reporter strains. Turk. J. Biol. 35: 313–324.

Kohajdová, Z., Karovièová, J. 2007. Fermentation cereals for specific purposes. J. Food Nutr. Res. 46: 51–57.

Kowalczyk, M., Mayo, B., Fernández, M., Aleksandrzak Piekarczyk, T. 2016. Updates on metabolism in lactic acid bacteria in light of "omic" technologies. In: Biotechnology of Lactic Acid Bacteria: Novel Applications. F. Mozzi, R.R. Raya, and G. M. Vignolo, eds. John Wiley, Oxford.

Kurtzman, C.P., Fell, J.W., Boekhout, T. 2011. Definition, classification and nomenclature of the yeasts. In: Kurtzman, C.P., J.W. Fell, and T. Boekhout, eds. The Yeasts, a Taxonomic Study. Volume 1, 5th edn. Elsevier, New York, NY, pp. 3–5.

LeBlanc, J.G., Todorov, S.D. 2011. Bactetiocin producing lactic acid bacteria isolated from boza, a traditional fermented beverage from Balkan Peninsula—from to application. pp. 1311–1320. *In*: A. Méndez-Vilas (ed.). Science against Microbial Pathogens: Communicating Current Research and Technological Advances. Formatex Research Center. Badajoz, Spain

Lee, H., Yoon Ji, H., Kim, H., Park, H. Lee, J. 2011. Functional properties of *Lactobacillus* strains from kimchi. Int. J. Food Microbiol. 145: 155–161.

Liu, M., Bayjanov, J.R., Renckens, B., Nauta, A., Siezen, R.J. 2010. The proteolytic system of lactic acid bacteria revisited: a genomic comparison. BMC Genomics.11: 36.

Mahasneh, A.M., Abbas, M.M. 2010. Probiotics and traditional fermented foods: The eternal connection (mini-review). Jordan J. Biol. Sci. 3: 133–140.

Maheswari, S.U., Shetty, P.H. 2013. Enhancement of functional properties of Indian fermented food idli through incorporation of *Moringa oleifera* leave. Int. J. Sci. Res. 21: 213–215.

Manickavasagan, A., Reicks, M., Singh, V., Sawsana, A., Intisar, A.M.M., Lakshmy. R. 2013.Acceptability of reformulates grain-based food: Implication for increasing whole grain consumption. Food Science and Human Welfare.2(3-4): 105-112.

Muyianja, C.M.B.K., Narvhus, J.A., Treimo, J., Langsrud, T. 2003. Isolation, characterization and identification of lactic acid bacteria from bushera: A Ugandan traditional fermented beverage. Int. J. Food Microbiol. 80: 201–210.

Naumov, G.I., James, S.A., Naumova, E.S., Louis, E.J., Roberts, I.N. 2000. Three new species in the *Saccharomyces sensu stricto* complex: *Saccharomyces cariocanus*, *Saccharomyces kudriavzevii* and *Saccharomyces mikatae.* Int. J. Syst. Evol. Microbiol. 50: 1931–1942.

Nout, M.J.R. 2009. Rich nutrition for the poorest—Cereal fermentations in Africa and Asia Food Microbiol. 26: 685–692.

Nyanzi, R., Jooste, P.J. 2012. Cereal based functional foods. pp. 161–197. *In*: E.C. Rigobelo (ed.). Probiotics. InTech. Rijeka.

Östrürk, I., Karaman, S., Törnük, F., Sagdic, O. 2013. Physico-chemical and rheological characteristics of alcohol-free probiotic boza prepared using *Lactobacillus casei Shirota*: estimation of the apparent viscosity of boza using nonlinear remodeling techniques. Turk. J. Agric. For. 37: 475–487.

Panda, S.H., Ray, R.C. 2007. β-Carotene-rich sweet potato lacto-juice. Plant Foods Hum. Nutr. 62: 65–70.

Panda, S.H., Parmanick, M., Ray, R.C. 2007. Lactic acid fermentation of sweet potato (*Ipomoea batatas* L.) into pickles. J. Food Process. Preserv. 31(1): 83–101.

Pedersen, M.B., Gaudu, P., Lechardeur, D., Petit, M.A., Gruss. A. 2012. Aerobic respiration metabolism in lactic acid bacteria and uses in biotechnology. Annu. Rev. Food Sci. Technol. 3: 37–58.

Prado, F.C., Parada, J.L., Pandey, A., Soccol, C.R. 2008. Trends in non-dairy probiotic beverages. Food Res. Int. 41: 111–123.

Rahman, *I.E.*A., Hamad, S.H., Osman, M.A., Dirar, H.A. 2010. Characterization and distribution of microorganisms associated with kisra bread preparation from three sorghum varieties in Sudan. Curr. Res. Bacteriol. 3: 138–147.

Ray, R.C., Panda, S.H. 2007. Lactic acid fermented fruits and vegetables: an overview. pp. 155–188. *In*: M.V. Palino (ed.). Food Microbiology Research Trends. Nova Science Publishers Inc., Hauppauge, New York, USA.

Rivera-Esponoza, Y., Gallardo-Navarro, Y. 2010. Non-dairy probiotic products. Food Microbiol. 27: 1–11.

Roberts, J.S., Kidd, D.R. 2005. Lactic acid fermentation of onions. LWT - Food Sci. Technol. 38: 185–190.

Sekar, S., Mariappan, S. 2007. Usage of traditional fermented products by Indian rural folks and IPR. Indian J. Tradit. Knowl. 6: 111–120

Steinkraus, K.H. 2002. Fermentations in world food processing, Compr. Rev. Food Sci. Food Saf. 1: 23–30.

Tamang, J.P., Tamang, B., Schillinger, U., Franz, C.M., Gores, M. Holzapfel, W.H. 2005. Identifi cation of predominant lactic acid bacteria isolated from traditionally fermented vegetable products of the Eastern Himalayas. Int. J. Food Microbiol. 105(3): 347–356.

Viander, B., Maki, M. Palva, A. 2003. Impact of low salt concentration, salt quality on natural large- scale sauerkraut fermentation. Food Microbiol. 20: 391–395.

Xiong, T., Guana, Q., Songa, S., Hao, M., Xie, M. 2012. Dynamic changes of lactic acid bacterial flora during Chinese sauerkraut fermentation. Food Control 26: 178–181.

Yoon, K.Y., Woodams, E.E., Hang, Y.D. 2006. Production of probiotic cabbage juice by lactic acid bacteria. Bioresour. Tech. 97(12): 1427–1430.

Zaunmüller, T., Eichert, M., Richter, H., Unden, G. 2006. Variations in the energy metabolism of biotechnologically relevant heterofermentative lactic acid bacteria during growth on sugars and organic acids. Appl. Microbiol. Biotechnol. 72: 421–429.

Zorba, M., Hancioglu, O., Genc, M., Karapinar, M., Ova, G. 2003. The use of starter cultures in the fermentation of boza a traditional Turkish beverage. Process Biochem. 38: 1405–1411.

Zubaidah, E., Nurcholis, M., Wulan, S.N., Kusuma, A. 2012. Comparative study on symbiotic effect of fermented rice bran by probiotic lactic acid bacteria *Lactobacillus casei* and newly isolated *Lactobacillus palntarum* B2 in wistar rats. APCBEE Procedia 2: 170–177.

Chapter 4

Bacterial Quorum Sensing: Its Role in Physiological Characterization of Bacteria

*Aayushi Sharma, Jasina Arora and Vijay Kumar**

[1]Department of Microbiology,
Kurukshetra University, Kurukshetra, Haryana
**e-mail: vijaykuk@kuk.ac.in*

Introduction

Bacteria being the most primitive unicellular microorganism in the universe, makes a significant contribution that shows either as beneficial impact like cynobacteria that helps in nitrogen fixation in plants or *Ruminococcus bromii* that is a residential micro flora in the large intestine of humans or as harmful impact such as citrus canker that effect mainly citrus fruits and meningitis caused by *Haemophillus influenzae* in humans. Now the question comes how can an organism function both as boon and ban on the living world? The answer is the communication that occurs between the bacterial cell that results into pathogenicity and cause several diseases and some time may be fatal if left untreated. Bacteria are the main part of the animal or plant micro biome that looks around for nutrients as well as to increase in their number. Now these bacteria in the appropriate micro biome may be associated *i.e.,* symbionts (beneficial to the host) or pathogens (harmful to the host) to meet their requirements. This association relied on the releasing and recognition of signal molecules with the help of bacterial communication.

Quorum Sensing

About 50 years ago, bacteria was thought to have protoplasm and with no community sense. However, this concept was soon blown out alone with discovery of bacterial cell to cell signaling a phenomenon which later termed as quorum sensing (QS) first discovered in *Vibrio fischeri,* a gram negative bacteria. The QS systems operates via a wide range of signals such as: (a) Oligopeptides, (b) N-acyl homoserine lactones (AHLs, AI-1), (c) Furanosyl borate (AI-2), (d) Hydroxyl-palmitic acid methyl ester, (e) Methyl dodecanoic acid, and (f) Diffusible signal factor (DSF); cis-11- methyl-2-dodecenoic acid. Quorum sensing signals are arbitrated by small different molecules called as Autoinducer which includes (i) AHLs, which can differ in their length and oxidation state of acyl side chains and are mainly produced by Gram-negative bacteria to check their population density in QS control of gene expression, and (ii) Lux I protein family along with the oligo-peptides (oligo peptie contains between 5 and 34 amino acid residues). Basically what oligo peptide does is functions as transporter and transport the substrate produce from the amino acid in gram negative and gram positive bacteria respectively. Quorum sensing, an intracellular form of communication used by bacterial community to communicate and coordinated in group behaviors such as bioluminescence, formation of biofilm, virulence factor expression, symbiosis, sporulation, pigment production, motility, toxin and antibiotic production *etc.* (Rutherford and Bassler, 2012; Bassler, 2016).

As described earlier QS was at first used to explain how bioluminescence is produced in marine bacterium *V. fischeri. V. fischeri*, a free-living, bioluminescent bacterium that can become a symbiont with the Hawaiian bobtail squid, *Euprymna scolopes*. The *V. fischeri* symbiosis is established when it travels by means of flagella based motility from surrounding sea water to epithelial cell structure of squid and name them as their light organ. Cell grows to the maximum density within the light organ persisting the infection over whole life time. The indication of lucrative symbiosis is the luminescence produce by *V. fischeri* that camouflages the squid at night by eliminating its shadow within water column.

Overview of Mechanism

- ☆ The gene cluster (lux) is majorly responsible for bioluminescence. LuxI is an enzyme that produces the autoinducer (AI) molecule N-(3-oxohexanoyl)-L-homoserine lactone.
- ☆ When the cells are at low population densities, the expression of the lux I operon is very less, which results in very slow production of the autoinducer.
- ☆ As the cell density increases, the autoinducer start accumulating in the medium. As the concentration of autoinducer increases, it will diffuse across the membrane back into the cell and bind the LuxR protein.
- ☆ The **LuxR–AI complex** then activates transcription of the divergent lux operon, leading to increase in the production of various enzymes responsible for bioluminescence.

Quorum Quenching

Variety of natural and synthetic compounds has been identified, which can deplete and also inactivate the QS molecules. The need to study these compounds is of major importance because of antibiotic resistance, which has increased in past few years. Due to the sudden upsurge in antibiotic resistance among the human bacterial pathogens and quo status poor effect of antibiotics to treat the bacterial infections, there was a need to find novel alternative to antimicrobial approach. Keeping that in mind a new method has paved its way in producing antimicrobial agents that control the synthesis of these QS molecules. The system of QS can be affected via various methods like: (a) the activity of the AHL synthase is decreased, (b) inhibition of the production of AHLs, (c) degradation of the AHLs, and (d) by use of various antagonistic compounds to the signaling molecules. This disruption in QS molecules is done by a process known as quorum quenching.

As explained above the pathogenicity rely on cell communication mechanism i.e, quorum sensing which is counter attack by a host self develop disease control tactics, i.e quorum quenching. Certain bacteria degrade the AHL molecule by quorum quenching. Quorum quenching is thought to be evolved either by the quorum sensing secreting microbes that tends to recycle it or want to clear the zone of sensing produce by themselves. Quorum quenching interferes with the quorum sensing process through different mode of action either by cleaving the QS signal molecules, or by competitive inhibition. The molecules act in cleavage or inhibition can either be enzymes or some other chemicals compounds, quorum quenching generally affect the main steps that carry the process of QS i.e, synthesis, diffusion, accumulation and perception of QS signal may be affected. Those enzymes which inactivate the QS signals are termed as quorum quenching enzymes, while the chemical compound that disturb the QS pathways are known as QS inhibitor. Also physical parameters can also influence the temperature or pH they reduce the half life of QS signals. This is done to attain the control of the synthesis of AIs and their contact with the receptors, as well as the raise of their disintegration. Many natural and synthetic compounds have been identified which can degrade as well as inactivate the QS molecules. Compounds of natural origin are always taken into account in medical fields as they can be biodegraded and normally have very fast in their mode of action, therefore they serve as a suitable option for inhibition of biological infections, they are assumed to be better than other QS inhibitors, and can be used more confidently for a longer period of time and can reach the situation of GRAS (generally recognized as safe). For instance, a gram negative bacteria, *Delftia tsuruhatensis* SJ01 isolated from rhizosphere, has activity to degrade the AHL also have a potential to inhibit biofilm. Similarly, *Vibrio alginolyticus* produces certain secondary metabolites which can inhibit the virulence factors of *Pseudomonas aeruginosa* PAO1 by down regulating the elastase activity, motility ability, and rhamnolipid production. In addition, the *V. alginolyticus* extract have anti-biofilm activity against *P. aeruginosa*.

Classification of Quorum Sensing Controlled Genes

Genes are the basic unit of information found in every living cell. Genes are inherited by one's parents which is like set of instruction and work as the blueprint, that are required in making up of essential proteins required by our body, that help in proper functioning. Proteins are the end product that is produce by genes. Many of characteristics and traits are influenced by genes. In case of bacteria the genetic material is DNA. Bacterial DNA mostly plays two vital functions they are: replication and expression. Genetic material must replicate accurately as a result progeny can inherit all of the specific genetic characteristics (the genotype) of the parental organism. Specific genetic material expression under a particular set of growth conditions regulates the observable characteristics (phenotype) of the organism. Bacteria have few structural or developmental traits which can be detect easily, but they have a vast array of biochemical capabilities and patterns of susceptibility to antimicrobial agents. Expression of the microbial genes is regulated by intracellular or extracellular environment situation. This expression of genes can be interrupt at any step, resulting into different phenotype and different genotype. Quorum sensing genes are also like any other genes their normal expression results into cell to cell communication or pathogenicity and their disruption at any step result in quorum quenching.

On the Basis of Nature

On the basis of nature, the quorum sensing genes are divided into following criteria:

1. Cell maintenance and cell proliferation.
2. Cell behavior.
3. Horizontal gene transfer.
4. Interaction with host and other microbes.

Cell Proliferation and Cell Maintenance

Usually the term proliferation is used to describe the process when a cell starts to divide into 2 daughter cells. In context of bacteria this proliferation term can be interchange with the replication. Their proliferation will depend upon the nutrient rich external environment that helps them to grow and divide. This can be termed as bacterial cell cycle. Till now it is observed that the bacterial cell cycle is divided into three stages: the first stage, which is in between the division and initiation of cell chromosome replication i.e; cell birth (known as the B period). The second stage of replication is known as C period. The third and last stage is the time between end replication and completion of division; this is known as D period. Cell maintenance is known as the process that promotes, sustain and preserve the cell viability. Bacteria try to maintain themselves with variety of processes; the major stress bearing component of bacterial cell is its cell wall specifically peptidogly can in either gram negative or gram positive bacteria. A number of the processes can be reduced partially or shut down entirely when there is decline in availability of resources. Generally in external environment, the bacteria are tremendously limited to the energy source that does not cause their growth, and their metabolism is also repressed and only most essential functions are performed. Such type of minimal energy requirements by the bacteria are known as the basal power requirement. Quorum sensing can be seen in cell maintenance and proliferating cells; many species of bacteria use this to their gene expression to maintain the density of their community population. Quorum sensing can also be viewed as one of the way to sense how 'open' the bacterial environment is (Redfield 2002), besides from being a density-dependent regulatory process that permits a synchronous expression of gene(s) in a bacterial population. This theory is known as diffusion sensing, and has been reunited with the QS principle to generate the efficiency sensing concept. This concept proposes that bacteria can only measure the combination of cell density, spatial distribution and limitations to autoinducer mass transfer via production of autoinducers (Hense *et al.*, 2007). According to this concept, signal molecules would be a way for bacteria to scrutinize the laudability of the production of molecules, like extracellular enzymes. On the basis of various ecological models, it is understandable that QS and diffusion sensing should not be considered as opposing theories (West *et al.*, 2012) and additional roles of QS signals may be suggested (Platt and Fuqua 2010), like cluster sensing (sensing the existence of a microcolony in an open environment) or position sensing (at the single cell level, possibility to sense its position with respect to the group), in order to proliferate and maintain itself in the surrounding.

Cell Behavior

The behavior of the cell is dependent upon internal organelles like its DNA packaging, Cytoskeleton, and their metabolic processes (gylcolytic pathway, pentose phosphate pathway, *etc.*) and along with this their local environment. Cell can receive and respond to various signals, this sort of cell to cell communication helps the cell to sense the environment and spread information and behave accordingly. Sometimes these signals may be specific to cells and sometimes cell may not respond these signals, depending upon the signals and type of cells. In, *in vivo* conditions, the microenvironment directs an array of cell behaviors which are important for development and regeneration of cells. Tissue engineering has lead to development of biomimetic design "instructive" scaffold materials that activate an array of preferred downstream cellular responses. One of the major challenge in this bioinstructive scaffold design is to control cell behavior, which specifically determines the signals or input instructions which is required to direct cells to assemble and maintain a new functional tissue for repair or replacement. But, recent researchers have showed that biophysical aspects of a cell's microenvironment may have equally important and analogous roles in regulating cell behavior. Two cellular structures, focal adhesions and adherens junctions, have been suggested as key mediators in sensing and responding to the biophysical microenvironment, playing critical roles in cell migration, force-sensitive gene regulation, differentiation, and ECM assembly.

Horizontal Gene Transfer

It is also known as lateral transfer of gene, and is a process in which microbe transfer its genetic material to another microorganism which is not its progeny. Bacteria generate its copy not only from replication but also with the help of gene transfer mechanism. Gene transfer led to development of a new genotype. Bacteria can adapts to new environments which is a part of bacterial evolution and this results mainly from acquiring new genes via horizontal gene transfer instead of any alteration in gene functions through mutations.

Horizontal gene transfer is of mainly 3 types:

1. **Transformation** involves uptake of short fragments of naked DNA by naturally transformable bacteria.
2. **Transduction** involves transfer of DNA from one bacterium into another via bacteriophages.
3. **Conjugation** involves transfer of DNA via sexual pilus and requires cell -to-cell contact.

Interaction with Host and other Microbes

Bacterial cell have develop number of ways through which the bacteria host interaction can takes place. On the basis of expression of pathogenic factors and virulence, bacteria have different ways to interact with their specific host, leading to adherence, invasion and migration across the cell membranes. If we specifically talk about the successful host and bacterial cell interaction then the first step is adhesion and subsequently requires colonizing of the host tissues and finally intercellular invasion oh host cell. After all this, in the host cell the bacteria multiply and cause pathogenicity, or can be in dormant stage and become activate when suitable conditions are there. In case with the bacterial interaction with other microbes, it occur mainly by the exchange of molecular and genetic information, and many other mechanisms can be involved in this exchange, such as secondary metabolites, quorum sensing system, biofilm formation, siderophores, and cellular transduction signaling.

Microbe - microbe interactions are important for a win full establishment and keeping the microbial population maintained with another type of microbial population. These interactions arise by the close environment which is followed by exchange of molecular and genetic information that include many mechanisms and classes of molecules. One of the extensively studied mechanisms for microbial interaction is quorum sensing which is a stimulus mediated response system related to cellular concentration. The signaling molecules (auto-inducers) are produced which allows cells to communicate as well as respond to the environment in a coordinated manner. These mechanisms can also allow microorganisms to set up a multi-trophic community, which further could result in highly diverse microbial population. These multiple interactions can result in pathogenic or beneficial outcome to the host. Main focus of these researches is on microbial interactions and communication in the human medical filed. The microbial interaction is critical for the successful establishment, maintenance of colonization and to cause the infection. Antimicrobial host defenses mechanisms together with environmental factors also play essential roles. Microorganism communication results in a coordinated response to host and other environmental signals in the microbial population, they collectively regulate tier expression to achieve successful pathogenic outcomes which could have not been accomplished if the cells have been individual. There is a great variety of types of microbial interaction with reference to the organisms involved. Bacteria–bacteria, fungus–fungus, bacteria–fungus, fungus–plant/animal, bacteria–plant/animal and bacteria–fungus–plant/animal interactions, including parasitic, mutualistic interactions involve many mechanisms (Table 4.1).

Table 4.1: Type of Interactions Involved among Various Organisms

Organisms Involved	***Types of Interaction***	***Compound and Mechanism Involved***	***Findings***
V. fischeri **and squids or fishes**	Symbiotic–fish	Quorum sensing	The autoinducer molecule reaches a threshold and luminescence genes are activated in the symbiotic association with fishes or squids.
Rhizobium leguminosarum **and plant**	Symbiotic–plant	Quorum sensing	The QS system in these bacteria responsible for various functions in plant like nodule formation efficiency, inhibition of growth, nitrogen fixation and plasmid transfer.
Xanthomonas **or** ***Xylella*** **and grapevines or citrus**	Pathogen–host	Quorum sensing	QS signaling molecules control the virulence factor expression and biofilm formation.
Pantoea stewartii **and** ***Zea mays***	Pathogen–host	Quorum sensing	QS mutants of *P. stewartii* were not able to move in the vasculature, resulting in decrease of disease.
Candida albicans **and** ***Pseudomonas -aeruginosa***	Microbial community	Quorum sensing	*C. albicans* produces Farnesol which downregulated the QS of *P. aeruginosa*. *P. aeruginosa* QS system may affect the dimorphic nature of *C. albicans* by activating the hypha to yeast reversion and blocking the yeast to hypha transition.

On the basis of Functions

Exoenzymes Production

Bacteria produce different types of exoenzymes that function outside the cell. These exoenzymes breakdown the bulky and complex molecules into simpler and make them available for the cell to uptake it. Decomposition is the perfect example of exoenzymes. The soil organic, matter decomposition occurs by catalyzation of exoenzymes produce by microorganism. Many of the exoenzymes are pathogenic in nature, for example most of the bacteria and fungi use exoenzymes as the first line mechanism to cause disease. Basically what exoenzymes do is breakdown the cell wall of host and takes entry into it and in case of complex animals by damaging the tissue known as Necrotizing. Few examples of exoenzymes as virulence factor are (Table 4.2).

Table 4.2: Examples of Various Exoenzyme Producing Bacteria

Exoenzymes	*Virulence Factor*	*Example*
Necrotizing Enzyme	Destroy cells and tissues	*Streptococcus pyogenes*
Coagulase	Binds to prothrombin, ultimately converting fibrinogen to fibrin.	*Staphylococcus aureus*
Kinases	Opposite of Coagulase, dissolve clot (staphylokinases)	*Staphylococcus aureus*
Hyaluronidase	Enables a pathogen to penetrate deep into tissues	*Clostridium* species

Siderophores Synthesis

Mostly seen in fungi and bacteria, it's a mechanism through which iron assimilation is done and made available to the cell in iron deficient nutrient. This siderophore system consists of low molecular weight ferric ion ligands i.e; siderophores and analogus cell bound receptors and utilization components. Siderophores are genetically controlled by the process of quorum sensing as described: the quorum-sensing regulator LasR controls the cell density-dependent gene expression in *Pseudomonas aeruginosa*. In LasR null mutants there was significant 2-fold decrease in production of the catecholate-hydroxamate siderophore pyoverdine when it was grown under iron-limiting conditions. When autoinducer PAI-1 defective LasI mutants was grown they also showed the decrease in the significant 2-fold decrease in production of pyoverdine production which was restored when PAI-1 autoinducer was provided from outside. This was one of the first reports of QS regulation for production of siderophore in bacteria and showed that cell density does enhance the production of siderophore rather than expression of essential signal for pyroverdine production.

Sporulation and Acid Resistance

In several Gram-positive bacteria, sporulation is controlled by quorum sensing, Sporulation, QS-dominate phenotype, shows when there is maximal stress and nutritional deficiency. An extend time period of nutrient deficiency drives the cell to enter the process of sporulation, which result in production of a metabolically dormant endospore, that regain growth once the favorable conditions are provided. Spore formation is a complex and tightly regulated phenomenon, which requires several hundred genes, directly and indirectly involved as described: Bacteria use quorum sensing for adaptation properties, the infectious cycle of *Bacillus thuringiensis* in the insect host shows the role of quorum sensing in natural habitat. It is maintained by reciprocator through regulator linked to the RNPP family and directly manages by re-internalized signaling peptides. One such RNPP regulator is NprR, which regulate the genes responsible for survival of bacteria in insect. NprR acts in the presence of its similar signaling peptide NprX as a transcription factor. If NprX is absent then NprR negatively controls sporulation and independently of its transcriptional activator function. Further, NprR ultimately inhibits the expression of Spo0A-regulated genes by preventing the KinA-dependent phosphorylation of the phosphotransferase Spo0F, which delays the initiation of sporulation process.

Biofilm Formation

Biofilm development and quorum sensing (QS) are closely linked processes. Biofilm formation is a mutual group behavior that involves bacterial populations living submerged in a self-produced extracellular matrix. QS is cell communication mechanisms that integrate gene expression in response to population cell density. When the population density reaches a threshold level, the QS gene will apparently switched on. Different bacterial species coincides the activation of QS occurs in the formed biofilm and activates the maturation of the biofilm in a coordinate manner.

Greenberg and his colleagues (1998) first described the role of the *Las* quorum sensing in biofilm formation of *Pseudomonas aeruginosa*, a Gram-negative bacterium. It can cause fatal systemic diseases in humans as it is one of the most common opportunistic pathogen mainly causing nosocomial infections. In case of cystic fibrosis patients this pathogen causes infection in lungs with formation of biofilm. Another example is Gram-positive bacteria, *Staphylococcus aureus* and certain other streptococci, which can also use signal peptide-mediated systems for quorum sensing. Like *S. aureus* causes nosocomial infections worldwide and can also cause mild form of skin infections and potentially fatal systemic disorders.

Virulence Factor

Ability of microbes to cause infection to the host is known as virulence. As described before, that bacteria produces diffusible signal molecules for their communication which are known as autoinducers. Initially, these molecules are produced at very low levels and they start accumulating during the growth, once their concentration reach a certain critical level these autoinducers can control the expression of various target genes. As this phenomenon is also called quorum sensing and controls the expression of virulence genes in number of microorganisms. This quorum sensing controlled virulence has now become an attractive target for the discovery and development of new therapeutic strategies. Both Gram-positive as well as Gram-negative bacteria use distinct kind of quorum sensing. Peptides, also known as auto inducing peptides (AIPs), are used as signaling molecules in Gram-positive bacteria. In case of Gram-negative bacteria communication done via small molecules as AIs. These may be either acyl-homoserine lactones (AHLs) or other molecules (Wei *et al.,* 2011).

These are extra polymeric substance of high molecular weight that is to be secreted by microbes either on their cell surface as glycocalyx layer or in their surrounding environment. *Sinorhizobium meliloti* is a soil bacterium, which can establish a symbiotic relationship by invading the alfalfa plants. This invasion process by *S. meliloti* requires synthesis of at least one of the two symbiotically important exopolysaccharides, succinoglycan and EPS II. The *S. meliloti* have a *sinRI* locus which encodes a quorum sensing system that plays a role in this symbiotic process. The *sinRI* locus controls the regulation of EPS II synthesis. Disruption of the autoinducer synthase gene, *sinI*, inihibts the production of EPS II as well as the expression of several other genes in the *exp* operon that are responsible for EPS II synthesis. This phenotype was complemented by the addition of acyl homoserine lactone (AHL) extracts from the wild-type strain but not from a *sinI* mutant, indicating that the *sinRI*-specified AHLs are required for *exp* gene expression. This was further confirmed by the observation that synthetic palmitoleyl homoserine lactone ($C_{16:1}$-HL), one of the previously identified *sinRI*-specified AHLs, specifically restored *exp* gene expression. Most importantly, the absence of symbiotically active EPS II in a *sinI* mutant was confirmed in plant nodulation assays, emphasizing the role of quorum sensing in symbiosis.

Bioluminescence

Quorum sensing was initially used to describe the mechanism of bioluminescence phenomenon in marine bacterium *V. fischeri*. *V. fischeri,* is free-living, bioluminescent bacteria which can become a symbiont with the Hawaiian bobtail squid, *Euprymna scolopes*. This symbiotic interaction is established when *V. fischeri* travels via flagella based motility from nearby sea water to epithelial cell structure of squid and name them as their light organ. Cell grows to the maximum density within the light organ persisting the infection during the whole life time. The indication of lucrative symbiosis is the luminescence produce by *V. fischeri* that camouflages the squid at night by eliminates its shadow within water column.

Antibiotics Resistance

Due to the extensive and uncontrolled use of antibiotics around the world, antibiotic resistance is increasing among microorganism from past 20 years. Since the introduction of antibiotics, now almost of the pathogenic bacteria have developed resistance to antibiotics which are commonly used for treatment of infections. Most antibiotics which are currently being used to kill the pathogenic bacteria act either by destroying the cell membranes or by interfering with protein synthesis. This "life or death" selection pressure has led to the evolution of resistance among microbes, and such a large-scale exploitation of antibiotics has brought some serious issues related to resistance. Quorum sensing can activate various cellular processes like virulence and drug resistance mechanisms, antibiotics tolerance, and damage the host. Some mechanisms of drug resistance are - chemical modifications, systemic elimination of antibiotics via efflux pump, and modification of drug-targeting genes. Also, many pathogenic bacteria can form a dense biofilm, which makes the bacteria highly resistant.

Host Colonizing Factors

Gram-positive bacteria uses common principles quorum sensing like production, detection, and response to AIs for host colonizing. In case of many Gram-positive bacteria they have AIs as oligopeptide AIPs; their size varies in range from 5 to 17 amino acids, they can be modified post translationally, and can be linear or cyclic (Havarstein *et al.,* 1995; Magnuson *et al.,* 1994). AIPs can be detected extracellular via membrane-bound two-component signal transduction sensor Kinases (Hoch and Silhavy 1995), and they a conserved histidines when bound by the AIP, there after the sensor Kinases is auto phosphorylated. The phosphoryl group is then passed from the histidine to a conserved aspartate on a cognate cytoplasmic response-regulator protein, and this phosphorylated response regulator controls the expression of quorum sensing target genes. The most well-studied system in this group of pathogens is the *S. aureus* Agr system. Gram-negative bacteria typically use LuxI/LuxR-type QS systems homologous to the first described QS system from the bioluminescent marine symbiotic bacterium *Vibrio fischeri* (Ruby 1996). In these systems, the LuxI homolog is an AI synthase that catalyzes a reaction between SAM and an acyl carrier protein (ACP) to produce a freely diffusible acyl homoserine lactone (AHL) AI. At high concentrations, AHL AIs bind to cognate cytoplasmic LuxR-like transcription factors. When not bound by AI, LuxR-type proteins are rapidly degraded, presumably to prevent bacteria from "short-circuiting" their QS systems. AI binding stabilizes the LuxR-type proteins, allowing them to fold, bind DNA, and activate transcription of target genes.

REFERENCES

Braga, Mesquite, Raissa., Dourado, Nóbrega, Manuella., LuizAraújo, Welington. 2016. Microbial interactions: ecology in a molecular perspective. Braz. J. Microbiol, DOI 10.1016/j.bjm.2016.10.005.

Grandclement, Catherine., Tannieres, Melanie., Morera, Solange., Dessaux, Yves., and Faure, Denis. 2016. Quorum Quenching: Role in nature and applied Developments. FEMS. Microbiol. DOI 10.1093/femsre/fuv038.

Hemmati, Fatemeh., Salehi, Roya., Ghotaslou, Reza., Kafi, Samadi, Hossein., Hasani, Alka., Gholizadeh, Pourya., Nouri, Roghayeh., Rezaee, Ahangarzadeh,

Mohammad. Quorum Quenching: A Potential Target for Antipseudomonal Therapy. Infect. Drug. Resist. DOI 10.2147/IDR.S263196.

Hense, A, Burkhard., Kuttler, Christina., Müller, Johannes., Rothballer, Michael., Anton Hartmann, Anton., Kreft, Ulrich, Jan. 2007. Does efficiency sensing unify diffusion and quorum sensing. Nat. Rev. Microbiol. DOI 10.1038/nrmicro1600.

Hossain, Tanim J., Chowdhury, Sumaiya I., Mozumder, Halima A., Chowdhury, Mohammad N. A., Ferdausi, Rahman Nabila., and Dey, Sujan. 2020. Hydrolytic Exoenzymes Produced by Bacteria Isolated and Identified From the Gastrointestinal Tract of Bombay Duck. Front. Microbiol. DOI 10.3389/fmicb.2020.02097.

Moat, G.Albert. Foster, W.John. Spector P. Micheal. Microbial physiology, Wiely-Lissa, John Wiely and Sons Inc, Publication, New York. pp 664-669.

Paluch, E., Soroczyñska, Rewak, J., Jêdrusik, I., Mazurkiewicz, E., Jermakow, K. 2020. Prevention of biofilm formation by quorum quenching. Appl. Microbiol. Biotechnol. DOI 10.1007/s00253-020-10349-w.

Platt, T.G., Fuqua, C.2010. What's in a Name? The Semantics of Quorum Sensing. Trends. Microbiol. DOI 10.1016/j.tim.2010.05.003.

Redfield, J, Rosemary.2002. Is Quorum sensing a side effect of diffusion sensing. Trends. Microbiol. DOI 10.1016/s0966-842x(02)02400-9.

Rutherford, T, steven., Bassler, L, Bonnie.2012. Bacterial Quorum Sensing: Its Role in Virulence and Possibilities for Its Control. Cold.Spring.Harb.Perspect.Med. DOI 10.1101/cshperspect.a012427.

Stintzi, A., Evans, K., Meyer, M.J., Poole, K. 1998. Quorum-sensing and siderophore biosynthesis in Pseudomonas aeruginosa: lasR/lasI mutants exhibit reduced pyoverdine biosynthesis. FEMS.Microbiol. DOI 10.1111/j.1574-6968.1998.tb13910.x.

Verma, C, Subhash., Miyashiro, Tim. 2013. Quorum Sensing in the Squid-Vibrio Symbiosis. Int. J. Mol. Sci. DOI 10.3390/ijms140816386.

West, A, Stuart., Winzer, Klaus., Gardner, Andy., Diggle, P, Stephen. 2012.Quorum sensing and the confusion about diffusion. Trends.in.Microbiol. DOI 10.1016/j.tim.2012.09.004.

Yi-Hu, Dong., Lian-Hui, Wang., Lian-Hui, Zhang. 2007. Quorum-quenching microbial infections: mechanisms and implications. Philos. Trans. R. Soc. Lond.B.Biol.Sci. DOI 10.1098/rstb.2007.2045.

Yi-Hu, Dong., Lian-Hui, Wang., Lian-Hui, Zhang. 2007. Quorum-quenching microbial infections: mechanisms and implications. Phil. Trans. R. Soc. DOI 10.1098/rstb.2007.2045.

Chapter 5

Applications of Probiotics in Gynecological Health: A Review

Vijay Kumar and Manpreet Kaur*

Department of Microbiology,
Kurukshetra University, Kurukshetra, Haryana
**e-mail: vijaykuk@kuk.ac.in*

ABSTRACT

The normal microflora of vagina is extremely important for maintenance of healthy vaginal state and consequently in prevention of urogenital infections. Since a healthy female vaginal microbiota is dominated by species of Lactobacillus, the perception of restoration and/or maintenance of a healthy vagina by exogenous administration of probiotic Lactobacillus seem feasible. Urogenital infections are treated using antibiotics. But use of antibiotics have several drawbacks, including a decrease in the numbers of beneficial bacteria and an increase of drug resistance; moreover, they are often ineffective and increase the risk of infection recurrence Unlike antibiotics, probiotics can be used over an extended period without adverse effects. Probiotics properly colonized in the vagina may help in reducing the recurrent infection. So, various strains of probiotics can be used in treatment and the load of probiotics may be used in the management of vaginal infections or its recurrence.

***Keywords**: Female fertility, Health, Infection, Infertility, Lactobacillus, Probiotic.*

Introduction

Infertility is a worldwide problem, and has steadily increased over the last 30 years, with a foremost financial burden on the international health care industry.

As the role of male in fertility is often overlooked, women are consistently blamed for infertility in a society where child bearing is of social significance. Infertility hence leads to emotional, social, and psychological problems in females. Thus, female infertility being a grave concern must be given utmost importance. The use of probiotics in genital tract infections was based on the reports of occurrence of Lactobacilli in the healthy vagina of women with no history of infection (Reid and Bruce, 2003). Lactobacilli may colonize the vagina of healthy women by traversing the perineum and block the entry of pathogens by forming a barrier. Probiotics use to treat genital tract infections could represent an alternative option to antibiotics since resistance to drugs; recurrent infections in addition to side effects are of major concern with antimicrobial treatment. *In vitro* studies and clinical trials in recent years have assembled evidence on the outcome of probiotics mainly Lactobacilli in the course of vaginal infection by counteraction of pathogens and competition (Mastromarino *et al.,* 2013; Verdenelli *et al.,* 2014). Strains generally often exploited as probiotics belong to the *Lactobacillus* and *Bifidobacteria.* Various species employed as probiotics are presented in Table 5.1.

Table 5.1: Commercially Exploited Probiotics Species

Lactic acid bacteria (LAB)	*Lactobacillus sp.*	*L. fermentum*
		L. acidophilus
		L. gasseri
		L. johnsonni
		L. rhamnosus
		L. salivarius
		L. caseii
		L. paracaseii
		L. lactis
		L. plantarum
		L. reuteri
	Bifidobacterium sp.	*B. bifidum*
		B. breve
		B. infantis
		B. lactis
	Pediococcus sp.	*P. pentosaceus*

Use of Probiotics in Genital Tract Infections

The use of probiotics in genital tract infections was based on the reports of occurrence of Lactobacilli in the healthy vagina of women with no history of infection (Reid and Bruce, 2003). Lactobacilli may colonize the vagina of healthy women by traversing the perineum and block the entry of pathogens by forming a barrier. Lactobacilli can defend against genital tract infection by: i) preventing the adherence of potentially pathogenic microorganisms to the vaginal epithelial cells, biolfilm formation, co-aggregation with microbial pathogens ii) producing antimicrobial products such as organic acid, H_2O_2 and bacteriocins, maintenance of low pH, competition for nutrients, production of surfactants with anti adhesive properties and iii) stimulating local immune response (innate as well as adaptive immune system) (Rose *et al.*, 2010; Wagner and Jhonson *et al.*, 2012; Machado *et al.*, 2013; Petrova *et al.*, 2013; Bisanz *et al.*, 2014; Doerflinger *et al.*, 2014; Gong *et al.*, 2014; Mastromarino *et al.*, 2014).

Coaggregation

Coaggregation of probiotic with pathogenic bacteria forms a physical-chemical hindrance that averts the pathogen entry, inhibits their growth, and prevents colonization of mucosal sites (Schachtsiek *et al.*, 2004; Schellenberg *et al.*, 2006; Nishiyama *et al.*, 2014). Pascual and colleagues (2008) reported coaggregation of *L. rhamnosus* and *L. fermentum* L23 with *C. albicans, G. vaginalis* and *E. coli.* Mastromarino *et al.* (2002) observed a reduction in the adhesion by 57.7 per cent upon coaggregation of *G. vaginalis* with *L. salivarius*FV2 and *L. gasseri.* Coaggregation might be an essential factor in preserving the health of vagina as the antimicrobial products generated by lactobacilli is highly concentrated in coaggregated area around the pathogen constituting a vital defense of host against the infection (Kotikalapudi, 2009; Taheri *et al.*, 2009) and reducing the viability of pathogen (Saunders *et al.*, 2007; Jones and Versalovic, 2009; McMillan *et al.*, 2011; Younes *et al.*, 2012; Gregorio *et al.*, 2014).

Production of Growth Inhibitory Substances

Various growth inhibitory substances produced by Lactobacilli are fatty acids, organic acids, biosurfactants, bacteriocins and hydrogen peroxide to facilitate the inhibition of both Gram +ve and Gram –ve microorganisms (Rolfe, 2000; Dunne *et al.*, 2001). Antimicrobial acids and H_2O_2 produced by Lactobacilli are linked with restriction and the management of pathogens. Main products of carbohydrate fermentation by lactic acid bacteria are lactic acid and acetic acid. These acids in the undissociated hydrophobic form disperse through the cell membrane of the targeted microorganism. Once inside the cytoplasm of the bacteria, acids are exposed to neutral pH, gets dissociated subsequently to anions and H^+ which in turn reduce the pH of the cytoplasm and the metabolic activities (Kotikalapudi, 2009; Dalie *et al.*, 2010). Low cytoplasmic pH interferes with signal transduction, inhibits glycolysis, and prevents active transport. Further, anionic protein accumulates inside the bacterial cell and cannot diffuse through the wall of the cell freely. This accumulation of anions in the bacterial cell leads to internal osmotic disorders. Kaewsrichan *et al.* (2006) showed that Lactobacilli produces hydrogen peroxide at the concentration that is inhibitory to *E. coli* and *G. vaginalis*. Production of H_2O_2 in the vaginal milieu presents a nonspecific mechanism-based defense against microbial infections. The specific inhibition mechanism is splitted into two portions:

(i) toxicity to microorganisms or (ii) acts as a peroxidase system catalyst (Boris and Barbes, 2000). In the vaginal fluid oxidation of halides into halogen or toxic hyphohalous acid is catalyzed by the peroxidase system through the use of H_2O_2.

Adherence and Competition to Pathogens

Major pre-requisite considered in the management of pathogenic species within the genital region to exert health promoting benefits is adherence of the Lactobacilli and to out-number the vaginal pathogens on Vaginal Epithelial Cells (VECs). Adherence is based on definitive interactions between the vaginal adhesions and complementary receptors. Several reports, show *in vitro*adherence of *Lactobacillus* to VECs (Andreu *et al.*, 1995; Kaewsrichan *et al.*, 2006). Moreover, blockage of the adhesion of pathogens of urogenital origin to the VEC's has been identified by some reporters (Chan *et al.*, 1985; Boris *et al.*, 1998). Some researchers consider that the stronger affinity of *Lactobacillus* to VEC receptors than pathogens result in the competitive exclusion manner of action (Osset *et al.*, 2001), others propose that exclusion does take place as a result of mechanical obstruction by the fragments of *Lactobacillus* (Chan *et al.*, 1985). Competition is also included by some studies in the displacement assay form (Boris *et al.*, 1998; Zarate and Nader Macias, 2006). In such assays, VECs are mixed with pathogens for 30 minutes, subsequently *Lactobacillus* were added on to thc mixture. Following the addition of *Lactobacillus* to the mixture a decrease was observed in the number of pathogens adhered per cell. The decrease is considered to take place because of a displacement of pathogens due to the higher affinity harboured by certain strains of *Lactobacillus* to the VEC receptors. Kwok *et al.*,(2006) and Butler *et al.* (2016) detected adherence of *L. crispatus*CTV-05 to VECs in large number of females. Although, little is known regarding the molecular mechanism involved in the adherence, different components, such as carbohydrates, lipoteichoic acids and glycoproteins, have been thought to be reasonably involved in the action (Chan *et al.*, 1985; Reid *et al.*, 1993; Boris *et al.*, 1998). Moreover, due to limited sources of nutrients, there is constant competition for the nutrients in this biological niche. Competition for nutrients and sites for binding are crucial in the management of pathogenic species in the genital tract of a woman.

Immunomodulation

Investigations regarding cytokine induction by probiotic bacteria present a great deal of interest. Regulating properties in *in vitro* and *ex vivo* models have been identified by several studies among the *Lactobacillus* strains when Lactobacilli were cultured along with immune cells (Christensen *et al.*, 2002; Morita *et al.*, 2002). Probiotics increases the generation of anti-inflammatory cytokines for instance IL-10, IgA, host defense peptides such as β defensin 2, dendritic cell maturation and reduces the generation of proinflammatory cytokines (by means of action on NFêB pathway) (Devine and Marsh, 2009). Macrophages are found to secrete immune mediators on stimulation by the lipoteichoic acid (LTA) present on the cell surface of Lactobacilli. Improved anti-inflammatory activity has also been observed *in vivo* in murine colitis models on LTA removal or substitution (Grangette *et al.*, 2005; Mohamadzadeh *et al.*, 2011). Anti-inflammatory properties in the supernatant of Lactobacilli have also been observed in cultured decidual cells, placental trophoblast cells, macrophages, and monocytes (Lin *et al.*, 2008; Yeganegi *et al.*, 2009; 2010; 2011; Li *et al.*, 2014). Induction in the generation of IL-12 and TNFα in primary splenocytes of mice was reported with internationally recognized *L. casei* shirota. *L. rhamnosus* CNCM I-4036 supernatant was more effective than live bacteria in decreasing the production of pro-inflammatory cytokines in human decidual cells upon *E. coli* challenge (Bermudez *et al.*, 2013).

Route of Administration

Probiotics have to reach intact and populate the vagina to bestow the benefits. The recognized methods of administration are orally and vaginally. Vaginal probiotic capsules or inclusion of Lactobacilli into vagina by means of a pessary have been widely used to directly introduce the probiotic bacteria into the vagina to overcome the pathogen or reduce their ability to proliferate. Probiotics are also administered orally to offer health benefits. A pioneering research held at the University of Western Ontario in 2001, reported the ability of specific strains of *Lactobacillus* to repopulate and maintain the normal urogenital flora on oral consumption for just 28 days (Reid *et al.*, 2001, 2002). Reid and colleagues (2002) found that bacterial and yeast pathogens in the vagina were reduced after combination of Lactobacilli strains were taken orally. It may be assumed that probiotic microbes should be competent of ascending to vagina *via* perineum following excretion from the rectum comparable to the pathogenic microorganisms of colonic origin that causes genital disorder. This is justified by the observation of clinical trials where vagina was colonized with the microflora of intestinal origin revealing microbial ascension is a normal process (Reid *et al.*, 2001). Strus and colleagues (2012) reported the recovery of orally administered Lactobacilli from the vagina. Thus, daily oral probiotic intake is effective at shifting pathogen friendly environments to a pathogen resistant vaginal microflora. However, the preferred route of delivery for probiotic lactobacilli in genital infections is intravaginal, as the time of vaginal colonization after oral consumption is longer compared to intravaginal administration. Further, the viability and survival in the gastric acid and bile salts and the Lactobacilli load that could be delivered orally to colonize the vagina is lower than direct vaginal administration. Moreover, the treatment duration with oral capsules is longer than the vaginal capsules (Bodean *et al.*, 2013). However, oral route has an advantage of reducing pathogen ascendance from the rectum to vagina through perineum, concerns with intravaginal approach is more invasive instillation of microbes.

Administration Vehicles

Probiotics can be administered in various forms comprising oral capsules, vaginal capsules and vaginal tablets. Positive effects of oral administration of capsules for treatment of vaginosis have been reported within clinical trials (Reid *et al.*, 2001; Reid *et al.*, 2002; Anukam *et al.*, 2006; Petricevic *et al.*, 2008). Alleviation of symptoms and decreased occurrence of Bacterial Vaginosis has been reported with vaginal probiotic tablets (Ozkinay *et al.*, 2005; Marcone *et al.*, 2008; Mastromarino *et al.*, 2009; Homayouni *et al.*, 2014). Some studies have also stated the efficacy of vaginal capsules in easing the symptoms of Bacterial Vaginosis (Anukam *et al.*, 2006; Ehrstrom *et al.*, 2010; Ya *et al.*, 2010). Limited studies support food as probiotic vehicles. Shalve and colleagues (1996) reported reduced BV episodes on use of fermented milk consisting of Lactobacilli.

Relation between Probiotics and Fertility

Reproduction process is simple and starts with conception upon contact of sperm with egg and failure is frequent. Infertility studies have provided useful insights into the details of reproduction success. One third of the infertility cases are of unexplained origin, while remaining is attributed to anatomical and physiological conditions. Complete and multiple system complications including reproductive potency is caused due to atrophy of cell layers or tissues of a morphological organization. Although detailed research is lacking on the relationship between reproductive morbidity (infertility) and reproductive tract microflora (vagina), studies have reported correlation between microbes (infections) and infertility. Kaur and Prabha (2013 and 2014) reported infertility in female BALB/c mice colonized with *Staphylococcus aureus* and *Escherichia coli* in vagina (frequent colonizers of genital tract). Many of the microorganisms present in vagina and cervix are not culturable and thus, not providing a types and number of bacteria present and which are beneficial and which are harmful that needs to be eradicated. It is hypothesized that microorganisms associated with vaginal infections might play a role in reduced ability of conception (Eckert *et al.*, 2003; Van oostrum *et al.*, 2013) as the vaginal disorder therapy improved the pregnancy rate. Presence of microorganisms in the reproductive tract might affect the transport of spermatozoa, result in immobilization/agglutination or loss of acrosome pertinent in successful conception. Impediments associated with microorganisms are measured as preventable and treatable. However, adverse effects associated with antimicrobial therapy are increasing, reduction in use of these agents should be reasonable.

There is growing evidence of the presence of certain species in the healthy genital tract defending the host against diseases by pathogenic microbes. Microbes in the body's natural orifices *e.g.*, vaginal cavity form a symbiotic host microorganism relationship through immune system regulation. Fluctuation in the microbial community leaves the host susceptible to bacterial over colonization and pathogen entry. It is proposed that the understanding and recognition of host microbe interactions will change much of medical views. Further, various pharmaceutical formulations enclosing probiotic Lactobacilli strains upon administration, being well-tolerated have been found to reduce, treat, or prevent genital infections and improve the microflora profile in vaginal milieu (Rossi *et al.*, 2010; Facchinetti *et al.*, 2013; Vicariotto *et al.*, 2014).

In contrast, despite the therapeutic efficacy of probiotics, there is no relevant evidence of their expression *in vivo* on the fertility aspect. Further, reproductive tract pathogenesis and potential causes of impaired fertility can be understood by illumination of host microbe interactions inside the vaginal cavity. Based on the positive impact on vaginal infection Borges *et al.* (2011) reported the prevention of colonization of *Listeria monocytogenes* in pregnant women by the use of a vaginal probiotic (*Pediococcus pentosaceus* SB83). Sirota *et al.* (2014) reported increase in implantation rates and reduction in immigration of pathogenic microbes upon inoculation with *L. crispatus* in uterine microbiome at the moment of embryo transfer (Braundmeier *et al.*, 2015). Although there is a lack of detailed research, further evidence is required in the field of association between the microbiota of the reproductive tract and the fertility rates. A meeting held in Aberdeen, Scotland in 2014 by the International Scientific Association for Probiotics and Prebiotics summarized the research on the role of microbes in conception (reproduction). Pelzer and colleagues (2013) have documented the presence of Lactobacilli in human follicular fluid and associated the Lactobacilli with embryo maturation and transfer while reporting correlation between microbes and infertility causes. Chimura *et al.*, 1992 has reported the presence of Lactobacilli in the ovaries that could have ascended from vagina and cervix and traversed all the way to the fallopian tubes. Functions of Lactobacilli in the healthy ovary or follicular fluid of fertile females can only be speculated at this point.

Bhandari *et al.* (2015) noted displacement of spermagglutinating *E. coli* by *Lactobacillus plantarum,* reinforcement of normal vaginal flora and finally restitution of fertility while evaluating the profertility effect of *L. plantarum.* In a study it was observed that the spermagglutinating strain of *E. coli* produced a factor with negative effects on the spermatozoa and rendered the female mice infertile on colonization in the vagina (Kaur and Prabha, 2013, 2014). Displacement of this sperm impairing *E. coli* led to the restoration of fertility in female mice. Blockage of the sperm impairing *E. coli* adherence to vaginal epithelial cells can be the possible explanation of the displacement of pathogens and restoration of normal vaginal microflora. Moreover, the possible harmful effects of the sperm impairing *E. coli* on the conception might be affected by the displacement of pathogens by *L. plantarum* that led to the restitution of fertility. Barbonetti and colleagues (2011) also reported a correlation between Lactobacilli and increased rate of fertilization potential. They found Lactobacilli can improve the fertilization potential of females by exerting a protective role on spermatozoa from free oxygen radical species present in the milieu in vaginal infections. The study of Lactobacilli and its effect on the reproductive potential is in its infancy due to the enduring influence of the focused paradigm on pathogen associated infertility rather than why Lactobacilli is associated with fertility and the difficulties of collecting samples from healthy pregnant women. Albeit, studies have not addressed the role of probiotics in fertility outcome, there i sevidence that probiotics are effective in vaginal infections associated with infertility. Thus, supplementation of probiotics aimed at improving reproductive health status and fertility makes sense. Further, more studies are required to confirm the concept of the role of probiotics and elucidation of the beneficial effects of probiotic therapy, strain composition, duration, route of administration, dosage and understanding of the mechanism involved, that could impact or improve conception, successful pregnancy and fertility outcome and result in efficacious intervention.

Probiotics Commercially available for Urogenital Tract

Although at present, a small number of probiotic products are commercially available for the urogenital tract, research is expanding in the area of urogenital probiotics, further it will possibly not be long prior the products for various genital conditions with different probiotic components may be offered in the market.

1. VSL#3: a combination consisting of 8 distinct strains (four lactobacilli strains (*L. plantarum, L. delbrueckii* subsp. *bulgaricus, L. acidophilus* and *L. casei*) *Streptococcus salivarius* subsp. *thermophilus*) and three bifidobacteria strains (*Bifidobacterium longum, B. infantis* and *B. breve*) was found to improve pregnancy outcome with some success in terms of modulation in immune parameters and vaginal microbiota on oral administration (Vitali *et al.,* 2012).
2. Probalac Vaginal: vaginal capsule containing *L. acdophilus, L. rhamnosus* and *S. thermophilus* resulted in lower recurrence rate of infection (Ya *et al.,* 2010).
3. Eco Vag R: vaginal capsule containing *L. rhamnosus Lbp PB01-DSM14870* and *L. gasseri Lba EB01- DSM14864* supplementation after clindamycin treatment significantly reduced the bacterial vaginosis recurrence rate (Larsson *et al.,* 2008).
4. Florisia vaginal tablets: a probiotic product for vaginal use comprises *L. salivarius FV2, L. brevis CD2* and *L. plantarum FV9.* These strains are chosen on the basis of characteristics such as mucosal colonization that is capability to attach the epithelial cells and to temporarily colonize human vagina, for production of antimicrobial compounds effective towards pathogenic microorganisms and its capacity to inhibit pathogen binding to cell membrane (Maggi *et al.,* 2000; Mastromarino *et al.,* 2013).
5. Fem-Dophilus™: available as vaginal capsules contain *Lactobacillus* RC-14 and *Lactobacillus* GR-1. With respect to their potential probiotic properties, these strains have been selected for use in the urogenital tract (Reid *et al.,* 2001, 2002, 2014).

Aspects of Safety

Lactobacilli, mainlyreferred to as "good" commensal microbes participate in the defense in opposition to pathogenic microorganisms in the gut and the reproductive tract. Diseases caused by *Lactobacillus* are uncommon and infections associated with the ingestion of probiotic *Lactobacillus* have occurred not more than in a few cases relating to *L. rhamnosus* GG (Reid and Hammond, 2005). These cases are related with patients suffering from chronic infection or devastating situations that present direct access from leaky gut to bloodstream. On the other hand, products consisting of probiotic *Lactobacillus* are considered safe corresponding to the approximate worldwide sale of 20 billion doses per annum (Reid, 2003). Over 1 billion doses administered annually for urogenital health have been well tolerated and bacteremia caused by *Lactobacillus* is extremely rare. Moreover, no reports of toxicity or infections as a consequence of genital probiotics (species of *Lactobacillus* and *Bifidobacteria*) have been described (Gupta and Garg, 2009). Probiotics are viable microorganisms, its use in healthy people is safe, but it should be used cautiously in immunocompromised persons. As a result of an amplified concern on the subject of conjugative transfer of antibiotic resistance genes, safety or risks must be considered besides benefits (Hammerman *et al.,* 2006). Several clinical trials that place efficiency corresponding to side effects or safety, infectivity in immunocompromised are therefore needed (Borriello *et al.,* 2003; Anukam *et al.,* 2006b; Marrazzo *et al.,* 2006; Sanders *et al.,* 2010).

Conclusions

As the vaginal microflora is frequently affected by various physiological factors of females, it can cause changes in the microbial ecosystems, thereby increasing the risk of infections. Lactobacilli are among the dominant microflora of vagina and it can contribute to control of growth of pathogens. So the use of probiotics can prevent or they can be used as alternative to use of antibiotics and many researchers have supported this idea. Probiotics can help in maintaining the health of vagina, as probiotics produces various natural antimicrobial substances that inhibit the growth of various pathogens and also promotes the growth of normal microflora. Authors conclude that once clinically tested, probiotics provide a very promising role in fertility of females, but its safety aspects need to be studied at a larger level. Also efficacy and cost effectiveness of these vaginally administered probiotics makes them a good choice treatment or prophylaxis of female urogenital infections.

Declarations

The authors have no conflict of interest and all the authors have contributed equally for the article.

REFERENCES

Andreu, A., Stapleton, A., Fennell, C., Hillier, S. and Stamm, E. 1995. Hemagglutination, adherence and surface properties of vaginal Lactobacillus species. *J. Infect. Dis.* 171: 1237-1243.

Anukam KC, Osazuwa E, Osemene GI, Ehigiagbe F, Bruce AWand Reid G. 2006. Clinical study comparing probiotic Lactobacillus GR-1 and RC-14 with metronidazole vaginal gel to treat symptomatic bacterial vaginosis. *Microbes Infect.* 12-13: 2772-6.

Barbonetti, A., Cinque, B., Vassallo, M. R. C., Mineo, S., Francavilla, S.and Cifone M. G. 2011. Effect of vaginal probiotic lactobacilli on in vitro-induced sperm lipid peroxidation and its impact on sperm motility and viability. *Fertil. Steril.* 95: 2485–2488.

Bermúdez-Humarán, L. G., Aubry, C., Motta, J. P., Deraison, C., Steidler, L., Vergnolle, N., Chatel, J. M. and Langella, P. 2013. Engineering lactococci and lactobacilli for human health. *Curr. Opin. Microbiol.* 16(3): 278–283.

Bhandari, P.and Prabha, V. 2015. Evaluation of profertility effect of probiotic *Lactobacillus plantarum* 2621 in a murine model. *The Ind. J. Med.Res.* 142(1): 79–84.

Bisanz, J.E., Enos, M.K., Mwanga, J.R., Changalucha, J., Burton, J.P., Gloor, G.B. and Reid, G. 2014. Randomized open-label pilot study of the influence of probiotics and the gut microbiome on toxic metal levels in Tanzanian pregnant women and school children. *mBio.* 5(5): e01580-14.

Bodean, O., Munteanu, O., Cirstoiu, C., Secara, D. and Cirstoiu, M. 2013. Probiotics—a helpful additional therapy for bacterial vaginosis. *J.Medicine Life* 6(4): 434–436.

Borges, S.F., Silva, J.G.L. and Teixeira, P.C.M. 2011. Survival and biofilm formation of *Listeria monocytogenes* in simulated vaginal fluid: influence of pH and strain origin. *FEMS Immunol. Med. Microbiol.* 62: 315–320.

Boris, S. and Barbes, C. 2000. Role played by lactobacilli in controlling the population of vaginal pathogens. *Microb. Infect.* 4: 543–546.

Boris, S., Suárez, J.E., Vázquez, F. and Barbés, C. 1998. Adherence of human vaginal lactobacilli to vaginal epithelial cells and interaction with uropathogens. *Infect. Immun.* 66(5): 1985-1989.

Borriello, S.P., Hammes, W.P., Holzapfel, W., Marteau, P., Schrezenmeir, J., Vaara, M.and Valtonen, V. 2003. Safety of probiotics that contain lactobacilli or bifidobacteria. *Clin. Infect. Dis.* 36(6): 775-80.

Braundmeier, A. G., Lenz, K. M., Inman, K. S., Chia, N., Jeraldo, P., Walther-António, M. R., Berg Miller, M. E., Yang, F., Creedon, D. J., Nelson, H.and White, B. A. 2015. Individualized medicine and the microbiome in reproductive tract. *Front. Physiol.* 6: 97.

Butler, D.S.C., Silvestroni, A. and Stapleton AE. 2016. Cytoprotective Effect of *Lactobacillus crispatus* CTV-05 against Uropathogenic *E. coli*. *Pathogens* 5(1): 27.

Chan, R.C., Reid, G., Irvin, R.T., Bruce, A.W. and Costerton, J.W. 1985. Competitive exclusion of uropathogens from human uroepithelial cells by Lactobacillus whole cells and cell wall fragments. *Infect. Immun.* 47: 84–89.

Chimura, T., Funayama, T., Murayama, K. and Numazaki, M.1995. Ecological treatment of bacterial vaginosis. *Japanese J. of Antibiotics* 48: 432–436.

Christensen, H. R., Frøkiaer, H.and Pestka, J. J. 2002. Lactobacilli differentially modulate expression of cytokines and maturation surface markers in murine dendritic cells. *J. Immunol.* 168(1): 171–178.

Dalié, D.K.D., Deschamps, A.M. and Richard-Forget, F. 2010. Lactic acid bacteria – Potential for control of mould growth and mycotoxins: A review. *Food control.* 21: 370-380.

De Gregorio, P.R., Juárez Tomás, M.S., Leccese Terraf, M.C. and Nader-Macías, M.E. 2014. *In vitro* and *in vivo* effects of beneficial vaginal lactobacilli on pathogens responsible for urogenital tract infections. *J. Med. Microbiol.* 63: 685–696

de Oliveira Leite, A.M., Miguel, M.A., Peixoto, R.S., Rosado, A.S., Silva, J.T. and Paschoalin, V.M. 2013. Microbiological, technological and therapeutic properties of kefir: a natural probiotic beverage. *Braz. J. Microbiol.* 44(2): 341-9.

Devine, D. A. and Marsh, P. D. 2009. Prospects for the development of probiotics and prebiotics for oral applications. *J.Oral Microbiol.* 1: 1,1949.

Doerflinger, S.Y., Throop, A.L. and Herbst-Kralovetz, M.M. 2014. Bacteria in the vaginal microbiome alter the innate immune response and barrier properties of the human vaginal epithelia in a species-specific manner. *J. Infect. Dis.* 209(12): 1989-99.

Dunne, C., O'mahony, L., Thornton, G., Feeney, M., Daly, C., O'Sullivan, G. and Collins, J.K. 2001. In vitro selection criteria for probiotic bacteria of human origin: Correlation with in vivo findings. *Am. J. Clin. Nutr.* 73: 386-392.

Eckert, L.O., Moore, D.E. and Patton, DL. 2003. Relationship of vaginal bacteria and inflammation with conception and early pregnancy loss following in vitro fertilization. *Infec. Dis. Obstet. Gynecol.* 11: 11–17.

Ehrström, S., Daroczy, K., Rylander, E., Samuelsson, C., Johannesson, U., Anzén, B.and Påhlson, C. 2010. Lactic acid bacteria colonization and clinical outcome after probiotic supplementation in conventionally treated bacterial vaginosis and vulvovaginal candidiasis. *Microbes Infect.*12(10): 691-9.

Facchinetti, F., Dante, G., Pedretti, L., Resasco, P., Annessi, E.and Dodero, D. 2013. Studio pilota sul ruolo dei probiotici nel trattamento della vaginosi batterica in gravidanza [The role of oral probiotic for bacterial vaginosis in pregnant women. A pilot study]. *Minerva Ginecol.* 65(2): 215-21.

Gong Z., Luna Y., Yu P. and Fan H. 2014. Lactobacilli inactivate *Chlamydia trachomatis* through lactic acid but not H_2O_2. *PLoS ONE* 9: e107758.

Grangette, C., Nutten, S., Palumbo, E., Morath, S., Hermann, C., Dewulf, J., Pot, B., Hartung, T., Hols, P. and Mercenier, A. 2005. Enhanced antiinflammatory capacity of a *Lactobacillus plantarum* mutant synthesizing modified teichoic acids. *Proc. Nat. Acad. Sci. USA* 102(29): 10321–10326.

Gupta, V.and Garg, R. 2009. Probiotics. *Ind. J. Med. Microbiol.* 27: 202-209.

Hammerman, C., Bin-Nun, A.and Kaplan, M. 2006. Safety of probiotics: comparison of two popular strains. *BMJ*333: 1006-8.

Homayouni, A., Bastani, P., Ziyadi, S., Mohammad-Alizadeh-Charandabi, S., Ghalibaf, M., Mortazavian, A.M. and Mehrabany, E.V. 2014. Effects of probiotics on the recurrence of bacterial vaginosis: a review. *J. Low. Genit. Tract Dis.* 18(1): 79-86.

Jones, S.E. and Versalovic, J. 2009. Probiotic *Lactobacillus reuteri* biofilms produce antimicrobial and anti-inflammatory factors. *BMC Microbiol.* 9: 35.

Kaewsrichan, J. Peeyananjarassri, K. and Kongprasertkit, J. 2006. Selection and identification of anaerobic lactobacilli producing inhibitory compounds against vaginal pathogens. *FEMS Immunol. Med. Microbiol.* 48(1): 75-83.

Kaur, K.and Prabha, V. 2013. Sperm Impairment by Sperm Agglutinating Factor Isolated from *Escherichia coli*: Receptor Specific Interactions. *Biomed Res. Int.* 2013: 548497.

Kaur, K.and Prabha, V. 2014. Immunocontraceptives: new approaches to fertility control. *Biomed Res. Int.* 2014: 868196.

Kotikalapudi, B. L. 2009. Characterization and encapsulation of probiotic bacteria using a pea-protein alginate matrix. Master thesis. University of Saskatchewan. Canada.

Kwok, L., Stapleton, A.E., Stamm, W.E., Hillier, S.L., Wobbe, C.L. and Gupta, K. Adherence of *Lactobacillus crispatus* to vaginal epithelial cells from women with or without a history of recurrent urinary tract infection. *J Urol.* 176(5): 2050-2054.

Larsson, P.G., Stray-Pedersen, B., Ryttig, K.R.and Larsen, S. 2008. Human lactobacilli as supplementation of clindamycin to patients with bacterial vaginosis reduce the recurrence rate; a 6-month, double-blind, randomized, placebo-controlled study. *BMC Womens Health.* 8: 3.

Lin, K. L., Suzuki, Y., Nakano, H., Ramsburg, E. and Gunn, M. D. 2008. $CCR2^+$ monocyte-derived dendritic cells and exudate macrophages produce influenza-induced pulmonary immune pathology and mortality. *J. Immunol.*180(4): 2562–2572.

Maggi, L., Mastromarino, P., Macchia, S., Brigidi, P., Pirovano, F., Matteuzzi, D.and Counte, U. 2000. Technological and biological evaluation of tablets containing different strains of lactobacilli for vaginal administration. *Eur. J. Pharm. Biopharm.* 50: 389–395.

Marcone. V., Calzolari, E.and Bertini, M. 2008. Effectiveness of vaginal administration of *Lactobacillus rhamnosus* following conventional metronidazole therapy: how to lower the rate of bacterial vaginosis recurrences. *New Microbiol.* 31(3): 429-33.

Marrazzo, J.M., Cook, R.L., Wiesenfeld, H.C., Murray, P.J., Busse, B.and Krohn, M. 2006. Women's satisfaction with an intravaginal Lactobacillus capsule for the treatment of bacterial vaginosis. *J Womens Health* (*Larchmt*) 15(9): 1053-60.

Mastromarino, P., Brigidi, P., Macchia, S., Maggi, L., Pirovano, F., Trinchieri, V., Conte, U. and Matteuzzi, D. 2002. Characterization and selection of vaginal Lactobacillus strains for the preparation of vaginal tablets. *J. Appl. Microbiol.* 93(5): 884-93.

Mastromarino, P., Di, P.M., Schiavoni, G., Nardis C., Gentile, M. and Sessa R. 2014. Effects of vaginal lactobacilli in *Chlamydia trachomatis* infection. *Int. J. Med. Microbiol.* 304: 654–661

Mastromarino, P., Macchia, S., Meggiorini, L., Trinchieri, V., Mosca, L., Perluigi, M.and Midulla, C. 2009. Effectiveness of Lactobacillus-containing vaginal tablets in the treatment of symptomatic bacterial vaginosis. *Clin. Microbiol. Infect.* 15(1): 67-74.

Mastromarino, P., Vitali, B.and Mosca, L. 2013. Bacterial vaginosis: a review on clinical trials with probiotics. *New Microbiol.* 36(3): 229-38.

McMillan, A., Dell, M., Zellar, M.P., Cribby, S., Martz, S., Hong, E., Fu, J., Abbas, A., Dang, T., Miller, W. and Reid, G. 2011. Disruption of urogenital biofilms by lactobacilli. *Colloids Surf. B. Biointerfaces* 86(1): 58-64.

Mohamadzadeh, M., Pfeiler, E. A., Brown, J. B., Zadeh, M., Gramarossa, M., Managlia, E., Bere, P., Sarraj, B., Khan, M. W., Pakanati, K. C., Ansari, M. J., O'Flaherty, S., Barrett, T. and Klaenhammer, T. R. 2011. Regulation of induced colonic inflammation by *Lactobacillus acidophilus* deficient in lipoteichoic acid. *Proc. Nat. Acad. Sci. USA*108 Suppl 1: 4623–4630.

Morita, H., He, F., Fuse, T., Ouwehand, A. C., Hashimoto, H., Hosoda, M., Mizumachi, K. and Kurisaki, J. 2002. Adhesion of lactic acid bacteria to caco-2 cells and their effect on cytokine secretion. *Microbiol.Immunol.* 46(4): 293–297.

Nishiyama, K., Seto, Y., Yoshioka, K., Kakuda, T., Takai, S., Yamamoto, Y. and Mukai, T. 2014. *Lactobacillus gasseri* SBT2055 reduces infection by and colonization of *Campylobacter jejuni. PLoS One* 9(9): e108827.

Osset, J., Bartolome, R.M., Garcia, E. and Andreu, A. 2001. Assessment of the capacity of Lactobacillus to inhibit the growth of uropathogens and block their adhesion to vaginal epithelial cells. *J. Infect. Dis.* 183: 485–491.

Ozkinay, E., Terek, M. C., Yayci, M., Kaiser, R., Grob, P. and Tuncay, G. 2005. The effectiveness of live lactobacilli in combination with low dose oestriol (Gynoflor) to restore the vaginal flora after treatment of vaginal infections. *BJOG: Int.J. Obstet. Gynecol.*112(2): 234–240.

Pascual, L.M., Daniele, M.B., Ruiz, F., Giordano, W., Pájaro, C. and Barberis, L. 2008. *Lactobacillus rhamnosus* L60, a potential probiotic isolated from the human vagina. *J. Gen. Appl. Microbiol.* 54(3): 141-8.

Pelzer, E.S., Allan, J.A., Waterhouse, M.A., Ross, T., Beagley, K.W. and Knox, C.L. 2013. Microorganisms within Human Follicular Fluid: Effects on IVF. *PLoSONE,* 8(3): e59062.

Petricevic, L., Unger, F.M., Viernstein, H.and Kiss, H. 2008. Randomized, double-blind, placebo-controlled study of oral lactobacilli to improve the vaginal flora of postmenopausal women. *Eur. J. Obstet. Gynecol. Reprod. Biol.* 141(1): 54-7.

Petrova, M.I., van den Broek, M., Balzarini, J., Vanderleyden, J. and Lebeer, S. 2013. Vaginal microbiota and its role in HIV transmission and infection. *FEMS Microbiol. Rev.* 37(5): 762-92.

Reid, G. 2014. Modulating the vaginal microbiome: the need for a bridge between science and practice. *Semin. Reprod. Med.* 32: 28–34.

Reid, G.and Bocking, A. 2003. The potential for probiotics to prevent bacterial vaginosis and preterm labor. *Am. J. Obstet. Gynecol.* 189: 1202–1208.

Reid, G.and Burton, J. 2002. Use of Lactobacillus to prevent infection by pathogenic bacteria. *Microbes Infect.* 4: 319–324.

Reid, G.and Hammond, J.A. 2005. Probiotics. Some evidence of their effectiveness. *Can. Fam. Physician.* 1(11): 1487-93.

Reid, G., Zalai, C.and Gardiner, G. 2001. Urogenital lactobacilli probiotics, reliability, and regulatory issues. *J. Dairy Sci.* 84: E164–E169.

Reid, G. and Bruce, A.W. 2003. Urogenital infections in women: can probiotics help? *Postgrad. Med. J.* 79: 428-432.

Rolfe, R.D. 2000. The role of probiotic cultures in the control of gastrointestinal health. *J. Nutr.* 130: 396S–402S.

Rose, M.A., Stieglitz, F., Köksal, A., Schubert, R., Schulze, J. and Zielen, S. 2010. Efficacy of probiotic Lactobacillus GG on allergic sensitization and asthma in infants at risk. *Clin. Exp. Allergy* 40(9): 1398-405.

Rossi, M.and Amaretti, A. 2010. Probiotic properties of bifidobacteria. B. Mayo, D. van Sinderen (Eds.), Bifidobacteria: Genomics and Molecular Aspects, Caister Academic Press,UK (2010), pp. 97-123.

Sanders, M.E.and Marco, M.L. 2010. Food formats for effective delivery of probiotics. *Annu. Rev. Food Sci. Technol.* 1: 65-85.

Saunders, S., Bocking, A., Challis, J. and Reid, G. 2007. Effect of *Lactobacillus* challenge on *Gardnerella vaginalis* biofilms. *Colloids Surf. B. Biointerfaces* 55: 138–142.

Schachtsiek, M., Hammes, W.P. and Hertel, C. 2004. Characterization of *Lactobacillus coryniformis* DSM 20001T surface protein Cpf mediating coaggregation with and aggregation among pathogens. *Appl. Environ. Microbiol.* 70(12): 7078-85.

Schellenberg, J., Smoragiewicz, W. and Karska-Wysocki, B. 2006. A rapid method combining immunofluorescence and flow cytometry for improved understanding of competitive interactions between lactic acid bacteria (LAB) and methicillin-resistant *S. aureus* (MRSA) in mixed culture. *J. Microbiol. Methods* 65(1): 1-9.

Shalev, E., Battino, S., Weiner, E., Colodner, R., and Keness, Y. 1996. Ingestion of yogurt containing *Lactobacillus acidophilus* compared with pasteurized yogurt as prophylaxis for recurrent candidal vaginitis and bacterial vaginosis. *Arch. Fam. Med.* (10): 593-6.

Sirota, I., Zarek, S.M.and Segars, J.H. 2014. Potential influence of the microbiome on infertility and assisted reproductive technology. *Semin. Reprod. Med.* 32: 35–42.

Strus, M., Helwich, E., Lauterbach, R., Rzepecka-Wêglarz, B., Nowicka, K., Wiliñska, M., Szczapa, J., Rudnicka, M., S³awska, H., Szczepañski, M., Waœko, A., Miko³ajczyk-Cichoñska, A., Tomusiak-Plebanek, A. and Heczko, P. B. 2018. Effects of oral probiotic supplementation on gut *Lactobacillus* and *Bifidobacterium* populations and the clinical status of low-birth-weight preterm neonates: a multicenter randomized, double-blind, placebo-controlled trial. *Infect.Drug Resistance,* 11: 1557–1571.

Taheri, H.R., Moravej, H., Tabandeh, F., Zaghari, M. and Shivazad, M. 2009. Screening of lactic acid bacteria toward their selection as a source of chicken probiotic. *Poultry Sci.* 88: 1586–1593.

van Oostrum, N., De Sutter, P., Meys, J.and Verstraelen, H. 2013. Risks associated with bacterial vaginosis in infertility patients: a systematic review and meta-analysis. *Hum. Reprod.* 28(7): 1809-15.

Verdenelli, M.C., Coman, M.M., Cecchini, C., Silvi, S., Orpianesi, C.and Cresci, A. 2014. Evaluation of antipathogenic activity and adherence properties of human Lactobacillus strains for vaginal formulations. *J. Appl. Microbiol.* 116: 1297–1307.

Vicariotto, F. 2014. Effectiveness of an association of a cranberry dry extract, D-mannose, and the two microorganisms *Lactobacillus plantarum* LP01 and *Lactobacillus paracasei* LPC09 in women affected by cystitis: a pilot study. *J. Clin. Gastroenterol.* 48(1): S96-101.

Vitali, B., Cruciani, F., Baldassarre, M.E., Capursi, T., Spisni, E., Valerii, M.C., Candela, M., Turroni, S. and Brigidi, P. 2012. Dietary supplementation with probiotics during late pregnancy: outcome on vaginal microbiota and cytokine secretion. *BMC Microbiol.* 12: 236.

Wagner, R.D. and Johnson, S.J. 2012. Erratum to: Probiotic *lactobacillus* and estrogen effects on vaginal epithelial gene expression responses to *Candida albicans*. *J. Biomed. Sci.* 19: 84.

Ya, W., Reifer, C.and Miller, L.E. 2010. Efficacy of vaginal probiotic capsules for recurrent bacterial vaginosis: a double-blind, randomized, placebo-controlled study. *Am. J. Obstet. Gynecol.* 203(2): 120.e1-6.

Yeganegi, M., Leung, C. G., Martins, A., Kim, S. O., Reid, G., Challis, J. R. and Bocking, A. D. 2010. *Lactobacillus rhamnosus* GR-1-induced IL-10 production in human placental trophoblast cells involves activation of JAK/STAT and MAPK pathways. *Reprod Sci.* 17(11): 1043–1051.

Yeganegi, M., Leung, C. H., Martins, A., Kim, S. O., Reid, G., Challis, J.and Bocking, A. D. 2011. *Lactobacillus rhamnosus* GR-1 Stimulates Colony-Stimulating Factor 3 (Granulocyte)(CSF3) Output in Placental Trophoblast Cells in a Fetal Sex-Dependent Manner. *Biol. Reprod. 84:* 18-25.

Yeganegi, M., Watson, C. S., Martins, A., Kim, S. O., Reid, G., Challis, J. R.and Bocking, A. D. 2009. Effect of *Lactobacillus rhamnosus* GR-1 supernatant and fetal sex on lipopolysaccharide-induced cytokine and prostaglandin-regulating enzymes in human placental trophoblast cells: implications for treatment of bacterial vaginosis and prevention of preterm labor. *Am. J. Obstet. Gynecol.* 200(5): 532.e1–532.e5328.

Younes, J.A., van der Mei, H.C., van den Heuvel, E., Busscher, H.J. and Reid, G. 2012. Adhesion Forces and Coaggregation between Vaginal Staphylococci and Lactobacilli. *PLOS ONE* 7(5): e36917.

Zarate, G. and Nader-Macias, M.E. 2006. Influence of probiotic vaginal lactobacilli on *in vitro* adhesion of urogenital pathogens to vaginal epithelial cells. *Lett. Appl. Microbiol.* 43: 174–180.

Chapter 6

Application of Synthetic Biology in Bioremediation

Akhlash P. Singh

Genomics and Proteomics Lab,
GGD SD College (Panjab University), Chandigarh
**e-mail: akhlash@ggdsd.ac.in*

ABSTRACT

The intense anthropological activities have caused a substantial level of environmental pollution that is a major threat to human existence on earth. The increasing pollution is not only affecting the animals and plant's health but also adversely affects the ecosystem and brought serious climatic changes. There are numerous sites of environmental degradation including the fresher water bodies, soil, and oceans that are deeply contaminated with extremely harmful non-degradable synthetic organic compounds produced from different industrial activities. To circumvent the problem various chemical, physical and biological strategies were adopted in the past, among them the bioremediation has proved the most potential and cost-effective and sustainable approach. However, bioremediation is not a new science to monitoring and cleaning hazardous chemicals and pollutants from the environment. But, after the emergence of systems, biological methodologies have created a wealth of knowledge in terms of genomics, proteomics, and metabolic data related to natural scavenging bacteria, plants, algae, and fungi. This allows us to rewrite or redesign the genetic programming of subject organisms by using synthetic biology tools and techniques. Therefore, these organisms can convert harmful organic compounds into less harmful or valuable products. The revolution in the area of synthetic genetic circuit construction can help us in the highly sensitive detection of pollutants but also their cleaning from the ecosystems. The main aim of the chapter is not to discuss the types and pollutants and their sites in detail. The major emphasis has been put on the recent advances made in the field of systems and synthetic biology in the area of bioremediations.

***Keywords**: Bioremediations, Phytoremediation, Synthetic biology, Systems biology.*

Introduction

From the very inception of human civilizations, anthropological activities particularly after the industrial revolution have created environmentally unfriendly situations on the earth. The byproducts of these activities are harmful to the earth's atmosphere and simultaneously hazards to the sustenance of human health too. Historically, mining for metals, hydrocarbons, and deforestation were the major activities that have created the large scale polluted sites world over that has adversely affected nearby human and animal population and natural microflora in ecosystems (Hong *et al.,* 1996). Moreover, after the industrial and green revolution, environmental pollutants have increased many folds due to the over-exploitation of natural resources in the virgin area of forests and oceans. Currently, a large number of diseases in man, plants, and animals have been caused by environmental pollutions and climatic changes, furthermore, several are on the rise, for example, respiratory diseases, allergy especially, cancers. Because synthetic chemicals are accumulated in tissues and thereby interfere with the normal physiological functions in living organisms, particularly the concerted effect of pollutants (mixture of pesticides) is extremely dangerous (Rhind, 2009). Hence, a very pertinent question arises: *what are pollutants*? The pollutants are unnatural or synthetic harmful compounds which originated from anthropological activities in the environment.

During the industrial revolution, excessive use of synthetic organic materials, petroleum-based products, and metal industries released polyaromatic hydrocarbon and heavy metal ions respectively, which contaminated the ecosystems in various countries of the world. Simultaneously, at the time of the green revolution intensified agriculture was practiced which involved the overuse of aromatic pesticides sprayed and high amounts of chemical fertilizers both have contaminated the large chunk of agricultural soils and freshwater bodies (Figure 6.1) (Davis *et al.,* 2002). The above-mentioned activities created various types of pollution such as degradation of forest, agricultural land, mineral, forest, sand, rocks, and the presence of toxic chemicals in the water. Now the question arises how we can overcome this problem. Scientists and researchers have exploited the various types of physical, chemical, and biological methodologies including exploited the metabolic cleaning ability of microorganisms and plants. The physical and chemical bioremediation methods are suffering some serious drawbacks like transport, contamination, excavations, expensive, temporary, and exposure to workers. The physical means of pollutants cleansing required a great deal of mechanical setup in terms of machinery and great cost. Current chapter is divided into two parts, first deals with a brief overview of bioremediation and the second describes the systems and synthetic biology approaches applied in bioremediation.

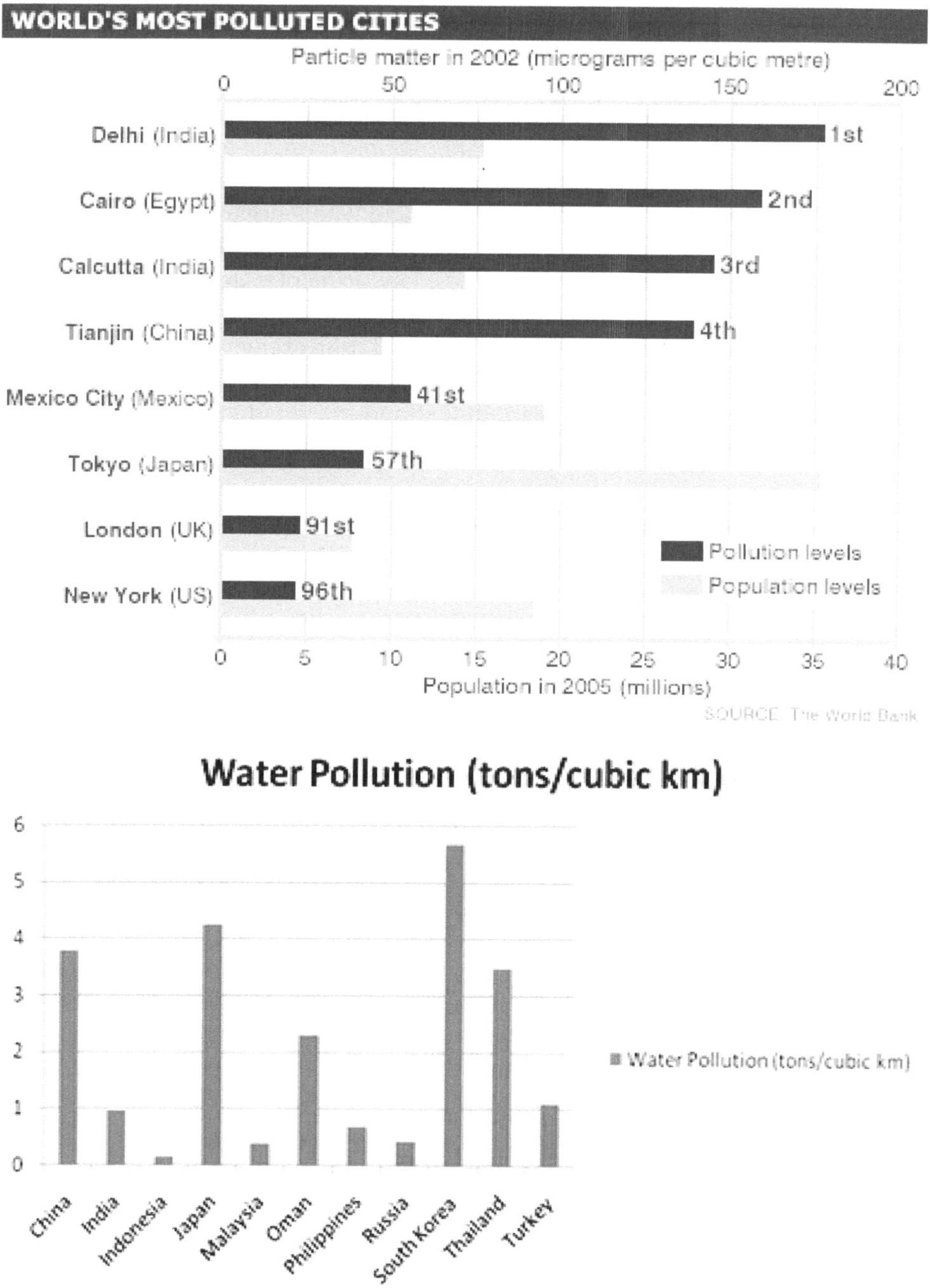

Figure 6.1: The Data Related to the Environmental Pollution Worldover.

Bioremediation: An Overview

In the view of above facts techniques of Bioremediation to eliminate or neutralize environmental pollutants from affected sites such as soils, water bodies, oceans, the air has been proved effective, sustainable, and free from drawbacks. But it requires a substantial amount of expertise and greater knowledge about the organisms involved in bioremediations. "*Bioremediation is the application of living organisms including microorganisms, plants, fungi, and algae to remove or degrade the contaminants from the pollutant sites*". Principally, bioremediation approaches generally use enzymes, metabolic pathways, and whole organisms to convert the highly harmful substances into less harmful or economically beneficial products. Under the overall bioremediation process, few principles are generally applied such as the use of mixed microbial communities, application of novel electron acceptor and donor, and sequestration or concentration of environmental pollutants inside the cells that are harvested at a later stage. Another major emphasis of all bioremediation-related studies is to understand the rapid growth of organisms under adverse environmental conditions like pollutant sites (Harms and Junca, 2014).

Currently, hydrocarbons from soils and spilling sites, polychlorinated biphenyls (PCBs), polyaromatic hydrocarbons (PAHs), heterocyclic compounds (such as pyridine or quinolone), pharmaceutical substances, radionuclides, and heavy metals have been remediated by using different advanced strains of microorganisms (Harms and Junca, 2014). Hence, likewise in other biological disciplines to understand a biological significant process we need a systems-level understanding of physiology, metabolisms of bacterial strains that are involved in bioremediations, therefore, whole-cell can be engineered by applying systems and synthetic biology principles. The synthetic biology and environmental biotechnology approach not affect the pattern of cleansing and removal part of the bioremediation process but it also improved and implement the efficient bio-sensing of hazard compounds at affected sites by inventing sophisticated biosensors and photo-sensors (Chakraborty *et al.*, 2012).

There are several methodologies in an application for bioremediations including, phytoremediation, bioventing, bioleaching, land farming, bioreactor, composting, bioaugmentation, rhizofiltration, and biostimulation, the total available bioremediation techniques have been divided into two broad categories *i.e. in situ* and *ex situ* processes.

Target Sites for Bioremediations

Water Bodies

The main water bodies include rivers, ponds, aquifers, lakes, and groundwater that are generally a source of fresh water supply for drinking purposes. The water can be used for agricultural activities too, so when contaminated water is consumed in human activities this can increase the biomagnification in animals and plants. Therefore, the consumption of animals and plants as food can cause toxicity in humans. The major sources of water bodies' pollution include discharge of untreated sewage, agriculture runoff, and effluents from small-scale factories. Fertilizers and pesticides used in agriculture, during monsoon due to flooding, many solid wastes from cities and industries and contaminated water washes away inside the rivers and lakes. The situation of water pollution is worse in developing countries (Figure 6.1) and according to a rough estimation, 14000 people killed or die from pollution-related ailments (http://environment.about.com/od/environmentalevents/a/waterdayqa.htm).

Oceans

The Sea occupies the third-fourth part of the earth's surface and closely affects the climate and lifeforms on the earth. A large number of sources have been identified which cause sea pollution such as humans dumping the trash, medical wastes, sludge dumping, oil spills, dumping of mercury, cadmium compounds, plastic, oil, and other dangerous wastes.

Mining Sites

There is a large number of mining sites worldwide which are generally exploited for energy products and extraction of coal, iron ore, manganese, mica, bauxite, copper, kyanite, chromite, beryl, apatite, *etc.* The mining sites' mediated pollution includes groundwater contamination and soils. The mining sites can cause a combination of air, water, and land pollution simultaneously. The impact of mining can be extremely dangerous; it can cause loss of diversity in affected regions. In some extreme cases, the chemicals have been percolated deep into the ground and make groundwater polluted or toxic.

Field Soils

As discussed above the fertilizers (ammonia; source of nitrogen, Urea; nitrogen source, Superphosphate; containing a certain amount of nitrogen and phosphate and Diammonium phosphate), and pesticides (organochlorine, organophosphate, organometallic, pyrethroids, and carbamates) are the major source of pollution in the arable land which has adversely affect agriculture soil promote the bio-magnification in consumers.

Municipal Sewages

Includes domestic waste, commercial waste, and industrial waste which include the amalgam of a wide spectrum of pollutants and their separation is an extremely troublesome process. Currently, technologies are available which can convert sewage into valuable or less harmful products (Owa, 2014).

Organisms Used in Bioremediation Methods

Currently among the scientific community, excitement prevails which promotes them to use bio-based methods to remove pollutants from the soil and water.

Therefore, they are using bacteria, fungi, plants, and algae for bioremediation processes in soil and aquatic systems.

Microorganisms

Recently with the emergence of modern high-throughput analytical methods such as genomics, proteomics, metagenomics, eco-genomics, and bioinformatics numerous studies have been conducted to investigate the effect of pollutants, hydrocarbons, heavy metals including high temperatures on the living systems including bacteria, plants algae, and fungus. These studies have identified a large number of genes, proteins, and metabolites responsible to respond to the entry and catabolism of pollutants/toxins and contaminants inside the living cells. However, a large number of bacterial species have been identified to decontaminate the soil, water, air, oceans and rivers, and mining sites, but have been proved effective. A few of them are P*seudomonas, Aeromonas, Moraxella, Beijerinckia, Flavobacteria, chlorobacteria, Nocardia, Corynebacteria, Atinetobacter, Mycobacteria, Modococc, Streptomyces, Bacilli, Arthrobacter, Aeromonas, Cyanobacteria Dehalococcoides, Dehalobacter, and Desulfitobacterium, etc. which can detoxify petroleum products* (Maphosa *et al.,* 2012). Recently Uranium degrading bacteria *Pseudomonas sp., Pantoea sp.* and *Enterobacter* sp., *Geobacter* and *Thermus scotoductus* have been identified from the sites of uranium mining sites.

Case Study of Uranium Bioremediation

This case study shows how extremely harmful radioactive substances can be removed from the soil or contaminated water. Uranium is a radioactive material that emits harmful radiation which can certainly cause cancer due to mutation in genetic material. This type of operation was carried out in 3 FRC in Oak Ridge in the USA. This area was contaminated with a 200um concentration of uranium, apart from this other harmful chemicals were also present there such as calcium and a high concentration of chlorinated solvents. The researchers applied their systems' biological approach for bioremediations. They have identified the various microorganism strains *Desulfovibrio, Geobacter, Anaeromyxobacte*r, *Desulfosporosinus, Acidovorax,* and *Geothrix* spp. and prepared the 16s clone library-based screening. The various molecular markersTh dsrAB, nirK, nirS, amoA, and pmoA were detected. At a later stage, a meta-genomics study was conducted where the two most potent bacterial species *i.e.* denitrifying b-*Proteobacteria* and *g-Proteobacteria* which possess the uranium degrading genes characterized by using the PhyloChip. During these microbial community-based studies it was observed that the growth of some bacteria (Geobacter or Geothrix spp) was hampered whereas the growth of others (Geobacter or Geothrix spp) improved, which provided the major clue for further studies.

Plants

The plant-mediated bioremediation process is also known as the phytoremediation process which is an *in situ* remediation technique.

Phytoremediation is a green technology powered by solar energy. It is the most suitable cost-effective technique whereby the low concentration of contaminants is present simultaneously, it can be used for many pollutants removal at the time. Like the bacteria plant also having the wonderful metabolic machinery which can convert a wide variety of substances into a wide spectrum of products. Hence, many plant species have been exploited for bioremediation purposes. The method includes firstly plants growing at the contaminated sites as a result they accumulate a large number of heavy metals and salts in its aerial parts. At a later stage, plants can be harvested. So by this method, we can remove the contaminants from the soils and water sites so far the various plant species have been used to remediate the metals, pesticides, solvents, explosives, crude oil and its derivatives, and various other contaminants from affected sites (Table 6.1). Some plant species have been identified that are well suitable for phytoremediation such as mustard plants; alpine pennycress, hemp, and pigweed. In the last two decades, the plants have been genetically manipulated for the excessive accumulation of heavy and toxic metals like lead, uranium, and arsenic. The technology of phytoremediation has been successfully applied to clear the soil from the sites such as groundwater, metal-mining sites, and polychlorinated biphenyls used in landfilling and coal mines. Shortly plants can be used for bioremediation at a large scale. Recently, the plant signal transduction pathway is a redesign to sense the TNT contamination in the soil (Antunes *et al.*, 2011).

Table 6.1: Various Common Plants Used in the Bioremediation Process

Sl.No.	*Name of plant species*	*Applications*
1	*Agropyron repens* (Wheatgrass)	Stabilization of lead in soil
2	*Agropyron smithii*	Rhizodegradation of hydrocarbons
3	*Agrostis castellana* (bent grass)	Metal accumulation
4	*Armoracia rusticana* (horseradish)	Source for peroxidaseMetal accumulation
5	*Avena fatua* (wild oat)	Metals hyperaccumulation
6	*Bambusa* (bamboo)	Wastewater management
7	*Bromus inermis* Bromegrass	TNT exposure and phytotoxicity
8	*Cannabis sativa (hemp)*	Waste management
9	*Hordeum vulgare* (Barley)	Fluoranthene transformation
10	*Linum usitatissimum* (flax)	Waste management
11	*Panicum virgatum* (Switchgrass)	TNT exposure, rhizodegradation of PAHs
12	*Trifolium* spp (clover)	Methyl bromide removal
13	*Sorghum vulgare* (sudangrass or broom corn)	Atrazine metabolism, hexahydro-I,3,5-trinitro-1,3,5-triazine, RDX; phytotoxicity,217 rhizodegradation of hydrocarbons
14	*Vallisneria Americana* (tape grass)	TCE transformation and metalsAccumulation
15	*Sorghastrum nutans* (Indian grass)	Rhizodegradation of PAHs*Sorghum bicolor* Rhizodegradation of petroleum hydrocarbons
16	*Nelumbo nucifera* (Indian lotus)	TNT transformation

Algae

The algae can be a major agent for bioremediation efforts. It has played an important role in controlling and biomonitoring of organic pollutants in aquatic ecosystems. But most of the efforts are dependent upon the types of species and strains used in this endeavor. The use of algae in bioremediation is a comparatively nascent field and developing as a cost-effective potential mode of phytoremediation for organic pollutants such as bisphenol from water (Gattullo *et al.,* 2012). The algae also proved extremely useful in hyperaccumulation of heavy metals and degradation of xenobiotics and source of feedstocks for biofuel production. The list of algae strains has been given in (Table 6.2) which is used for bioremediation purposes.

Table 6.2: Different Algae Strains and Species Used in the Bioremediation Process

Sl.No.	*Name of Algae Species and Strains*	*Name of Organic Pollutant*
1	*Monoraphidium braunii*	Bisphenol
2	*Agmenellum quadruplicatum*	Bisphenol
3	*Selenastrum capricornutum*	Benzene, toluene, naphthalene, phenanthrene, pyren
4	A consortium of *Chlorella sorokiniana* and *Pseudomonas migulae*	Phenanthrene
5	*Chlamydomonas reinhardtii*	Herbicide (prometryne)
6	*Chlamydomonas reinhardtii*	Herbicide (fluroxypyr)
7	*Monoraphidium braunii*	Bisphenol
8	*Pediastrum tetras, Ankistrodesmus Fusiformis, Amphora coffeaeformis*	Herbicide (mesotrione)

Source: Chekroun, KB, Sánchez, B and Baghour, K. (2014) The role of algae in bioremediation of organic pollutants. *International Research Journal of Public and Environmental Health* Vol.1 (2), pp. 19-32).

Fungi or Mycoremediation

Like microorganisms and plants, fungi can also be used for bioremediation purposes. The fungus is the best candidate for this work because the fungus has already been a good cleaner of the environment; they are the best sources of enzymes for a breakdown of Cellulose and lignin that propose the best model for the breakdown of organic compounds. Several species of fungus have been used for the bioremediation of nerve gas and sarin gas (polycyclic aromatic hydrocarbons) currently Ecuadorian origin fungus species *Pestalotiopsis* can degrade the polyurethane in both environments aerobic as well as in anaerobic conditions, therefore can be used in the landfill environment. *Aspergillus* and different varieties of molds play a valuable role in the degradation of the starches, hemicelluloses, celluloses, pectin, and other sugar polymers in the favorable condition *i.e.* availability of nutrients. The fungus is also a very suitable source of industrial enzymes that can be further used for bioremediation purposes (Kües *et al.*, 2015).

Table 6.3: Important Industrial Enzymes Identified and Characterized Generally Used in Bioremediations

Sl.No.	Name of Enzymes	Target Environmental Pollutants	Name Reaction Catalysed
1	Mono-oxygenase	Various types of Alkane, steroids, fatty acids, and aromatic compounds	Introduction of an oxygen atom to substrate and utilize substrate as reducing agent. Desulfurization, dehalogenation, denitrification, ammonification, and hydroxylation of substrate
2	Dioxygenase	Benzene ring containing compounds (Aromatic compounds)	Incorporate of two oxygen atoms into the substrate resultsin intradiol cleaving and extradiol cleaving as result aliphatic product formation
3	Laccase	Ortho and paradiphenols, aminophenol, polyphenols, polyamines, lignin, and arylamines	Oxidation, decarboxylation, and demethylation of a substrate.
4	Lignin peroxidase	Halogenated phenolic compounds, polycyclic aromatic compounds	Oxidation of substrate in the presence of H_2O_2
5	Manganese peroxidase	Lignin and other phenolic compounds	Oxidizes the phenolic compounds in presence of Mn ion
6	Peroxidase	Phenolic aromatic compounds	The enzyme catalyses the electron transfer from an oxidizable substrate, as a result of reduction of by-products
7	Lipase	Lipids available in spill oils	The hydrolysis of triacylglycerol to glycerol and free-fatty acids
8	Cellulase	Lignocellulosic compounds	Hydrolysis polymers into simple sugars
9	Protease	Proteins	Peptide hydrolysis in presence of water

Advantages of Bioremediations

1. Highly suitable for low polluted areas and cost-effective green methods.
2. No need for excavation even can be applied to landfills and groundwater cleaning in anaerobic conditions.
3. No need for heavy machinery and transport methods.
4. The pollutants can be easily harvested or separated from plants, fungi, and microorganisms.

5. The extremely safe method in terms of exposure to workers.
6. Another very important factor is that organisms used in bioremediations are generally taken from the local environment because they are acclimatized to local conditions.
7. Comparatively less amount of effluent is produced which can easily be removed.
8. Being in *situ* and publically more acceptable and less ethical issues are associated.

Systems Biology Approach for Bioremediation

Genomics

With the emergence of genetic engineering a large number of genes involved in the protection against the adverse environmental conditions have been characterized and isolated from the bacteria, plants and various organisms (Urgun-Demirtas *et al.,* 2006). However, as we know that the isolations and characterization of the individual gene and protein are not sufficient to understand the complex biological process like bioremediations. However, with the advent of next-generation sequencing complete techniques, the whole genome has been sequenced of various organisms including several species of bacteria, fungi, and algae (Maphosa *et al.,* 2012). The genome sequence information will help to understand the pathway and enzymes related to detoxification of environmental pollutants at the molecular level. With the availability of fast sequencing technology, the genome of bacterial strain EbN1 has been sequenced. This bacterium is capable of degrading the aromatic environmental pollutants (chloroethene, chloroethanes, and polychlorinated benzenes) even in absence of oxygen, and pathways for these functions have been elucidated. It is a well-known fact that some bacteria cannot grow in certain conditions due to a lack of suitable nutrients, electron acceptors, and donors. Hence, the purpose of cellular structure investigation of bacteria is to identify the respiratory complexes and proteins which can perform this type of function. The protein profile of facultative denitrifying bacteria Aromatoleum *aromaticum* is studied when it grows with toluene and ethylbenzene (Trautwein *et al.,* 2012). The genome sequence of lessons can be learned from the recent genomic studies of *Burkholderia xenovorans* LB400 and *Rhodococcus* sp. strain RHA1, which have recently been sequenced provide opportunities to understand the catabolism oforganic compounds and moreover to understand long term evolutionary understanding. The genome pieces of information can also provide help to comparisons of bacterial genomes may also help to understand the crucial enzymes and metabolites (Chain *et al.,* 2006). Two major catabolic pathways namely, b-ketoadipate and PAA pathways reveal the catabolism of non-oxygenolytic ring-cleavage. This type of information may help to degrade the aromatic pesticides that prevailed in the soil and water. The availability of genome sequence data also provides information related to bacterial adaptation in the local niche.

The genome analysis of PCB-degrading bacterium *Burkholderia xenovorans* LB400 reveals that it can digest aromatic compounds because it contains the eleven such metabolic pathways which can be utilized in the digestion of homogentisate, homoprotocatechuate, hydroxyphenylacetate, and 4-hydroxyphenylacetate catabolism (Méndez *et al.*, 2011). The other bacterial *Cupriavidus necator* JMP134 strains degrade the 2, 4-D, halobenzenes, chlorophenols, and nitrophenol, in this experiment a metabolic construct has been designed which makes bacteria capable of bio-remediate the aromatic ring compounds. Later on, it has been grown on the media which contains 140 aromatic compounds tested, among them 60 act as a carbon and energy source for JMP134 strains (Pérez-Pantoja *et al.*, 2008). Systems-level studies have been carried out in *Corynebacterium glutamicum* to test its aromatic degradation capability. It degrades the aromatic compounds by using the primary metabolic pathway *i.e.* glycolysis pathways it provides enough preface for synthetic biology to manipulate bacterial genetic circuits for the degradation of the ring compounds (Shen *et al.*, 2012). The various genes and their regulation provide great opportunities for modification of the genetic program of bacteria. In a significant metabolic engineering experiment researchers constructed the whole integrated pyrene degradation pathways in *Mycobacterium vanbaalenii* PYR-1 Shewanella *oneidensis* MR-1 by using the genomics, proteomics, and metabolic engineering approaches. A marine bacteria *Novosphingobium pentaromativorans* US6-1, have exploited for the bioremediation of multiple polycyclic aromatic hydrocarbons and several crucial enzymes have been identified such as 4-hydroxybenzoate 3-monooxygenase, salicylaldehyde dehydrogenase, and PAH ring-hydroxylation dioxygenase. It generally uses the two pathways identified *i.e.* o-phthalate pathway or salicylate pathway, later on, can enter into the tricarboxylic acid (TCA) cycle (Lyu *et al.*, 2014). To exploit the full potential of bio-remedial capability is required the harness of transporters, genes, and proteins, which are responsible particularly in a local environment. Simultaneously need to identify various post-translational modifications of proteins carried out during the signaling process.

Proteomics

The proteins are the real molecular horses of the biological system and they are involved in virtually all biological processes. According to the central dogma of molecular biology, the information encoded in the gene is translated into the proteins, during the course a single gene can give rise to many proteins hence the investigation at the gene level does not provide exact insight. The study at the direct protein level is more appropriate to engineer organisms that can efficiently convert toxic agents into less toxic. Proteomics is a comparatively new discipline of science that deals with the expression, identification, modifications, interactions, and annotation of protein functions at a particular moment in the specific biological system. Proteomics is a combination of the 2D gel electrophoresis and MALDI-TOF that can provide us with the profiles of proteins under various conditions (Visioli and Marmiroli, 2012). The technologies have been also applied to bioremediation studies and revolutionized bioremediation in plants and microorganisms.

The global view of protein species expression which is involved in the bioremediation processes can offer a better opportunity to understand the mechanisms of detoxification in cells in a better way. Moreover, a combination of genomics and proteomics which is a part of functional genomics endeavor will elucidate the exact metabolic pathways and regulatory mechanisms ultimately gene functions. Therefore, the functional genomics approach will provide a good preface for synthetic biology in bacteria and plants. According to the principles of bioremediation, bacteria utilized various strategies such as antioxidant enzymes, a transporter for drainage of toxic chemicals from the cell; increase the rigidity of the cell wall, a strong vesicle that can accumulate the heavy metal ions in excess amount. Other methods are also using xenobiotics as nutrients and using xenobiotics chemicals as electron acceptors and electron donors. Simultaneously, heavy metals, xenobiotics also adversely affect the cellular structure of organisms. In the last two decades, proteomics technology has been used to identify the proteins and enzymes and metabolic pathways both utilizing the carbon and energy inside the bacterial strains must be identified under the environmental condition *i.e.* stress conditions such as hydrocarbon and heavy metals. These types of studies will provide help to synthetic biology to construct the synthetic genetic circuits in scavenging the organisms that can behave on our expected line. Given a deep understanding of genomics and proteomics, allow us the manipulation of the genome at a large scale even the metabolic pathway can transplant in scavenging bacteria. Which cannot be possible with the conventional genetic engineering experiment, which is reductionist by nature. The systems-level study carried out on*Mycobacterium vanbaalenii* PYR-1 with help of 1D and LC-MS/MS has identified the 27 enzymes, which are responsible for pyrene degradation through o-phthalate and the b-ketoadipate pathway (Kim *et al.*, 2007). To test the resistance of the heavy metal such as lead, copper, and cobalt stress in the *Pseudomonas fluorescens,* several proteins, and enzymes have been identified (Xu *et al.*, 2015). The proteomic changes related to naphthalene degradation in *P. fluorescens* ATCC 17483 were carried out to identify different proteins quantitatively to understand the role of proteins during the bioremediations (Herbst *et al.*, 2013). Understanding the complete mechanisms of bioremediation by bacteria required different post-translational modifications that responsible regulate the activity of various proteins and enzymes. The 2D-based study is best suitable for PTM in bacteria in response to abiotic stress including heavy metals. The combination of genome sequencing and proteomics was used in the investigation of protein complexes and their encoding genes in *Acinetobacter sp. KS-1* unravels the role of enzymes, proteins, and enzyme complexes involved in benzoate degradation pathways. In a significant study to know the quantitative analysis of genes and proteins in response to phenol-containing media in *P. putida* KT2440 bacteria, a SILAC based study found that proteins related to energy pathways, fatty acid biosynthesis, transcription regulation, and transport were up regulated. The same approach was also applied to *Herminiimonas arsenicoxydans* and monitored its growth in arsenic-rich growth media that shows it is a powerful microorganism that can be used for arsenic bioremediations.

Metabolomics

It is also a high-throughput technique of metabolites profiling analysis under a particular condition in an organism. The metabolomics is carried out with the help of gas chromatography coupled with mass spectrometry (GC-MS). The metabolites profiling is carried out in microorganisms, plants, and fungus (Schauer and Fernie, 2006). Currently, technology has been utilized to create a relationship between genes and metabolites so the genome level of a network could have produced that are extremely useful in systems and synthetic biology endeavors. The combination of LC-MS or LC-TOF MS (time-of-flight mass spectrometry) creates wonderful metabolic fingerprints, which are applied to toxicology. Currently, the fungus has been utilized to bioremediate an herbicide Alachlor by using the *Paecilomyces marquandii*. During the degradation process, it has been observed that the rate of glycolysis, TCA cycle, and level of antioxidants enzymes increased (Szewczyk *et al.*, 2015). So, metabolic studies can provide the major clue in the field of bioremediation.

Systems Biology

In the recent past, the reductionist approach that is limited to a single gene and protein investigation in various organisms could not provide the expected result in the past. Hence, the worse and ever-increasing problem of pollution cannot be surmounted. The recent focus of scientists has been shifted from reductionist to a systems approach. Systems biology is the multidisciplinary approach that comprises the contribution of genomics, transcriptomics, proteomics, metabolomics, phenomics, and lipidomics. Simultaneously, it also seeks help from physical, chemical, mathematical, and computational sciences. In the context of detoxification or bioremediation, systems biology offers the study at a larger level *i.e.* starting from single-molecule to systems levels (ecosystem) in a variety of areas or contaminated sites (Figure 6.2). Systems biology exploits the crucial interactions ranging from single molecules to even available microbial communities, which can be proved the efficient strategy to understand the catabolic process of toxic environmental products. Systems biology also provides simulation opportunities to understand the *in situ* perturbation in terms of toxic agents in cellular metabolism. Systems biology is aimed to elucidate the key points and their optimized survival chances of plants and microbial cells in adverse conditions that prevailed in a specific ecosystem. Soon systems biology will mature and provide opportunities for bioremediation of radioactive, hydrocarbon, metals, xenobiotics, and chlorinated solvents from soils and static water.

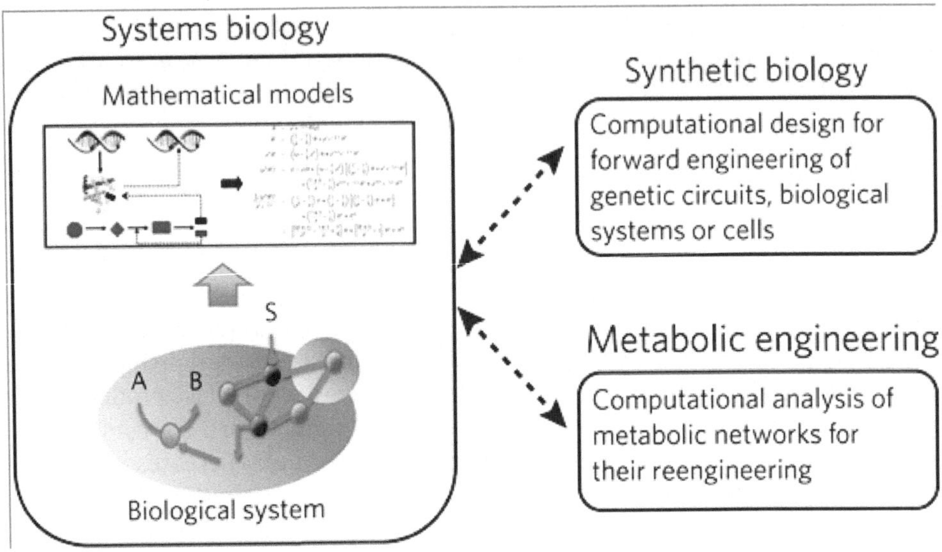

Figure 6.2: The Systems Biology has Empowered the Synthetic Biology Extremely Helpful in the Process of Bioremediation.

(*Source*: http://www.katyanovablog.com/picsgevs/water-pollution-graph Figure No 6.2 Interrelationship between systems, synthetic biology, and metabolic engineering

(*Source of figure*: Nielsen J, Fussenegger M, Keasling J, Lee SY, Liao JC, Prather K, Palsson B (2014) Engineering synergy in biotechnology. *Nat Chem Biol.* 10(5):319-22. DOI: 10.1038/nchembio.1519).

The latest approach of systems and synthetic biology which operate at a very large scale provides the help to overcome the bioremediation problems via identifying and design the crucial electron donor, acceptor, and crucial enzymes and their activities which are useful in cleaning processes. However, to reap the maximum benefits of multi-omics sciences we need to create databases in the area of bioremediation and a common standard language hence, data can be exchanged among the scientists.

Synthetic Biology

This is a newly emerging area of interdisciplinary science that applied the engineering principles to biology to design or redesign the biological systems to inculcate new functionality like bioremediations and detoxification (Dvorak *et al.*, 2015). In synthetic biology, we design artificial genetic networks by using the basic genetic parts such as promoter, transcription factors, ribosome binding sites, and proteins (activator and repressors). These synthetic gene circuits can be put into a microbial cell known as the chassis. If we draw a parallelism between electronics and synthetic gene circuits, they follow the same principles. Similar structural levels are prevailed in electrical circuits as in natural genetic circuits the classic example of circuitry is bacteria is *Lac* and *Trp* operon discovered by Jacob and Monad in 1961. The best example of the biological switch is the bacteriophage lysis–lysogenic switch (Khalil and Collins, 2010; Bhattacharjee *et al.,* 2020). Therefore, by engineering these circuits and switches a new function can be introduced in the cells.

To construct a synthetic circuit, we need highly efficient genome manipulating agents such as Megnuclease, ZFN, TALEN and CRISPR/Cas9. These are highly specific engineered proteins and RNA-based endonucleases that can manipulate the genome with pinpoint accuracy. In this manner, we can manipulate the genome of bacteria, plants, and fungus (Esvelt and Wang, 2013). But to design a synthetic bacteria or circuit we follow the design cycle of synthetic biology which includes fundamental engineering concepts like, standardization, functional decoupling, abstraction, modeling, formalized design process, and construction of the new biological circuit, module systems. However, there is a great challenge to engineer biology due to incomplete knowledge of biological systems, standard biological parts, and manipulating tools (Way *et al.,* 2014; Rylott *et al.,* 2020). The synthetic biology approaches have been applied to microorganisms and manipulated to overcome the problems of food, energy, and the environment. A large number of synthetic genetic circuits are designed in microorganisms and successfully applied in biosensing, medicine, and biofuel production. However, synthetic biology is in its nascent stage in the case of plants, algae, and fungi.

Role of Synthetic Biology in Bioremediation

Detection Process

The environmental pollutants released in the environment are highly toxic to the biological systems in the nano and picogram hence their detection is very difficult with the normal calorimetric methods. Additionally, traditional detection methods are quite expensive, inefficient, and labor-intensive, and time-consuming. To overcome these problems recently biosensors are developed which are most suitable for detection methods. Biosensors are devices based on biological principles and organisms comprise the receptor part to detect the subjected molecule, amplification unit, and signal part as in the case of the electronic device. The receptor bind with a subject molecule, this binding event is amplified with the help of the signal transduction part which makes it measurable in the form of fluorescence or chemiluminescence biosensors that detect a nano or pico amount of molecules. However, biosensors have been used in the medicals, army operations, food industry, diagnostics but their use in bioremediations is comparatively new. But biosensors must pass some desirable characteristics such as being highly specific, efficient, rapid, inexpensive, sensitive, operational at a variety of conditions, need a minute amount of sample, small in size, and able to measure in real-time. As mentioned above they are bio-based and have been constituted from enzymes, DNA, RNA, and proteins (Khalil and Collins, 2010; Bhattacharjee *et al.,* 2020). There are the following types of biosensors used in environmental-related issues.

- **Enzyme-based biosensors:** The enzyme-based biosensor is generally based upon the quantification of the inhibition process against a particular enzyme (Mross *et al.*, 2015). The oxidoreductase and hydrolases classes of enzymes have been exploited for this purpose.
- **Immuno-biosensors:** This type of biosensor is based upon antibody and antigens interaction.
- **DNA-RNA or Nucleic acid-based biosensors:** The principal component of these markers is specific oligonucleotides which act as sensing components. These nucleotides can hybridize with DNA/RNA and polluting agents as a target molecule. This is the best method of in situ detection of pollutants.
- **RNA-based biosensors:** The RNA-based biosensors are used to sense temperature, ligands, and tertiary structure caused by environmental conditions.
- **Quorum sensing-based biosensors**: In this type of biosensors, the transcription process is tamed. The transcription process is programmed in such a manner it can detect the environmental signal.
- **RNA aptamer-based biosensors**: This type of biosensor comprises fused aptamer and RNA regulatory elements, it blocks the translation process after binding the target molecule with the aptamer part.
- **Peptide-based biosensors**: The principle is based upon the fact that the target molecule binds with the receptor part, consequently the detection process is amplified via signal transduction process.
- **Whole-cell biosensors:** Whole-cell of microorganisms can be used as biosensors in which a reporter gene is fused with an inducible gene promoter to sense the target molecule. For example, *E.coli* based biosensors have been used in the detection of salmonella in food material. TNT and Atrazine, a major pollutant, can also be detected with this type of biosensors. Under the iGEM project, an arsenic biosensor has been developed. The future development of future generation biosensors will be based on multitasking and multi-detection systems in single biosensors. Synthetic biology will play a greater role in the development of biosensors via modifying the pathways and receptors, amplification, and signal all three components of the biosensors by modifying the pathways, enzyme, and signal transduction process in the organisms.

Pathways Engineering

Most recently, the pathways have been designed by using the synthetic biology approaches to degrade the various heterocyclic nitrogen, sulfur, and oxygen (NSO) compounds in different microbial consortia. The synthetic genetic circuits were designed on the line of Monod inhibition mode (Oberoi *et al.*, 2015; Bhattacharjee *et al.*, 2020). Currently, the combine approaches of synthetic biology and metabolic engineering have been applied to overcome the environmental pollution through genomics, proteomics and metabolomics of microorganisms. This method has provided deep insight in the complex nature of microbes. However, bioremediation in natural environment is slow but can be enhanced via metabolic engineering. In future, we need a single computational platform equipped with all bioinformatics tools and databases for data analysis and metabolic pathways, hence will help to reconstruction of new metabolic pathways in the suitable community of microbes. However, it requires collaboration of scientists from different laboratories and share data, and updates current databases. Therefore, need new technology for metabolic engineering and pathway reconstruction in view genome editing protocols like gene shuffling and molecular biological tools used in productive microbial strains. Moreover, coming synthetic biology and metabolic engineering approaches like cybernetics, nanobioremediation technology can significantly contribute to the bioremediation processes (Dangi *et al.*, 2019).

Conclusion

With the advent of high throughput methods like genomics, proteomics and other multi-omics techniques will provide a comprehensive view of molecular structure in the organisms. It will support the synthetic biology efforts and we would like to design the new bacterial strains, which can remediate the pollutants with more efficiency. However, it will also depend upon the cost efficiency of the available technology and more knowledge in the area of basic sciences. In the coming era, many scavenger organisms will be identified which will prompt bioremediation research.

REFERENCES

Antunes MS, Morey KJ, Smith JJ, Albrecht KD, Bowen TA, Zdunek JK, Troupe JF, Cuneo MJ, Webb CT, Hellinga HW, Medford JI. (2011) Programmable ligand detection system in plants through a synthetic signal transduction pathway PLoS One. 25; 6(1): e16292.

Barrington K, Chopin T, Robinson B (2009) Integrated multi-trophic aquaculture (IMTA) in marine temperate waters. In: Soto D, editor. Integrated mariculture: a global review FAO Fisheries and Aquaculture Technical Paper No 529. Rome: FAO. 7–46y

Bhattacharjee, Gargi, Nisarg Gohil, and Vijai Singh. "Synthetic biology approaches for bioremediation." Bioremediation of Pollutants. Elsevier, 2020. 303-312.

Chain PS, Denef VJ, Konstantinidis KT, Vergez LM, Agulló L, Reyes VL, Hauser L, Córdova M, Gómez L, González M, Land M, Lao V, Larimer F, LiPuma JJ, Mahenthiralingam E, Malfatti SA, Marx CJ, Parnell JJ, Ramette A, Richardson P, Seeger M, Smith D, *et al.* (2006) Burkholderia xenovorans LB400 harbors a multi-replicon, 9.73-Mbp genome shaped for versatility. Proc Natl Acad Sci U S A. 17; 103(42): 15280-7.

Chekroun, KB, Sánchez, B and Baghour, K.(2014) The role of algae in bioremediation of organic pollutants. International Research Journal of Public and Environmental Health Vol.1 (2), pp. 19-32.

Dangi, Arun Kumar, *et al.,* "Bioremediation through microbes: systems biology and metabolic engineering approach." Critical reviews in biotechnology 39.1 (2019): 79-98.

Davis, D. L., Bell, M. L. and Fletcher, T. (2002) A look back at the London smog of 1952 and the half-century since. Environ. Health Perspect. 110, 734–735.

Dvorak P, Bidmanova S, Damborsky J, Prokop Z (2014) Immobilized synthetic pathway for biodegradation of toxic recalcitrant pollutant 1,2,3-trichloropropane. Environ Sci Technol. 17; 48(12): 6859-66.

Esvelt KM, Wang HH. (2013) Genome-scale engineering for systems and synthetic biology. Mol Syst Biol. 2013; 9: 641

Gattullo CE, Bährs H, Steinberg CE, Loffredo E (2012) Removal of bisphenol A by the freshwater green alga Monoraphidium braunii and the role of natural organic matter. Sci Total Environ; 416: 501-6.

Harms H, Junca H.(2014) Editorial overview: environmental biotechnology. Curr Opin Biotechnol; 27: 7-9

Herbst FA, Taubert M, Jehmlich N, Behr T, Schmidt F, von Bergen M, Seifert J.(2013) Sulfur-34S stable isotope labeling of amino acids for quantification (SULAQ34) of proteomic changes in Pseudomonas fluorescens during naphthalene degradation. Mol Cell Proteomics 12(8): 2060-9.

Hong, S., Candelone, J.-P., Patterson, C. C. and Boutron, C. F. 1996 History of ancient copper smelting pollution during Roman and medieval times recorded in Greenland ice. Science 272, 246 –249. (doi: 10.1126/science.272. 5259.246)

Karigar CS, Rao SS. (2011) Role of microbial enzymes in the bioremediation of pollutants: a review. Enzyme Res 2011: 805187.

Khalil AS, Collins JJ (2010) Synthetic biology: applications come of age. Nat Rev Genet. 11(5): 367-79

Kim SJ, Kweon O, Jones RC, Freeman JP, Edmondson RD, Cerniglia CE (2007) Complete and integrated pyrene degradation pathway in Mycobacterium vanbaalenii PYR-1 based on systems biology. J Bacteriol. 189(2): 464-72.

Kües U. (2015) Fungal enzymes for environmental management. Curr Opin Biotechnol. 33: 268-278.

Lyu Y, Zheng W, Zheng T, Tian Y.(2014) Biodegradation of polycyclic aromatic hydrocarbons by Novosphingobium pentaromativorans US6-1. PLoS One; 9(7): e101438.

Maphosa F, Lieten SH, Dinkla I, Stams AJ, Smidt H, Fennell DE.(2012) Ecogenomics of microbial communities in bioremediation of chlorinated contaminated sites. Front Microbiol.; 3: 351.

Méndez V, Agulló L, González M, Seeger M (2011) The homogentisate andhomo protocatechuate central pathways are involved in 3- and 4-hydroxyphenylacetate degradation by Burkholderia xenovorans LB400. PLoS One.; 6(3): e17583

Mross S, Pierrat S, Zimmermann T, Kraft M. (2015) Microfluidic enzymatic biosensing systems: A review. Biosens Bioelectron.; 70: 376-391

Oberoi AS, Philip L, Bhallamudi SM. (2015) Biodegradation of Various Aromatic Compounds by Enriched Bacterial Cultures: Part B-Nitrogen-, Sulfur, and Oxygen-Containing Heterocyclic Aromatic Compounds. Appl Biochem Biotechnol.

Owa, JK (2014) Water pollution: sources, effects, control and management. International Letters of Natural Sciences Vol. 8 pp 1-6

Pérez-Pantoja D, De la Iglesia R, Pieper DH, González B (2008) Metabolic reconstruction of aromatic compounds degradation from the genome of the amazing pollutant-degrading bacterium Cupriavidus necator JMP134. FEMS Microbiol Rev; 32(5): 736-94.

Rhind SM. (2009) Anthropogenic pollutants: a threat to ecosystem sustainability? Philos Trans R Soc Lond B Biol Sci.; 364(1534): 3391-40

Rylott, Elizabeth L., and Neil C. Bruce. "How synthetic biology can help bioremediation." Current Opinion in Chemical Biology 58 (2020): 86-95.

Shen XH, Zhou NY, Liu SJ.(2012) Degradation and assimilation of aromatic compounds by Corynebacterium glutamicum: another potential for applications for this bacterium? Appl Microbiol Biotechnol; 95(1): 77-89.

Szewczyk R, Soboñ A, S^{3}aba M, D^{3}ugoñski J.(2015) Mechanism study of alachlor biodegradation by Paecilomyces marquandii with proteomic and metabolomic methods. J Hazard Mater.; 291: 52-64.

Trautwein K, Lahme S, Wöhlbrand L, Feenders C, Mangelsdorf K, Harder J, Steinbüchel A, Blasius B, Reinhardt R, Rabus R. (2012) Physiological and proteomic adaptation of "Aromatoleum aromaticum" EbN1 to low growth rates in benzoate-limited, anoxic chemostats. J Bacteriol. 194(9): 2165-80

Urgun-Demirtas M, Stark B, Pagilla K.(2006) Use of genetically engineered microorganisms (GEMs) for the bioremediation of contaminants. Crit Rev Biotechnol; 26(3): 145-64

Visioli G, Marmiroli N.(2013) The proteomics of heavy metal hyperaccumulation by plants. J Proteomics; 79: 133-45.

Way JC, Collins JJ, Keasling JD, Silver PA.(2014) Integrating biological redesign: where synthetic biology came from and where it needs to go. Cell; 157(1): 151-61

Xu YB, Zhou Y, Ruan JJ, Xu SH, Gu JD, Huang SS, Zheng L, Yuan BH, Wen LH.(2015) Endogenous nitric oxide in *Pseudomonas fluorescens* ZY2 as mediator against the combined exposure to zinc and cefradine. Ecotoxicology; 24(4): 835-843.

Chapter 7

E. coli as an Indicator of Organic Pollution in Water: A Review

*Anjali Malan and Hardeep Rai Sharma**

Institute of Environmental Studies,
Kurukshetra University, Kurukshetra – 136 119, Haryana
**e-mail: sharmahardeeprai@gmail.com*

ABSTRACT

The present review chapter is focused on the presence of E. coli in different surface and groundwater sources and its impact on human health. The physiology and types of E. coli are discussed here. E. coli is the main indicator organism for microbial pollution in water as stated by World Health Organisation (WHO, 2011). This bacterium has many strains which can be found in soil and water and the most commonly found species of E. coli in water is Shiga toxin-producing Escherichia coli (STEC). Its presence in any water resource depicts the organic pollution in water and also indicates the probable existence of other microbes as well. Many disease outbreaks have also been occurred due to consumption of water contaminated with E. coli. The most common infections reported due to E. coli contamination in water are diarrhea, urinary tract infections (UTI), meningitis and pneumonia etc. Some treatment measures can also be adopted at household level for the removal of coliforms from water for example, boiling, chlorination and ultrafiltration methods etc.

Introduction

A German microbiologist Theodor Escherich in 1884 carried out research on infant's intestinal microbes and their role in digestive system and causing diseases. He discovered fast-growing bacteria, which he named as the *E. coli* community, but it is now called as *Escherichia coli* bacteria (Zimmer, 2008; Shulman *et al.*, 2007; Escherich, 1988).

Escherichia coli is a gram-negative bacillus, with a length of only about 1 μm and a width of 0.35 μm, but it will vary greatly depending on the species and conditions- even if it is enlarged, it looks like a rod like small sausages (Figure 7.1). It may have whip like flagella, which is used to navigate its surroundings, or a hair like pili that enables it to attach itself to the surfaces or with other cells. Physiologically, it is facultative aerobic, which means that it can grow successfully in the presence or absence of oxygen, but it cannot grow at extreme temperatures or pH values, nor can it decompose pollutants and perform photosynthesis. In terms of phylogeny, it belongs to the Enterobacteriaceae family, and is closely related to pathogens such as *Salmonella, Klebsiella, Serratia* and the notorious *Yersinia pestis* that cause plague (Brenner and Farmer, 2007).

Figure 7.1: *Escherichia coli*
[*Source*: U.S. Environmental Protection Agency (USEPA) 2018].

Commonly, bacteria are single-celled organisms and naturally found in lakes, streams and rivers. A major group of these bacteria does not cause any harm to humans; however, some bacteria, which are commonly found in the intestines of warm-blooded animals, can cause human diseases and *E. coli* bacteria is one of them. It is from faecal coliform group, commonly found in the guts of humans and animals. In water, it is a powerful indicator of pollution from sewage or animal waste which may contain many pathogens. The drinking of such contaminated water can cause serious illness especially in children under five, people with weakened immune systems, and the elderly are particularly susceptible to infection [United States Geological Survey (USGS), 2018].

Total Coliform (TC)

The TC are gram-negative bacteria, which do not form spores and are rod shaped. They can be aerobic or facultative anaerobes. Initially, people thought that these bacteria were a manifestation of faecal contamination; however, it has been found that TC are widespread in nature and are not always related to the guts of warm-blooded animals.

Faecal Coliform (FC)

The FC are a subset of coliforms and are used to set up the first standard of microbial water quality. The ability to grow at high temperatures (44.5°C) separates these bacteria from all coliforms, making it a more accurate index of faecal contamination in warm-blooded animals. Faecal coliforms are detected by counting the number of dark to blue-grey colonies growing on a 0.65 μm filter, placed on mFC agar and incubated in an oven at 44.5°C for 22-24 hours. The presence of faecal coliforms in the water depicts that the water has been contaminated by the feces of warm-blooded animals (USGS, 2018).

Escherichia coli (E. coli)

E. coli belongs to the FC group which is characterized by its inability to decompose urease. It is a rod-shaped bacterium that usually exists in the guts and the feces of warm-blooded animals. The fresh water count is determined by counting the colonies growing on a 0.45-micron filter having light brown and yellow color, incubated at 35.0°C for 22-24 hours by placing in m-TEC medium. The presence of *Escherichia coli* can be confirmed by adding urea substrate. The presence of this bacteria is a direct evidence of contamination by the faecal matter of warm-blooded animals. The *E. coli* bacteria can cause severe diseases such as sepsis, meningitis, pneumonia, diarrhea, urinary tract infections (UTI) and some intestinal and respiratory infections *etc.* The total amount of *E. coli* in the environment is still widely used as an indicator of good quality drinking water in the United States and many other countries.

E. coli exists in the environment, food, and intestines of human and animals. This is a very large and varied group of bacteria. Though maximum strains of *E. coli* are harmless and prevent the growth of harmful bacterial species and also helpful in synthesizing necessary vitamins [SDWF (Safe Drinking Water Foundation), 2017]. However, some strains may be opportunistic pathogens, while other strains can cause gastrointestinal diseases if ingested by healthy people. Opportunistic pathogens are organisms that usually live in the host without causing harm, but can cause infections in people with weak immunity (SDWF, 2017). The *E. coli* 0157: H7A is a newly discovered strain of *E. coli* and it can cause serious illness and death in children and the elderly people (USGS, 2018).

Types of *E. coli*

Escherichia coli is composed of many kinds of bacteria. The pathogenic *E. coli* can be classified according to serogroup, virulence factors, pathogenic mechanism, and clinical symptoms (Kaper *et al.*, 2004; Nataro and Kaper, 1998). The pathogenic strains of *Escherichia coli* are classified into pathogenic types and six of them are related to diarrhea, collectively referred to as *Escherichia coli*.

- ☆ STEC can also be called verocytotoxin-producing *Escherichia coli* (VTEC) or enterohemorrhagic *Escherichia coli* (EHEC).
- ☆ Enterotoxigenic *Escherichia coli* (ETEC)
- ☆ Enteropathogenic *Escherichia coli* (EPEC)
- ☆ Enteroaggregative *Escherichia coli* (EAEC)

Shiga Toxin-Producing *Escherichia coli* (STEC)

This can cause serious foodborne illness. The STEC outbreaks sources include mainly faecal contamination of raw and half-cooked meat or other products, vegetables and raw milk. The disease will go away on its own, but it can lead to dangerous impacts, like hemolytic uremic syndrome (HUS), especially in old age people and young children. STEC is sensitive to heat. When cooking at home, one must be sure to observe food hygiene habits, such as "cooking properly" (WHO, 2018)

STEC produces toxins called Shiga toxins because they are similar toxins are produced by *Shigella dysenteriae*. STEC can grow in a temperature ranging from 7°C-50°C, with an optimal temperature of 37°C. A few species of STECs can grow in foods with acidic nature at a pH as high as 4.4 and also in foods with water activity <0.95. Symptoms of STEC disease are associated with diarrhea and abdominal cramps, in some cases can lead to bloody diarrhea along with fever and vomiting. The incubation period can range from 3 to 8 days, with an average of 3 to 4 days. Most patients recover within 10 days, but in a number of patients with weak immunity, the infections can be life-threatening such as HUS, hemolytic anemia and thrombocytopenia (low platelet count) (WHO, 2018).

Most information about STEC is related to serotype 0157: H7 because it is biochemically distinguishable from other *E. coli* strains having cattle as its main host. Moreover, other ruminants like goat, sheep, and deer are also categorised as its important hosts, while dogs, cats, pigs, rabbits, and horses and birds (like chickens) are also infected. *E. coli* 0157: H7 is mainly transmitted to humans by the consumption of contaminated food and water and other foods contaminated by feces and cross-contamination during cooking (WHO, 2018).

The strain of *Escherichia coli* O104: H4 produced Shiga toxin, which caused a major outbreak in Europe in 2011, commonly referred to as EHEC. The reported cases of HUS were 855 and non-HUS cases were 2987 which caused around 53 deaths in different European countries. The *E. coli* O157: H7 (usually abbreviated as *E. coli* O157 or "O157" for short) is usually associated with the *E. coli* outbreaks [Centers for Disease Control and Prevention (CDC), 2014].

Besides, *E. coli* O157, some other types of STEC can also spread diseases. Those serogroups in the group of STEC, along with *E. coli* O145, are referred to as "non-O157 STEC". Current surveillance data is limited regarding public health on non-O157 STEC incidence rates; many infections associated with STEC O145 cannot be diagnosed or reported. As compared to STEC-O157 infection, it is more difficult to identify STEC infections other than O157. First, clinical laboratories must detect the presence of Shiga toxin in stool samples. Positive samples must be sent to public health laboratory for identification of STEC O157. The non-O157 STEC cannot be recognized usually in clinical laboratories (CDC, 2014).

E. coli O157:H7

Although *E. coli* was identified in 1885, it's strain O157: H7 was not discovered in the University Centers for Disease Control (CDC) until 1975. It was in 1982, seven years later, that this strain of *E. coli* was identified as the cause of intestinal disease in an outbreak related to contaminated hamburgers. The bacterium was found associated with several cases of bloody diarrhea. Since then, it has been thought that most infections are caused by eating undercooked ground beef. This association led to the designation of strain O157:H7 as enterohemorrhagic *E. coli* as it causes damage to the intestinal epithelium. Despite infection of *Escherichia coli* O157: H7 could occur to anyone, however children under five, the elderly, and people with weakened immune systems have an increased risk of serious illness and are more likely to develop long-term infections due to infection complications (SDWF, 2017).

The O157: H7 strain of *E. coli* is commonly found in a few livestock farms, and its main habitat is the intestines of healthy cattle. Infected people or animals can release millions of microorganisms through the intestine. The source of O157: H7 are faecal contaminated water sources. The waste can enter in the water sources through sewer overflows, failures in sewage system, and contaminated rainwater (CDC, 2015).

Sources of Organic Water Pollution in Water Bodies

The water bodies may get polluted by various sources as described below:

Surface Water Pollution

Surface water accounts for about 70 per cent of the earth including oceans, rivers, and lakes. In freshwater sources, the surface water accounts for more than 60 per cent of household water use, but most of it are threatened. According to the latest national water quality survey conducted by the United Nations Environmental Protection Agency (USEPA), about half of rivers and around one-third of our lakes are polluted and are not suitable for purposes like drinking, swimming and fishing (USEPA, 2018). Pollution from nitrate and phosphate is the main type of pollution for the freshwater resources. Although these nutrients are vital for the growth of plants and animals, agricultural waste and fertilizer runoff have made them major pollutants. Municipal and industrial wastes also have a significant impact on toxins-random waste that is dumped directly into water bodies by industry and people (NRDC, 2021). In surface water, the major pathogen sources include municipal sewage and agricultural runoff, and on-site septic systems and pit latrines (Suthar *et al.,* 2009).

Groundwater Pollution

The rainwater percolates deep underground and fills up the cracks, and pores of aquifers (mainly underground reservoirs), and then it becomes groundwater, which is one of the most inconspicuous and important natural resource. Sometimes this water is the only source of fresh water for people living in the rural areas. When pollutants (from sources like pesticides and fertilizers, leaching waste from landfills and sewage treatment plants) enter the aquifer, groundwater is contaminated, making it unsafe for human consumption. Removal of pollution from groundwater can be difficult, impossible, and costly. Sometimes the aquifer remains unused from many years or decades. The polluted groundwater can also affect other sources which are at a long distance from the original source of pollution because it can seep into lakes, streams, and oceans [Natural Resources Defense Council (NRDC), 2021]. Safe disposal of human excreta is essential to reduce the spread of diseases. As the leakage from septic tanks and pit latrines can contaminate the groundwater as well as surface water bodies which may cause many water-borne diseases. Moreover, according to some studies, the wells and drinking water sources located nearby latrines were found contaminated with *E. coli* and nitrates (Verheyen *et al.,* 2009; Banerjee, 2011; Vinger *et al.,* 2012).

Wastewater and Sewage

The water which is utilized for some purpose is sewage. Its source can be our showers, sinks, toilets and some industrial, commercial, and agricultural activities (NRDC, 2021). According to the United States (U.S.), > 80 per cent of the wastewater in the world returns to the environment without treatment; in some less developed countries, this proportion exceeds 95 per cent. In the U.S., sewage treatment plants treat approximately 34 billion gallons of wastewater each day. These systems can reduce pathogens in wastewater, can remove pollutants like nitrogen and phosphorus along with heavy metals and toxic chemicals from the industrial waste before the purified water is discharged back into the water. The aging and malfunctioning wastewater treatment systems will also discharge >850 billion gallons of untreated sewage (NRDC, 2021). In 1978 -79 the generation of sewage was 7,067 million litres per day (MLD) in Indian cities which increased to 62,000 MLD in 2014-15. Whereas the treatment capacity increased from 2,758 MLD to 23,277 MLD only (CPCB, 2021). The discharge of wastewater and sewage is responsible for polluting the surface and groundwater bodies. Some agricultural processes such as uncontrolled spread of manure, refusal to bathe sheep, tilling the soil, ploughing the land, and using pesticides and fertilizers can cause water pollution. Accidental spills of dairy products can also affect water quality (UK environmental law association, 2017).

E. coli Contamination in Water: Case Studies from India

Faecal water pollution is a major problem in India. Although faecal pollution can cause various pathogens in the water, these pathogens can cause diseases such as cholera, salmonellosis and shigellosis, but their concentration is relatively low. It is impractical to check every water sample for pathogens. Instead, the presence of pathogens is determined by testing "indicator" organisms (such as *E. coli*). The coliform has the same source as the pathogen. Coliform bacteria are relatively easy to identify. They are usually more numerous than the most dangerous pathogens and are similar to the response of many pathogens to the environment, sewage and water treatment. Therefore, the detection of coliform bacteria is considered a reasonable indicator of the presence of other pathogens (India Water Portal 2020).

E. coli or other faecal coliforms in water comes from the excrement of humans and animals. When it rains, melts snow, or other forms of precipitation, *E. coli* can be washed into streams, rivers, creeks, lakes, or groundwater. *E. coli* may enter the drinking water with improper treatment of the water. In India many studies reported the presence of *E. coli* in major rivers, ground water, and in piped water for distribution (Sharma *et al.*, 2000; Suthar *et al.*, 2009; Rai *et al.*, 2010; Dimri *et al.*, 2019; Rayasam *et al.*, 2019 and Malan *et al.*, 2020).

The river Ganges originated in the Himalayas and is considered the most sacred river in Hinduism. Human activities in the Ganges deteriorate the water quality and raise serious doubts about the purity of the Ganges water and its use for drinking and other purposes. In a study by the Ganga river water samples were collected from different *ghats* (steps leading down to a river) located in Badarinath, Devprayag, Uttarkashi, Rishikesh and Haridwar in Uttarakhand state. All the sites were found infected with gut bacteria, and the most important species identified were *E. coli, Klebsiella* and *Proteus*. In a study by Rai *et al.* (2010), the highest microbial count in the Ganga River was recorded in rainy season as compared to the winter. The FC count ranged from 3.2×10^3 to 4.2×10^7 per 100 ml. The main bacterial genera isolated were *Salmonella, Vibrio, E. coli,* and *Clostridium*.

According to CPCB (2017), TC and FC count ranged from 5 to >1800 MPN per 100 ml and the E. coli ranged from 1.8 to >1800 MPN per 100ml at 40 different locations of river Ganga and minimum count was found at Gangotri. According to the CPCB, the water in river Ganga is not suitable for direct drinking and is potable only after disinfection from the seven places that it passes through. The polluted water of the river is having Faecal coliforms count ranging >50,000 MPN (The

Economic Times, 2019). The water samples from river Ganges in Kolkata were analysed for TC count from April 2008 to April 2020 during in the pre-monsoon season. There were significant spatial differences in the TC matrix of the study area, and the Babughat value is relatively high. (10,271.92 ± 3050.35 MPN per 100 ml) as compared to the second Hooghly Bridge (8 485.58 ± 2589.40 MPN per 100 ml). Interestingly, the TC value also drops sharply during the COVID19 lockdown period (April 2020), regardless of the location of the study.This sudden decrease in *E. coli* can be attributed to the reduction of non-working industries, waste disposal and fishing activities, tourism, transportation, lack of bathing and religious ceremonies on the banks of the Ganges in the middle of the quarantine phase caused by COVID19 (Mukherjee *et al.,* 2020).

River Yamuna is the longest tributary in India and also the largest tributary of river Ganga. The TC and FC were reported in river Yamuna water at village Palla, the entrance point of river Yamuna into Delhi (the capital of India) which were found to increase several fold times at Okhla barrage, the downstream of Yamuna in Delhi during 1995. Between the two seasons, the value of TC and FC were both found to be increased manifold times. During monsoon season, the TC range was observed 3950 per 100ml at Palla and 27000 per 100 ml at Okhla barrage; and the FC count was found to be 255 per 100 ml at Palla and 1335 per 100 ml at Okhla barrage. The main reason behind heavy load of microbes in river Yamuna was dumping of untreated sewage into the river because of unavailability of some sewage treatment in New Delhi (Sharma *et al.,* 2000). However, the trends from 2007 to 2016 have shown that the TC and FC count were extremely high even after implementation of Yamuna Action Plan (YAP) in 1993 in the Delhi segment of Yamuna River. The observed values of TC in the Delhi stretch during 2007 to 2016, ranged from 10000 to 92×10^7 MPN/100 mL. These higher levels of TC and FC in the river Yamuna indicates that there is contamination from the faecal matter and some other sources like sewage and industrial discharge (Mazhar *et al.,* 2021). In another report of CPCB (2017), 23×10^4–16×10^7 MPN/100 ml coliform content were observed in 2016 along with 0.1–1.1 mg/l DO, and 29–67 mg/l BOD at a site of river Yamuna in New Delhi.

Compared with the control sample, the physico-chemical properties and bacterial counts in the water samples collected at the most critical point of the of River Cauvery in Karnataka showed higher biological oxygen demand and coliform bacteria ($p \leq 0.01$) (Skariyachan *et al.,* 2015). The report of CPCB revealed FC in river Cauvery ranged from 38,400 to 7.8 lakh MPN per 100 ml (Deccan Herald News, 2016). Moreover, in a study from different station of river Cauvery and Bhawani, severe contamination was found for TC at Vairapalayam site of river Cauvery (1800 MPN/100ml) and Bhavani Sagar of Bhawani River (920 MPN/100ml). The main reason observed behind the higher concentration of coliforms in these rivers was open defecation and barrage sedimentation (Sivaraja and Nagarajan, 2014).

In a study from rural areas of Northern Rajasthan, drinking water samples from groundwater were found contaminated with microbes and the concentration calculated by standard plate count (SPC) was varied between 8.3×10^4 ?28.3×10^4 in different sites. Total 10 bacterial species were isolated from contaminated water samples including *E. coli, Enterobacter, Pseudomonas, Klebsiella, Bacillus, Staphylococcus* species *etc.* Drinking water contaminated with such a huge load of microbes can cause serious health problems (Suthar *et al.,* 2009).The microbial count in groundwater samples from the Brij region of Uttar Pradesh was found high among 40 per cent samples with high SPC values, and 3500 colony-forming unit/ ml. In these samples, coliforms (MPN index) were also higher than the permissible limits of Bureau of Indian Standards (BIS), (2012) except two sites. In 26.67 per cent samples, the *E. coli* bacteria were isolated and 18.75 per cent were producing red colonies on congo red agar indicating their pathogenic nature(Sharma *et al.,* 2017).

Around 80 per cent of the drinking water samples (n = 29) from peri-urban areas of Bengaluru were found contaminated as per WHO standards (2011) of Nil *E. coli*/100 ml of sample. About 40 per cent samples were having very high *E. coli* counts which indicated that the households were at the risk of diarrheal diseases. The bacterial population (*E. coli* count) were observed higher in the months of December, May and September (>150 CFU/100 ml) as compared to January and March (<30 CFU/100 ml). The chemically and microbiologically contaminated water poses threat to the consumer health (Sheeba *et al.,* 2017). The *E. coli* was isolated from domestic water from the piped distribution system in Alibag (Maharashtra). The isolates were analyzed for drug resistance and genotyped by multilocus sequence typing. Around 35 per cent samples were found contaminated with coliforms and 13 per cent were biochemically confirmed to have *E. coli* (Rayasam *et al.,* 2019).

The groundwater samples from open-defecation free villages of Kurukshetra district were having gram-negative, rod-shaped bacteria. The TC count ranged from 5 to 45, 19 to 110, and 13 to 52 MPN per 100 ml in the pre-monsoon, monsoon and post-monsoon seasons, respectively. In certain seasons, the total number of coliform was decreasing with increase in the depth in all the groundwater samples. The FC count were found in all samples, and the MPN indices for pre-monsoon, monsoon and post-season were 2 to 11, 4 to 13, and 3.6 to 11MPN per 100ml, respectively. Six bacterial genera were isolated from the groundwater samples namely *Leminorella, Citrobacter, Klebsiella, Cedecea, Acinetobacter,* and *Salmonella.* Although FC were found in the groundwater however, *E. coli* was not detected. The Multiple Antibiotic Resistance Index (MAR) score confirms that contamination in the groundwater is mainly from sources with non-human origin, such as cow dung or excreta from other animals (Malan *et al.,* 2020).

E. coli in Soil

E. coli exists in the gastrointestinal system and in the environment (Ksoll *et al.,* 2007). After separation from the mammalian host, it is believed that the *E. coli* population is in trouble for long-term survival in response to stress (Winfield and Groisman, 2003). It is confirmed by the reduction in *E. coli* O157:H7 content in manure (Looper *et al.*, 2009) and in the soil (Berry and Miller, 2005) as compared to their content in the host. Some species of *E. coli* isolated from subtropical environments, like the bottom of river banks and river sediments (Desmarais *et al.*, 2002). Some resistant populations of *E. coli* were reported in pastures (NandaKafle *et al.*, 2017), in temperate forests and river basins also (Ishii *et al.*, 2006).

The strain O157:H7 of *Escherichia coli* can survive and survive in a variety of environments like water, soil, food, and animal water tanks (Figures 7.2 and 7.3). *E. coli* O157:H7 can remain alive for one year in soil treated with faecal matter and 21 months in untreated, uncomposted feces (Jiang *et al.*, 2002). When the temperature is kept >50°C for six days, manure composting can effectively kill *E. coli* O157:H7. It can stay in water for a long time, especially at low temperatures. Sediments from water and contaminated cow manure can be used as long-term storage for *E. coli* O157: H7 (> 8 months), and live bacteria in feed animals are the source of infection (LeJejune *et al.*, 2001).

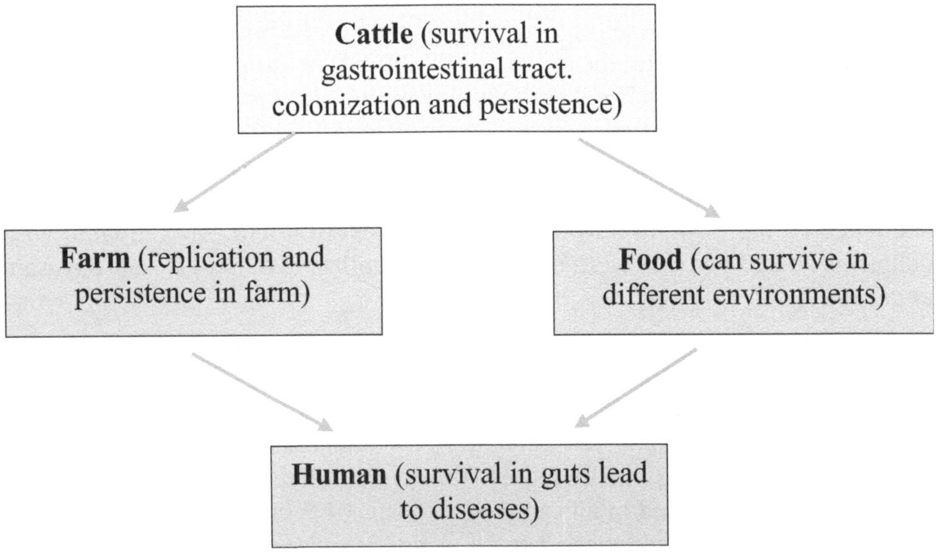

Figure 7.2: Ecological Scheme of *E. coli* O157:H7 (Derived from: Lim *et al.*, 2010).

In an experiment at Purdue University, Indiana, *E. coli* was added to liquid manure in the soil, shown that the bacteria in the rhizosphere or plant root zone of lettuce and radish were vigorous and active. Later, *E. coli* eventually reaches to the plant, where it survived for several weeks (Wallheimer, 2010).The *E. coli* is able to grow in the food crops such as in the tissues of spinach. Although faecal contamination is a major suspect, soil is underestimated as its potential reservoir. *E. coli* can be a soil symbiont, maintaining a stable population in the soil without predators and competition (NandaKafle *et al.*, 2018).

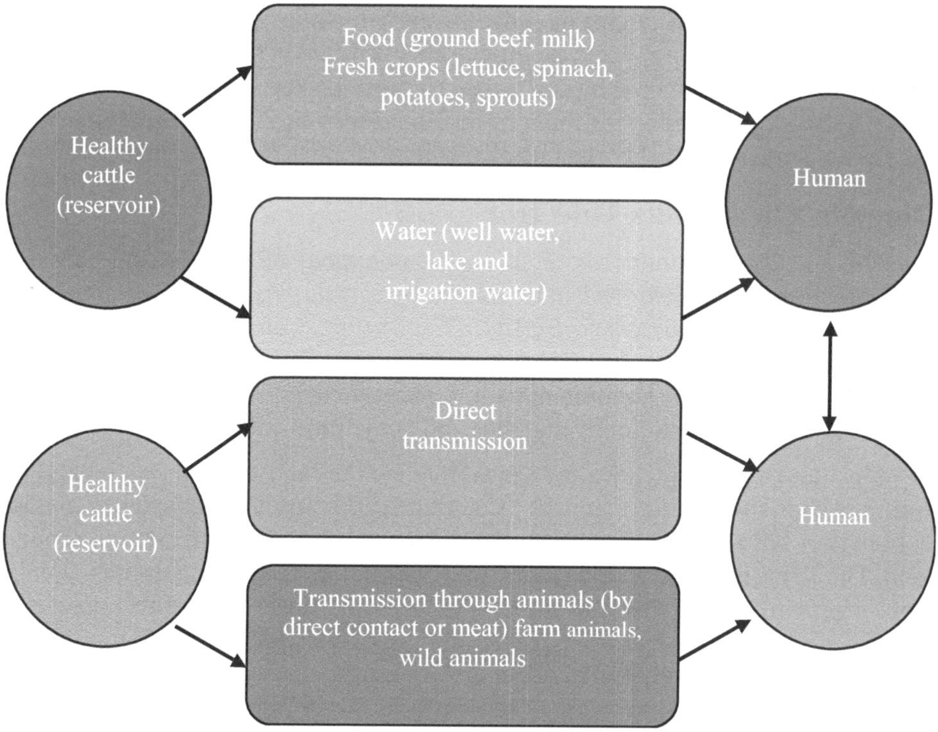

Figure 7.3: Transmission of *E. coli* O157:H7 (derived from: Lim *et al.*, 2010).

The O157: H7 strain of *E. coli* has been found related to the gut of cattle, especially the cecum (Wang *et al.*, 2017). It is currently the main source through which the bacteria enter into the food chain. Spinach, tomatoes, lettuce, and fresh fruits were identified as its sources (Denis *et al.*, 2016). Initially, these foods were believed to be contaminated with feces, but according to recent reports *E. coli* O157: H7 has grown in lettuce and tomato tissues also (Wright *et al.*, 2013, 2017). After inoculation from unknown sources, intestinal bacteria reproduce in the growing plants which are difficult to remove with the help of surface treatment like washing.

One of the biggest issues with the food safety is that the pathogens can spread from to the edible parts of plants from contaminated soil. This causes pathogens to survive longer on leaf surfaces, especially on sandy soils than on fumigated soils. Therefore, the use of fumigants may help reduce the spread of *E. coli* O157:H7 from the soil to the leaves. The *E. coli* O157:H7 is able to survive for a long time in the environment even under adverse conditions. Moreover, this pathogen can also remain alive in the soil >90 days (Ibekwe *et al.*, 2007).

In a study by Ibekwe, *et al.* (2011), the microbial biodiversity of *E. coli* determined by Denaturing gradient gel electrophoresis (DGGE) is significantly higher in clay than in sand, which causes the initial population decline in clay to be greater than that in sand. It indicates that after contamination of soil with *E. coli*

O157:H7, the pathogen cannot be killed by spraying alone, but it can reduce the microbial diversity, thereby improving the survival rate of the pathogen. Survival of *E. coli* is strongly affected by soil types, as these bacteria can easily survive in sandy soils as compared to the loamy soils (Ibekwe *et al.*, 2006). However, in clay, due to the properties of clay, the bacteria will last longer than in sand.

Disease Outbreaks Due to *E. coli*

The *E. coli* is common inhabitant in the intestines of humans and animals; however, intestinal infections can be caused by certain types of *E. coli*, such as *E. coli* O157: H7. The *Escherichia coli* O157:H7 and other strains that cause intestinal diseases are called STEC because they produce toxins. The intestinal infection may have symptoms such as abdominal pain, fever associated with diarrhea. In more severe cases it can cause bloody diarrhea and dehydration also (Healthline, 2019).

More and more epidemics are related to the intake of fruits and vegetables like sprouted beans, lettuce, and spinach). Contamination can also be caused by contact with pets or wild animal feces at a certain stage of planting or processing canned food. The STEC is also isolated from water bodies like lakes and streams, ponds, wells and drinking troughs, and survives in feces and water sediments for several months. According to reports, there is water transmission in both contaminated drinking water and recreational water (WHO, 2018).

The presence of *E. coli* can lead to potential disease outbreaks and shows the importance of understanding how these organisms survive in secondary environments such as soil. The biological and abiotic properties of soil play a very big part in the survival of *E. coli*. It is not yet cleared how these two aspects interact with survival and whether this is related to toxin genes. Non-virulence O157: H7 survives longer than carriers of virulence genes. Antibiotic resistance is more pronounced in soils with higher pH, indicating the existence of biotic and abiotic interactions. Bacterial soil diversity promotes the survival of O157:H7 through competition for resources and niches (Xing *et al.*, 2019).

The incidence of *E. coli* outbreaks varied between different countries. In the European Union and Norway, the confirmed cases in the laboratory were 1.3 cases per 1,00,000 population in 2004, while in the USA, it was 0.9 cases per 1,00,000 people in the same year. In Argentina, the frequency of EHEC was measured as 22 cases of HUS per 1,00,000 children between the age of 6 to 48 months (WHO, 2020). About 265,000 STEC infections were reported in the U.S. annually. Moreover, the STEC O157 was responsible for about 36 per cent of them. The estimates of CDC stated that in each year STEC can cause 3,600 hospitalizations and 30 deaths in U.S. (CDC, 2016).In India, disease burden includes around 1700 million diarrheal cases and 5,000,000 deaths per year among children (WHO, 2017), and high incidence of diarrhea is also reported in adults (Acree and Davis, 2017). One of the most common bacterial agents of infective diarrhea is *Escherichia coli*. In India, it has been documented that diarrhoeagenic *Escherichia coli* (DEC) is more prevalent among children and adults (Gomes *et al.*, 2016) and many outbreaks of diarrhea have been reported in India which were mainly foodborne (Pai *et al.*, 1997; Scavia *et al.* 2008).

Health Impacts of *E. coli* O157:H7

E. coli O157: H7 is one of the infective strains of *E. coli*. It produces potent toxins and can cause serious illness like diarrhea and abdominal cramps; sometimes infection can cause bloodless diarrhea. *E. coli* causes of many common bacterial infections such as bacteremia, UTI, cholecystitis, traveller's diarrhea, cholangitis, and other clinical infections such as neonatal meningitis and pneumonia (Medscape, 2019). People with very weak immunity, young children, pregnant women, and the elder people are at higher risk of these complications. Symptoms due to *E. coli* intestinal infection were usually between one to ten days which is known as the incubation period and may last for about 5-10 days (Healthline, 2019). Usually, *E. coli* causes urinary tract infections, and most commonly ascending infections (that is, from the perineum to the urethra). The pelvic inflammatory disease (PID) and prostatitis can also be caused by *E. coli*. It usually lives in the guts of animals and humans; however, intestinal infections can be caused by some strains having acquired genes. The diarrhea can be caused by the following strains if swallowed:

- ☆ **Enterohemorrhagic** strains (such as serotype O157:H7 and others) produce various neurotoxins, cytotoxins, and enterotoxins, including verotoxin, and can spread bloody diarrhea; 2-7 per cent of cases may have HUS. These strains are mainly derived from undercooked ground beef, but may also come from people who have infected their feces due to poor hygiene.
- ☆ **Enterotoxigenic** strains can lead to watery diarrhea.
- ☆ **Enteroinvasive** strains can spread inflammatory diarrhea.
- ☆ **Enteropathogenic** strains can lead to watery diarrhea.
- ☆ **Enteroaggregative:** Persistent diarrhea can occur in AIDS patients and children in tropical areas by some strains (MSD Manual, 2020).

Enterohemorrhagic *Escherichia coli* (EHEC) is a pathogenic strain of *E. coli*, which produces Shiga toxin, cause HUS, and hemorrhagic colitis (HC) and poses a threat to humans. The EHEC serotype such as O157: H7, O26: H11, O157: NM, O111: H8, and O91: H21, can cause many diseases in humans (Melton-Celsa *et al.*, 1996; Paton and Paton, 1999). *Escherichia coli* O157: H7 is the EHEC serotype which is most frequently isolated from patients in the Japan, United Kingdom and the United States.

Treatment Measures for *E. coli* Removal

The water should be boiled for one minute for killing or inactivating *E. coli* O157:H7, and at higher altitudes the boiling time should be three minutes. After sterilization the water must be cooled and stored in a clean container with a lid and refrigerated. The household methods such as boiling, chlorination, reverse osmosis and solar disinfection are effective in removal of microbes from drinking water (Malan, 2019). At present, no certified filter is available to remove *E. coli* from water. One can also disinfect the water well; and should consult their local health department for best practices. After disinfecting the well water, it should be checked regularly to reduce the chances of reoccurrence of the problem (CDC, 2015).

Preventive measures for *E. coli* infection include careful hand hygiene and isolation of the infected people, especially ESBL infections. To prevent infection with *E. coli* O157: H7, all ground beef must be thoroughly cooked and raw milk must be avoided. Because infections may be transmitted through water, patients infected with *E. coli* O157:H7 or having diarrhea should not use swimming pools and water slides and other recreational facilities within 2 weeks after symptoms subsided. Symptoms of O157: H7 must be diagnosed, and all outbreaks must be reported to health authorities, especially in nurseries. If there is an outbreak of *E. coli* in the media, the doctor should advise family members to check whether there are recalled foods in their refrigerator. People must also abide by food safety regulations and avoid eating raw or undercooked beef; wash their hands regularly, should clean kitchen counter tops and tableware immediately after touching raw minced meat; and avoid cross-contamination with other products in the refrigerator. If they think they may be suffering from disease due to the recalled food, and symptoms usually appear between 2 to 7 days after ingestion, they should also be encouraged to contact their healthcare provider (Makvana and Krilov, 2020).

The infection can be prevented by applying control measures at any stages of the food chain, that is from farms to food processing, production and preparation in commercial sector and kitchens at home. The WHO (2018) has described *Five keys* of a Safe Diet, that reduces the spread of pathogens and diseases caused by STEC.

The five keys to safer food as recommended by WHO (2018) are:

1. Cleanliness.
2. Separation of raw and cooked food.
3. Thorough cooking.
4. Keeping food at safe temperature.
5. Using safe raw materials and water

The *E. coli* bacteria is normally found in the intestine of warm-blooded animals including human beings. This bacterium is human friendly, however, if it is present in the environment in soil or in water, it can also become a source of severe diseases. Hence, one should maintain a good hygiene in households and at work places to avoid any contamination with microbes. It will be helpful in reducing the disease burden of the country as well as will maintain the good health among people.

REFERENCES

Acree, M., Davis. A.M. 2017. Acute diarrhoeal infections in adults. *J. Am. Med. Assoc.* 318: 957–8.

APEC water, 2020. Drinking Water Contaminants - *Escherichia coli, E. coli*. Available at: https: //www.freedrinkingwater.com/water-contamination/ecoli-bacteria-removal-water.htm (Accessed on 9 June 2021).

Banerjee, G. 2011. Underground pollution travel from leach pits of on-site sanitation facilities: a case study. *Clean Technol. Environ. Policy* 13: 489–497.

Berry, E.D. and Miller, D.N., 2005. Cattle feedlot soil moisture and manure content: II. Impact on Escherichia coli O157. *J. Environ. Qual.* 34: 656-663.

Brenner, D.J., Farmer, J.J. 2007. Family I. Enterobacteriaceae, In: Garrity, G., Brenner, D.J., Krieg, N.R., Staley, J.R. (Eds.). Bergey's Manual of Systematic Bacteriology. Volume 2, Springer, New York, pp. 587–850.

Brian Wallheimer, 2010. E. coli thrives near plant roots, can contaminate young produce crops. Purdue University News. Available at: https: //www.purdue.edu/newsroom/research/2010/101103TurcoEcoli.html#:~: text=coli%20thrives%20near%20plant%20roots%2C%20can%20contaminate%20young%20produce%20crops,November%203%2C%202010 and text=Purdue%20scientists%20added%20E.,roots%2C%20of%20lettuce%20and%20radishes (on 6 June 2021).

CDC, 2016. Escherichia coli (E. coli). Available at: https: //www.cdc.gov/ecoli/pdfs/CDC-E.-coli-Factsheet.pdf (Accessed on 11 June 2021).

Centers for Disease Control and Prevention (CDC), 2014. Shiga toxin-producing E. coli (STEC), National Center for Emerging and Zoonotic Infectious Diseases (NCEZID), Division of Foodborne, Waterborne, and Environmental Diseases (DFWED). U.S. Department of Health and Human Services. Available at: https: //www.cdc.gov/ecoli/general/index.html (Accessed on 5 June 2021).

Centers for Disease Control and Prevention (CDC), 2015. E. coli O157: H7 and Drinking Water from Private Wells. National Center for Emerging and Zoonotic Infectious Diseases (NCEZID), Division of Foodborne, Waterborne, and Environmental Diseases (DFWED), U.S. Department of Health and Human Services. Available at: https: //www.cdc.gov/healthywater/drinking/private/wells/disease/e_coli.html#: ~: text=human%20or%20animal.-,E.,water%20runoff%2C%20and%20agricultural%20runoff (Accessed on 5 June 2021).

CPCB, 2017. Microbial characterization of Ganga River. Available at: https: //cpcb.nic.in/openpdffile.php?id=UmVwb3J0RmlsZXMvNzk zXzE1NDc3MjUzMDlfbWVkaWFwaG90bzY0NTYucGRm (Accessed on 30 June 2021).

CPCB, 2021. Status of STPs. Available at: https: //cpcb.nic.in/status-of-stps/ (Accessed on 15 June 2021).

Deccan Herald, 2016. Cauvery samples reveal prevalence of fecal coliform: TN minister. Available at: https: //www.deccanherald.com/content/562256/cauvery-samples-reveal-prevalance-fecal.html (Accessed on 11 June 2021).

Desmarais, T.R., Solo-Gabriele, H.M. and Palmer, C.J., 2002. Influence of soil on fecal indicator organisms in a tidally influenced subtropical environment. *Appl. Environ. Microbiol.* 68: 1165-1172.

Dimri, A., Awasthi, C., Uniyal, S., Nautiyal, A. and Singh, K.P., 2019. Isolation and characterization of coliform bacteria and bacteriophages from Ganga River in Northern Himalayan regions. *Int. J. Curr. Microbiol. App. Sci*, 8: 1582-1592.

Escherich, T., 1988. The intestinal bacteria of the neonate and breast-fed infant. *Clin. Infect. Dis.* 10: 220-1225.

Gomes, T.A., Elias, W.P., Scaletsky, I.C., Guth, B.E., Rodrigues, J.F., Piazza, R.M., *et al.*, 2016. Diarrhoeagenic Escherichia coli. *Braz. J. Microbiol.* 47: 3–30.

Helathline, 2019. *E. coli* infection. Available at: https: //www.healthline.com/ health/e-coli-infection (Accessed on 6 June 2021).

Ibekwe, A.M., Grieve, C.M. and Yang, C.H., 2007. Survival of *Escherichia coli* O157: H7 in soil and on lettuce after soil fumigation. *Can. J. Microbiol.* 53: 623-635.

Ibekwe, A.M., Papiernik, S.K., Grieve, C.M. and Yang, C.H., 2011. Quantification of persistence of *Escherichia coli* O157: H7 in contrasting soils. *Int. J. Microbiol.* doi: 10.1155/2011/421379

Ibekwe, A.M., Shouse, P.J. and Grieve, C.M., 2006. Quantification of survival of *Escherichia coli* O157: H7 on plants affected by contaminated irrigation water. *Eng. Life Sci.* 6: 566-572.

Ishii, S., Ksoll, W.B., Hicks, R.E. and Sadowsky, M.J., 2006. Presence and growth of naturalized Escherichia coli in temperate soils from Lake Superior watersheds. *Appl. Environ. Microbiol. 72*: 612-621.

Jiang, X., Morgan, J. and Doyle, M.P., 2002. Fate of *Escherichia coli* O157: H7 in manure-amended soil. *Appl. Environ. Microbiol.* 68: 2605-2609.

Kaper, J.B., Nataro, J.P. and Mobley, H.L., 2004. Pathogenic *Escherichia coli. Nat. Rev. Microbiol.* 2: 123-140.

Ksoll, W.B., Ishii, S., Sadowsky, M.J. and Hicks, R.E., 2007. Presence and sources of fecal coliform bacteria in epilithic periphyton communities of Lake Superior. *Appl. Environ. Microbiol. 73*: 3771-3778.

LeJeune, J.T., Besser, T.E. and Hancock, D.D., 2001. Cattle water troughs as reservoirs of Escherichia coli O157. *Appl. Environ. Microbiol.* 67: 3053-3057.

Lim, J.Y., Yoon, J.W. and Hovde, C.J., 2010. A brief overview of *Escherichia coli* O157: H7 and its plasmid O157. *J. Microbiol. Biotechnol.* 20: 5-14.

Looper, M.L., Edrington, T.S., Callaway, T.R. and Rosenkrans Jr, C.F., 2009. Fate of *Escherichia coli* O157: H7 and Salmonella from contaminated manure slurry applied to soil surrounding tall fescue. *Lett. Appl. Microbiol.* 48: 513-516.

Mahmud, Z.H., Islam, M.S., Imran, K.M., Hakim, S.A.I., Worth, M., Ahmed, A., Hossan, S., Haider, M., Islam, M.R., Hossain, F. and Johnston, D., 2019. Occurrence of

Escherichia coli and faecal coliforms in drinking water at source and household point-of-use in Rohingya camps, Bangladesh. *Gut Pathog.* 11: 1-11.

Makvana, S. and Krilov, L.R., 2015. *Escherichia coli* infections. *Pediatr Rev.* 36: 167-70.

Malan, A. 2019. Assessment of groundwater quality and related health impacts in Nirmal Grams of Kurukshetra district of Haryana. Ph.D. Thesis. Institute of Environmental Studies, Kurukshetra University, Kurukshetra, Haryana, India.

Malan, A., Kumar, V. and Sharma, H.R., 2020. Bacteriological evaluation of groundwater in open-defecation-free villages of Kurukshetra district, Haryana, India. *Int. J. Environ. Stud.* 77: 928-941.

Mazhar, M.A., Ahmed, S., Husain, A. and Khan, N.A., 2021. Trends in organic contamination in river Yamuna: A case study of Delhi stretch (2007-2016). *Research Square.* DOI: https: //doi.org/10.21203/rs.3.rs-254013/v1

Medscape, 2019. *Escherichia coli (E. coli)* Infections. Available at: https: //emedicine.medscape.com/article/217485-overview#: ~: text=Escherichia%20coli%20is%20one%20of,as%20neonatal%20meningitis%20and%20pneumonia (Accessed on 2 June 2021).

Melton-Celsa, A.R., Darnell, S.C. and O'Brien, A.D., 1996. Activation of Shiga-like toxins by mouse and human intestinal mucus correlates with virulence of enterohemorrhagic *Escherichia coli* O91: H21 isolates in orally infected, streptomycin-treated mice. *Infect. Immun.* 64: 1569-1576.

MSD Manual Professional Version, 2020. *Escherichia coli* Infections (E. coli). Available at: https: //www.msdmanuals.com/en-in/professional/infectious-diseases/gram-negative-bacilli/escherichia-coli-infections (Accessed on 6 June 2021).

Mukherjee, P., Pramanick, P., Zaman, S. and Mitra, A., 2020. Eco-restoration of River Ganga water quality during COVID-19 lockdown period using Total Coliform (TC) as proxy. *NUJS J. Regul. Stud.* Special Issue.

NandaKafle, G., Christie, A.A., Vilain, S. and Brözel, V.S., 2018. Growth and extended survival of *Escherichia coli* O157: H7 in soil organic matter. *Front. Microbiol.* 9, p.762.

NandaKafle, G., Seale, T., Flint, T., Nepal, M., Venter, S.N. and Brözel, V.S., 2017. Distribution of diverse *Escherichia coli* between cattle and pasture. *Microbes Environ.* 32: 226-233.

Nataro, J.P. and Kaper, J.B., 1998. Diarrheagenic *Escherichia coli. Clin. Microbiol. Rev.* 11: 142-201.

Natural Resources Defense Council (NRDC), 2021. Water Pollution: Everything you need to know. Available at: https: //www.nrdc.org/stories/water-pollution-everything-you-need-know (Accessed on 5 June 2021).

Pai, M., Kang, G., Ramakrishna, B.S., Venkataraman, A., Muliyil, J. 1997. An epidemic of diarrhoea in south India caused by enteroaggregative *Escherichia coli*. *Indian J. Med. Res.*106: 7–12.

Paton, A.W. and Paton, J.C., 1999. Direct detection of Shiga toxigenic *Escherichia coli* strains belonging to serogroups O111, O157, and O113 by multiplex PCR. *J. Clin. Microbiol.* 37: 3362-3365.

Rai, P.K., Mishra, A. and Tripathi, B.D. 2010. Heavy metal and microbial pollution of the River Ganga: A case study of water quality at Varanasi. *Aquat. Ecosyst. Health Manag.* 13: 352-361.

Rayasam, S.D., Ray, I., Smith, K.R. and Riley, L.W., 2019. Extraintestinal pathogenic Escherichia coli and antimicrobial drug resistance in a Maharashtrian drinking water system. *Am. J. Trop. Med. Hyg.* 100: 1101-1104.

Safe Drinking Water Foundation (SDWF), 2017. Detailed *Escherichia coli*. Available at: https: //www.safewater.org/fact-sheets-1/2017/1/23/detailed-escherichia-coli (Accessed on 10 June 2021).

Scavia, G., Staffolani, M., Fisichella, S., *et al.,* 2008. Enteroaggregative *Escherichia coli* associated with a foodborne outbreak of gastroenteritis. *J. Med. Microbiol.* 57: 1141–1146.

Sharma, B., Parul, A.K., Jain, U., Yadav, J.K., Singh, R. and Mishra, R., 2017. Occurrence of multidrug resistant *Escherichia coli* in groundwater of Brij region (Uttar Pradesh) and its public health implications. *Vet. World.* 10: 293-301.

Sharma, H.R., Chhetry, D., Kaushik, A. and Trivedi, R.C., 2000. Variability in organic pollution of River Yamuna in Delhi. *Int. J. Environ. Pollut.* 7: 185-188.

Sheeba, G., Jalagam, A. and Venkatasubramanian, P., 2017. Drinking water contamination from peri-urban Bengaluru, India. *Curr. Sci.*113: 1702-1709.

Shukla, S.C., Tripathi, B.D., Kant, R., Kumari, D. and Panday, V.S., 1989. Physico-chemical and biological characteristics of river Ganga from Mirzapur to Ballia. *Indian J. Environ. Health* 31: 218-227.

Shulman, S.T., Friedmann, H.C. and Sims, R.H., 2007. Theodor Escherich: the first pediatric infectious diseases physician? *Clin. Infect. Dis.* 45: 1025-1029.

Sivaraja, R. and Nagarajan, K. 2014. Levels of indicator microorganisms (total and fecal coliforms) in surface waters of rivers Cauvery and Bhavani for circuitously predicting the pollution load and pathogenic risks. *Int. J. Pharmtech Res.* 6: 455-461.

Skariyachan, S., Mahajanakatti, A.B., Grandhi, N.J., Prasanna, A., Sen, B., Sharma, N., Vasist, K.S. and Narayanappa, R., 2015. Environmental monitoring of bacterial contamination and antibiotic resistance patterns of the fecal coliforms isolated from Cauvery River, a major drinking water source in Karnataka, India. *Environ. Monit. Assess.* 187: 1-13.

Suthar, S., Chhimpa, V. and Singh, S., 2009. Bacterial contamination in drinking water: a case study in rural areas of northern Rajasthan, India. *Environ. Monit. Assess.* 159: 43-50.

The Economic Times, 2019. Ganga River water unfit for direct drinking, bathing: CPCB. Available at:

https: //economictimes.indiatimes.com/news/politics-and-nation/ganga-river-water-unfit-for-direct-drinking-bathing-cpcb/articleshow/69573401.cms?utm_source=contentofinterest and utm_medium=text and utm_campaign=cppst (Accessed on 10 June 2021).

United Kingdom environmental law association, 2017. Sources of water pollution. Available at: http: //www.environmentlaw.org.uk/rte.asp?id=90 (Accessed on 10 June 2021).

United States Environmental Protection Agency (USEPA), *2018.* Bacteria and *E. coli* in water. Available at: https: //www.usgs.gov/special-topic/water-science-school/science/bacteria-and-e-coli-water?qt-science_center_objects=0#qt-science_center_objects *(*Accessed *on 15 June 2021).*

United States Geological Survey (USGS), 2018. Bacteria and *E. coli* in Water. Available at: https: //www.usgs.gov/special-topic/water-science-school/science/bacteria-and-e-coli-water?qt-science_center_objects=0#qt-science_center_objects (Accessed on 31 May 2021).

Verheyen, J., Timmen-Wego. M., Laudien, R., Boussaad, I., Sen, S., Koc, A. *et al.*, 2009. Detection of adenoviruses and rotaviruses in drinking water sources used in rural areas of Benin, West Africa. *Appl. Environ. Microbiol.* 75: 2798–2801.

Vinger, B., Hlophe, M., Selvaratnam, M. 2012. Relationship between nitrogenous pollution of borehole waters and distances separating them from pit latrines and fertilized fields. *Life Sci. J.* 9: 402–407.

Wang, Y., 2002. The function of OmpA in Escherichia coli. *Biochem. Biophys. Res. Commun.* 292: 396-401.

WHO, 2020. Factsheet on *E. coli*. Available at: https: //www.who.int/foodsafety/foodborne_disease/ecoli.pdf (Accessed on 15 June 2021).

WHO. Diarrhoeal disease. 2017. Available at: http: //www.who.int/mediacentre/factsheets/fs330/en/(Accessed on 16 June 2021).

Winfield, M.D. and Groisman, E.A., 2003. Role of nonhost environments in the lifestyles of *Salmonella* and *Escherichia coli*. *Appl. Environ. Microbiol.* 69: 3687-3694.

World Health Organization (WHO), 2018. *E. coli*. Available at: https: //www.who.int/news-room/fact-sheets/detail/e-coli (Accessed on 9 June 2021).

World Health Organization (WHO), 2018. *E. coli*. Available at: https: //www.who.int/news-room/fact-sheets/detail/e-coli (Accessed on 2 June 2021).

Wright, K.M., Chapman, S., McGeachy, K., Humphris, S., Campbell, E., Toth, I.K. and Holden, N.J., 2013. The endophytic lifestyle of *Escherichia coli* O157: H7: quantification and internal localization in roots. *Phytopathology*, 103: 333-340.

Wright, K.M., Crozier, L., Marshall, J., Merget, B., Holmes, A. and Holden, N.J., 2017. Differences in internalization and growth of *Escherichia coli* O157: H7 within the apoplast of edible plants, spinach and lettuce, compared with the model species N icotiana benthamiana. *Microb. Biotechnol.* 10: 555-569.

Xing, J., Wang, H., Brookes, P.C., Salles, J.F. and Xu, J., 2019. Soil pH and microbial diversity constrain the survival of *E. coli* in soil. *Soil Biol. Biochem.*128: 139-149.

Zimmer C. 2008. *Microcosm.* Pantheon Books, New York.

Chapter 8

Current Status of Combined Veterinary Vaccines

Vijay, Ojasvita, Swati Dahiya and N.K. Kakker*

*Department of Veterinary Microbiology,
Lala Lajpat Rai University of Veterinary and Animal Sciences,
Hisar – 125 004, Haryana
e-mail: swatidahiya@luvas.edu.in

ABSTRACT

Veterinary vaccines have played an important role in protecting animal and public health. Decrease in disease incidence and cost of treatment, efficient production of milk and meat for feeding the ever-increasing human population, reduction in the use of antibiotics are few of their numerous positive effects. Scientific and technological advancements in biotechnology, immunology and related domains have led to the development of many novel vaccines and improvements to existing ones. However, this came with several drawbacks: increased needle pricks, poor compliance, improper storage and in efficient handling are a few. Combined vaccines provide a way that reduces and overcomes some of these drawbacks and encourages the use of a wide range of vaccines. A combined vaccine is a single preparation containing multiple vaccines (against different types or groups of pathogens or antigens) used in vivo. By combining vaccines, the overall number of injections needed are reduced resulting in fewer needle pricks and less trauma; increased vaccine coverage, timely completion of vaccination schedule, reduced storage space requirement, easier shipping and reduced overall cost of immunization programme. Here we have discussed different types of combined vaccines, methods of preparation and success stories in bovines, ovines, caprines, poultry, canines and felines as well as their disadvantages. It is also emphasized here that the development of combined vaccines and their acceptance is crucial to address the immunization needs of the world and help in achieving the 'One Health' goal thus improving, protecting and defending the health of all the species including animals, humans, plants and environment.

Keywords: *Combined veterinary vaccines, Bovine, Canine, Ovine, Caprine, Feline, Poultry*

Introduction

Veterinary Vaccines

Animals have a natural defense mechanism against invading infections known as the immune system. During infection, immune system responds to eliminate the infection, but this response is not always fast enough to prevent a disease from causing major harm or even death to the animals. Using a vaccine can give the immune system an edge against the infection. A vaccine is an immunogenic preparation of a pathogen that provides the animals with the ability to fight infections caused by pathogenic bacteria, viruses, parasites, or a combination thereof, by acting as an infection and stimulating an immune response.

Historically, vaccination probably had the greatest impact on human/animal health by any medical/veterinary intervention technique. Ancient Chinese, who practiced the art of variolation, Benjamin Jesty and Edward Jenner's findings to prevent the smallpox infection (Fenner *et al.*, 1988) pioneered the science of vaccination. Later, in the 19th century, Louis Pasteur demonstrated that attenuated *Pasteurella multocida* could protect chickens from fowl cholera and paved the way for the development of veterinary vaccines (Pasteur *et al.*, 1881). Since then, the research and development of veterinary vaccines has paralleled the human medicine (Mandado, 2019) and today numerous vaccines are available which protect the animals from various diseases, some of which are zoonotic and pose a significant risk to humans as well.

Importance of Veterinary Vaccines

Veterinary vaccines are critical components in disease prevention and control programmes in animals, food production, and the public health sector. Global eradication of smallpox from humans and rinderpest from cattle attests to the effectiveness of vaccines in preventing and controlling infectious diseases. Innovations in the research and development have made the vaccines safer and more effective. The most recent example is the worldwide usage of COVID vaccinations in both humans and animals (Tizard, 2020; Mint, 2021; Zimmer *et al.*, 2021). Quality vaccines are now essential for modern society and ensure that our pets and farm animals are benefitted by preventing the occurrence of disease (Roth, 2011).

1. **Escherichia colif zoonotic diseases:** The use of vaccines has helped to control zoonotic diseases among food animals, companion animals,

and even wildlife. As a result, human exposure to zoonotic diseases has decreased as well. In the absence of certain vaccines, it is unlikely that some families would want to keep dogs or cats as pets because they fear being infected with zoonotic diseases such as rabies or toxoplasmosis (Roth, 2011).

2. **Control of newly emerging and exotic diseases of animals:** Emerging and exotic animal diseases are posing an increasing threat to animal health and putting food security in jeopardy. Increased human and animal populations, as well as environmental deterioration and worldwide trade - travel have increased the potential for pathogenic transmission within and between species. The recent pandemic caused by coronaviruses (SARS-CoV2 and MERS-CoV) which originated from bats is a perfect example befitting the situation (El-Sayed and Kamel, 2021). In the decades to come, emerging diseases will continue to plague the world. These emerging, re-emerging animal or trans-boundary animal diseases, can only be controlled through rapid development of animal vaccines.
3. **Reduction in the need for antibiotics:** There are increasing concerns about pathogens developing antibiotic resistance because of the extensive use of antibiotics (Singer *et al.,* 2003). The availability of affordable and efficacious vaccines will reduce the dependency on antibiotics in both veterinary and human medicine.
4. **Safe and efficient food production:** United Nations Department of Economic and Social Affairs Population Division estimates that by 2050, the world population will reach over 9.1 billion. The food production will need to be increased up to 70 per cent from 2009 to 2050 in order to adequately feed such population (Roth *et al.,* 2010). Vaccines preserve animal health, improve production, increase the life span and therefore are important components to fulfill the aforementioned needs.

Combined Vaccines in Animals

Scientific and technological advancements in biotechnology, immunology and related domains have led to the development of many novel vaccines and improvements to existing ones. Morbidity and death from a variety of diseases could be reduced dramatically, if these newer combination vaccines are fully utilized in at-risk groups. However, as more and more vaccines are introduced, there is an increase in the number of needle pricks causing animal trauma and increased visits to the veterinarian and thus, owners may fail to comply the resulting complex immunization regimen. Furthermore, storage needs and record-keeping associated with delivering separate injections will negatively impact the efficacy of health-care professionals and the increased overall cost of vaccine administration. As a result, the incidence of vaccine-preventable diseases may not decline. To overcome these problems, combined vaccines have been developed.

A combined vaccine is defined as a single preparation of multiple vaccines (against different types or groups of pathogens or antigens)used *in vivo* (Ellis and Brown, 1997).The groundwork for the development of combined vaccines was laid when three individual vaccines diphtheria, tetanus, and pertussis were combined into a single DTaP vaccine (Edwards and Decker, 2008). Since then, combined vaccines have revolutionized the framework of immunization programs and have become an essential tool in the healthcare sector.

Gustav Ramon of Pasteur Institute, considered as the father of combined vaccines, began the clinical studies in 1930s by combining diphtheria and tetanus antigens (Vitek and Wharton, 2008) disregarding the perception that antigenic competition would hinder the joint immunization. By the end of the decade, inactivated *Bordetella pertussis* bacteria was also added to DT combination, giving rise to the DTaP combined vaccine. At that time antibody response was taken as proof of efficacy and the interference between the antigenic components was not explored.

The Need for Combined Vaccines

The immunization schedule is already complex, making the process of incorporating new vaccines ever more difficult. However, it can be simplified by combining multiple vaccines to achieve numerous positive outcomes, *viz.*:

1. Reduction in the number of needle pricks by combining vaccines, thus reducing the trauma to the animals
2. Higher rate of compliance among animal/pet owners because of the reduction in the number of injections as well as the number of visits to veterinary professionals
3. Increased vaccination coverage, timely completion of vaccination schedules, thus lowering the overall administration costs (Weston and Klein, 2008)
4. Reduction in the storage space requirement
5. Reduced shipping costs
6. Allows incorporation of new vaccines into already existing immunization schedules

Types of Combined Vaccines

Vaccines are classified into three types: live or live-attenuated, non-live, and modern vaccines, and till now only the first two have been developed into combination products.

1. Live or live-attenuated vaccines are made up of viral or bacterial strains that have been cultured *in vitro* in a way that reduces their infectivity and pathogenicity to the target population. These strains retain their infectivity and replication ability to limited extent, but are not pathogenic (Alexandersen, 1996).

2. Non-live vaccines can be divided into inactivated/killed or subunit vaccines. The antigenic component in these vaccines can either be killed whole organisms (virus or bacteria), proteins purified from the organism, recombinant proteins or polysaccharides (Pollard and Bijker, 2021). To increase immunogenicity, non-live vaccines are frequently coupled with an adjuvant.

The combination vaccines can either have live or non-live components, successful combination of live and non-live components together is not yet recorded. The third type of vaccine, known as modern vaccines, is RNA/DNA-based; however, because vaccines of this type have only recently entered clinical trials, combinations of these vaccines are currently not available. The combined vaccines can further be divided into multi-disease or multivalent vaccines based on antigens combined.

Methods of Combinations

A number of methods can be used to incorporate complex mixtures of microbes and/or antigens to form a single preparation (Committee for Medicinal Products for Veterinary Use Guidelines, 2013). These methods include:

1. **Combined vaccines**: Combined or multi-disease vaccines are those that protect against more than one disease or pathogen and are approved by a single marketing authorization. These combined vaccines can either be delivered in a single container or in different ones
2. **Where contents are blended before administration.**
3. **Multivalent vaccines**: Multivalent combinations defend against only one disease by mixing the different antigens relevant for protection. These vaccines are used for immunization against different types of strains (serotypes or serogroups) of a single pathogen and the number of strains designates the valency of the vaccine (trivalent, tetravalent, *etc.*). The most common example is the use of single foot-and-mouth disease (FMD) trivalent vaccine against FMD virus serotypes O, A and Asia-1 in cattle, buffalo, sheep and goats in India. Generally, a multivalent combination vaccine is developed as a whole rather than developing each serotype vaccine separately.
4. **Association vaccines:** The use of two or more vaccines, each of which has its own marketing authorization, is regarded as association vaccines. The following association vaccines are possible:
    a. Combining two or more vaccine preparations for administration at a single site.
    b. Two or more vaccines are administered at the same time but at different sites.

Examples of Combined Vaccines for different Species

Combined Vaccines for Bovines

FMD caused by FMDV, Haemorrhagic septicemia (HS) caused by a bacteria *Pasteurella multocida* and black quarter (BQ) caused by *Clostridium chauvoei* bacteria are important diseases of bovines which cause significant losses to national economy. Among these the co-occurrence of FMD and HS has been recorded for the last two decades because the FMDV produces a transient immunosuppression that provides an opportunity to predispose the animals to *P. multocida* infection resulting into the death of the affected animals. This pattern of combined outbreak of FMD and HS has been recorded in Punjab in 1998 and Haryana since 2001 (Chhabra *et al.,* 2004; RRC Hisar, Annual Report 2016-17, 2018). This demands tremendous efforts to prevent the livestock from scourge of dual infection. And in such a situation the use of a combination vaccine would be extremely advantageous, as in case of Thailand where the combination vaccine for FMD and HS is routinely practiced and strong antibody and cellular immune responses have been recorded with no antigenic competition between the two antigens (Muenthaisong *et al.,* 2021). Vaccine manufacturers in India have come out with combined FMD+HS oil adjuvant vaccine as well as FMD+HS+BQ or HS+BQ combined vaccines that confer a long-lasting immunity to combat these infections.

Besides number of advantages of combined vaccines over single vaccines, questions have been raised about inter antigenic competition impairing the protective response to each of the constituent antigens. Several studies have been conducted where the combined vaccines were found safe, well-tolerated and immunogenic with no antigenic competition between the components, revealing their potential for the success (Chhabra *et al.,* 2004; Prasad *et al.,* 2019). The combined FMD+HS vaccine was found to be effective in eliciting an excellent immune response against both FMD and HS and duration of protection (upto 6 months) against both diseases was also similar to single vaccine (Reddy *et al.,* 2001).

Recently, the Dept. of Animal Husbandry and Dairying (DAH and D), Government of India, New Delhi has constituted a committee for "Recommendations on usage of FMD+HS combined vaccine". The scientific data generated by the Scientists of LUVAS, Hisar was presented in series of meetings held by DAH and D and discussed at length by the members from different state veterinary universities and Indian Council of Agricultural Research (ICAR) institutes. Based on this data, the Haryana state was allowed (first in the country) to use FMD+HS combined oil-adjuvant vaccine for mass vaccination of eligible bovine population, at a six-monthly interval on trial basis (RKVY, 2019). Presently, the Regional Research Centre on FMD, Department of Veterinary Microbiology, LUVAS, Hisar is participating actively in implementation of FMD+HS combined vaccine programme initiated by DAHD, Government of Haryana since April 2019 by providing technical and scientific support in the form of sero-monitoring and sero-surveillance work in all the 22 districts of Haryana. The data thus generated provided evidence that sufficient protection at herd level had been generated despite the fact that the disease was present in the neighboring states. As a result, there has been no incidence of FMD and HS in Haryana since the start of FMD+HS combined vaccination.

The type of combination in a combined vaccine is also influenced by the epidemiological status and immunization regimen practiced in a given area. Like, in several of the South and South-East Asian Countries, the antigen for Black Quarter *i.e. Clostridium chauvoei*is also included along with FMD virus and *P. multocida*in combination form. Reddy *et al.* (2001) studied the immune response of this FMD+HS+BQ combined vaccine in cattle and found that both routes (subcutaneous and intramuscular) of administration of combined vaccine produced satisfactory level of immune response on day-21post vaccination.

Rabies vaccination is also recommended for cattle in developing countries (Singh *et al.*, 1995). Palanisamy *et al.* (1992) found that vaccinating animals with FMD+ Rabies combined vaccine induces satisfactory serum antibody titers and the components of vaccine showed protective titers similar to when given individually. Srinivasan *et al.* (2001) also reached at the same conclusion with FMD+ HS+ BQ+ Rabies combined vaccine.

Combined Vaccines for Ovines and Caprines

In case of small ruminants, the two most contagious viral diseases are *Peste des petits ruminants* (PPR) and sheep/goat pox (SGP). They are found in the same geographic areas and cause huge economic losses. Generally, vaccination against both the diseases is done by individual monovalent vaccines. However, broad spatial distribution of animal population with difficult access due to hilly terrain/forests leads to low vaccine coverage and thus helps in the spread and maintenance of the diseases. The combined PPR/SGP vaccine confers a good protection against both infections in a single shot (Fakri *et al.*, 2015). Chaudhary *et al.* (2009) combined PPR and sheep pox vaccines in recommended doses and then studied its safety and immunogenicity along with challenge studies in sheep. The study found that the combined vaccine was safe and potent, also there was no interference between vaccine components and it was concluded that this vaccine can be used for cost effective immunization strategies in target population.

Other important combined vaccine routinely given to sheep and lambs are to protect them against Clostridial diseases like the CD-T toxoid that protects from enterotoxaemia and tetanus caused by *Cl. perfringens* types C/D and *Cl. tetani*, respectively. A 3, 7, or 8-way combos are also available including *Cl. perfringens* (type-B, C and D) along with *Cl. tetani, Cl. septicum, Cl. novyi, Cl. sordellii or Cl. haemolyticum* (Tizard, 2021). In such complex mixtures, the animals show individual variations in response and some antigenic competition also occurs. The strongest antibody response is produced by *Cl. tetani* and *Cl. novyi* type-B, whereas *Cl. septicum* induces the weakest.

The first pentavalent bluetongue virus (BTV) vaccine marketed in India to control the arthropod borne infection in sheep is an inactivated vaccine containing BTV antigens of polyvalent serotypes 1, 2, 10, 16 and 23 (Saminathan *et al.*, 2020).

Combined Vaccines for Poultry

Avian respiratory diseases caused by mixed infection are more common and severe than single infections (Kleven and Glisson, 1997). Therefore, combined vaccine should be preferred over a single-agent vaccine and in recent years a lot of combination vaccines have been developed for poultry. Most of these combination products are live attenuated vaccines which produce low antibody titers with shorter immunologically effective period than that of a live single-virus vaccine. The reason being interference, a phenomenon demonstrated among viruses (Birhane and Fesseha, 2020). Therefore, antigen load may be increased for mixed vaccines to get higher antibody titers.

Among combined live-attenuated vaccines, Newcastle disease and Infectious bronchitis (ND+IB) vaccines are being used for years now but their performance record is still conflicting. The main criticism being that the IB component may interfere with the chicken's ability to respond to the ND component (Thornton and Muskett, 1975). The interference pattern was not detected in inactivated ND+IB combined vaccines as satisfactory immune response was observed for both the components (Zanella, 1970; Uchinuno *et al.* 1973). Another combined vaccine containing *Haemophilus gallinarum* along with inactivated ND+IB viruses is also available and which is reported to induce superior immune response to IB as compared to the individual IB vaccine (Otsuki and Iritani, 1974).

Coccidiosis, a protozoan disease is also a significant disease of poultry affecting the intestinal tracts and the recent commercialized vaccines introduced for coccidiosis consist of live, sporulated oocysts of multiple *Eimeria* species (Soutter *et al.*, 2020).

Combined Vaccines for Canines and Felines

There are two broad categories of the vaccines available for dogs and cats: core and non-core vaccines (Day *et al.*, 2016). Core vaccines protect the animals against severe, life threatening diseases. The core vaccines for dogs are those that protect against canine distemper, canine adenovirus and the canine parvovirus type 2 while for cats it includes vaccines against feline parvovirus, feline calcivirus and feline herpesvirus-1. Vaccination against rabies should also be considered core for both species in geographical areas where it is endemic. Non-core vaccines are against specific infections that are based on the location, environment, or lifestyle of the animals.

In India, combined vaccines for dogs include core vaccines, rabies and leptospira antigens while for cats, mainly 'tricat' combination vaccines are used, consisting of a combination of core vaccines. Reddy *et al.* (2002) compared the efficacy of combined vaccine containing with individual components of monovalent vaccines and found that there was no significant variation when compared with the efficacy of monovalent vaccines. Also, no antigenic competition amongst the different valencies of the combined vaccine was detected and the vaccine was found to be safe. The duration of immunity conferred by combined vaccine was also studied (Reddy *et al.*, 2003) and it was reported that pups can possess protective antibody titers for a period of two years with a primary vaccination and a booster dose after one month.

Schetters *et al.* (2007) tested a combination of soluble parasite antigens of *Babesia canis* and *B. rossi* and found that the combination can protect the dogs against both of the infections.

Combined Vaccines: Issues to Consider

There are numerous problems in the development of combined vaccines that should be addressed as soon as feasible in order to maximize the likelihood of generating the final vaccine product in a shorter time frame.

Technical Issues

To obtain a formulation with a long enough shelf-life, the final combined vaccine product must be pharmaceutically stable and physical interactions among vaccine components and excipients must be at acceptable level (Paradiso *et al.*, 1993). Ensuring these is relatively easy for vaccines with similar components, like multivalent combinations but it becomes a challenge for multi-disease combinations where different components and excipients are brought together and optimization of formulations have to be carried out in such cases.

Clinical Issues

The clinical trials for multi-disease combined vaccines usually compare the combination directly with its component vaccines. The goal of these trials is to demonstrate the safety and effectiveness of combined vaccines.

1. **Safety:** Safety is the utmost important issue in designing and formulation of combined vaccines because mixing different antigenic components and/or adjuvants might elicit adverse reactions such discomfort at the site of injection, redness, irritability, and so on. Therefore, a combined vaccination should only be considered safe after comparing its reactogenicity profile to that of its individual components.
2. **Efficacy:** For combining multi-disease vaccines care should be taken to ensure that the protective effects of the combination are equivalent to that of the individual vaccines, and their adverse effects should not exceed the additive effect. The same is not true for multivalent combinations as individual component vaccines are not available for comparison. In such cases, the efficacy should be proven by serological monitoring and the combination should be shown to prevent the disease significantly

against all vaccine serotypes in aggregate and against at least one serotype individually.

Regulatory Issues

Combined vaccines are subject to regulatory scrutiny that encompasses both component antigens and the combination product as a whole. The demonstration of consistency at the manufacturing and clinical levels is a fundamental regulatory requirement during production of combined vaccines and regulatory agencies present all over the world have set instructions for the same.

Combination Issues

The combinations of the vaccines should never be random because one component may dominate or interfere with the response of the others and also the addition of vaccines against organisms that are not causing problems at the given geographical site can be a waste of time and resources. Additionally, combination of vaccines having different immunization schedules is a little tricky and a lot of factors need to be considered to develop a combined vaccine immunization schedule.

Disadvantages of Combined Vaccines

1. Combining vaccines may introduce negative outcomes such as decreased immunogenicity and/or efficacy, as well as increased side effects. Competition or interference between antigens may occur when different antigens administered at the same time and proper clinical trials are required to properly assess the risk of such phenomenon. Combined vaccine manufacturers should adjust their components in accordance with this.
2. Immunity generated by one vaccine componentmay last for years and other components may offer short term immunity. When such components are packaged together the component providing long-term immunity has to be given repeatedly to keep the immunity for the short-term component strong which exposes the animal repeatedly and may cause adverse reactions like skin or autoimmune disease, allergies and even death.
3. If there is any particular adverse event in animals due to combination, it may not always be clear which component is responsible for it.
4. There have been numerous outcries that the large number of pathogens and chemicals that the animals are exposed take a huge toll on the immune system and may lead to adverse consequences.

Conclusion

The ultimate and immediate goal of immunization is prevention and eradication of infectious diseases and in order to reach this goal vaccine coverage and compliance to vaccination is a must. Combined vaccines can help to achieve this goal by overcoming some of the key challenges by simplifying the immunization regimen and reducing the number of shots. However, there are many technical, clinical and regulatory issues that must be overcome to maintain the immunogenicity and safety of the combined vaccine. Keeping the future prospects in mind, the need for the use of combined vaccines is inevitable to not only protect the animal health but also affect the human health directly or indirectly by maintaining the food safety and security. This will ultimately help in achieving the 'One Health' goal which involves improving, protecting and defending the health of all the species including animals, humans, plants and environment.

REFERENCES

Alexandersen, S. 1996. Advantages and disadvantages of using live vaccines risks and control measures. *Acta. Vet. Scand. Suppl.* 90: 89-100.

Birhane, N. and Fesseha, H. 2020. Vaccine failure in poultry production and its control methods: A Review. *Biomed. J. Sci. Tech. Res.*DOI: 10.26717/BJSTR.2020.29.004827.

Chaudhary, S.S., Pandey, K.D., Singh, R.P., Verma, P.C. and Gupta, P.K. 2009. A vero cell derived combined vaccine against sheep pox and peste des petits ruminants for sheep. *Vaccine.* 27: 2548-2553.

Chhabra, R., Sharma, R. and Kakker, N.K. 2004. Comparative immunogenicity of foot and mouth disease virus antigens in FMD-haemorrhagic septicaemia combined vaccine and FMD vaccine alone in buffalo calves. *Indian J. Exp. Biol.* 42: 259–64.

Committee for Medicinal Products for Veterinary Use Guidelines. 2013. Guideline on the requirements for combined vaccines and associations of immunological veterinary medicinal products (IVMPs). European Medicines Agency, EMA/CVMP/IWP/594618/2010.

Day, M.J., Horzinek, M.C., Schultz, R.D. and Squires, R.A. 2016. WSAVA Guidelines for the vaccination of dogs and cats. *J. Small Anim. Pract.* 57: E1-E45.

Edwards, K.M. and Decker, M.D. 2008. Pertussis vaccine. In: Plotkin, S.A., Orenstein, W.A., Offitt, P. (Eds.). Vaccines. 5th edn. Saunders, PA, USA. pp. 471-528.

Ellis, R.W. and Brown, K.R. 1997. Combination Vaccines. *Adv. Pharmacol.* 39: 393-423.

El-Sayed, A. and Kamel, M. 2021. Coronaviruses in humans and animals: the role of bats in viral evolution. *Environ. Sci. Pollut. Res.* **28:** 19589-19600. https: //doi.org/10.1007/s11356-021-12553-1

Fakri, F., Ghzal, F., Daouam, S., Elarkam, A. and Douieb, L. 2015. Field application of a combined vaccine against peste des petits ruminants and sheep pox.*J. Vaccines Vaccin.*6: 301.

Fenner, F., Henderson, D.A., Arrita, L., Jezek, Z. and Ladnyi, M. 1988.*Smallpox and its Eradication*. World Health Organization, Geneva, p. 4.

Kleven, S.H. and Glisson, J.R. 1997. Multicausal respiratory disease. In: Calnek, B.W., Barnes, H.J. and Beard C.W. (Eds.). Diseases of Poultry. 10[th]edn, Iowa State University Press, Ames, IA. pp. 1008-1012.

Mandado, T. 2019. The current status of veterinary vaccinology: A Review. *J. Adv. Med. Pharm. Sci.* 20(2): 1-10.

Mint. 2021. Russia registers World's 1[st]Covid vaccine Carnivac-Cov for animals. https: //www.livemint.com/news/world/russia-registers-world-s-1st-covid-vaccine-for-animals-says-report-11617176371144.html Accessed on June 22, 2021.

Muenthaisong, A., Rittipornlertrak, A. and Nambooppha, B. 2021.Immune response in dairy cattle against combined foot and mouth disease and haemorrhagic septicemia vaccine under field conditions. *BMC Vet. Res.* 17: 186.

Otsuki, K. and Iritani, Y. 1974. Preparation and immunological response to a new mixed vaccine composed of inactivated Newcastle disease virus, inactivated infectious bronchitis virus and inactivated *Hemophilus gallinarum*. *Avian Dis.* 18: 297-303.

Palanisamy, R., Ramanna, B., Rao, K. and Srinivasan, V. 1992. Combined vaccination of cattle against FMD and rabies. *Microbiologica.*15(1): 45-9.

Paradiso, P. R., Hogerman, D. A., Madore, D. V., Keyserling, H., King, J., Reisinger, K. S., Blatter, M. M., Rothstein, E., Bernstein, H.H., Pennridge Pediatric Associates and Hackell, J. 1993. Safety and immunogenicity of a combined diphtheria, tetanus, pertussis and *Haemophilus influenzae* type b vaccine in young infants. *Pediatrics.* 92: 827-832.

Pasteur, L., Chamberland, C.E. and Roux, E. 1881.Sur la vaccination charbonne use. *CR Acad. Sci. Paris.*92: 1378–83.

Pollard, A.J. and Bijker, E.M. 2021. A guide to vaccinology: from basic principles to new developments. *Nat. Rev. Immunol.* 21: 83-100.

Prasad, V.S., Patel, S.B., Patel, K., Srinivas, K. and Thesia, M. 2019. Serological response to foot and mouth disease and Haemorrhagicsepticaemia combined vaccine in cattle. *J. Agri. Vet. Sci.* 12: 57–61.

Reddy, G.S., Ananda, K. and Srinivasan, V.A. 2001. Performance of oil adjuvant combined vaccine containing FMD, rabies, *Pasteurella multocida* and *Clostridium chauvoei* antigens. *Indian Vet. J.* 78(11): 990-993.

Reddy, S., Guddeti, S., Srinivasan, G. and Alwar, V. 2003. Duration of immunity conferred by combined vaccine containing canine distemper, canine hepatitis, canine parvo, Rabies and leptospira antigens. *Ind. Vet. J.* 80: 604-607.

Reddy, S., Srinivasan, G. and Alwar, V. 2002. Comparative efficacy of combined vaccine for canines containing canine distemper, hepatitis, parvovirus, rabies and leptospira antigens and monovalent vaccines. *Ind. Vet. J.*79: 435-439.

RKVY. 2019. Proceedings of the meeting of the State Level Sanctioning Committee (SLSC) of Rashtirya Krishi Vikas Yojna (RKVY-RAFTAAR) held on 13.03.20219. Agenda No. 5.9. Implementation of (HS+FMD) dual vaccine in the State of Haryana under RKVY for the year 2019-20, p: 25. SLSC_HR_13.03.2019.pdf (rkvy.nic.in)

Roth, J. A., Galyon, J. and Stumbaugh, A. 2010. Causes and consequences of emerging and exotic diseases of animals: Role of the veterinarian. In: Emerging and exotic diseases of animals, 4thedn, Iowa State University, Ames, IA. p. 10-17.

Roth, J.A. 2011. Veterinary vaccines and their importance to animal health and public health. *Procedia Vaccinol.* 5: 127 – 136.

RRC Hisar, Annual Report 2016-17. 2018. Dahiya, S. and Kakker, N.K. (Eds.) Department of Veterinary Microbiology, Lala Lajpat Rai University of Veterinary and Animal Sciences, Hisar, Haryana, India.

Saminathan, M., Singh, K.P., Khorajiya, J.H., Dinesh,M., Vineetha,S., Maity, M., Rahman,AT.F., Misri, J., Malik, Y.P.S., Gupta, V.K., Singh, R.K. and Dhama, K. 2020. An updated review on bluetongue virus: epidemiology, pathobiology, and advances in diagnosis and control with special reference to India. *Vet. Q.*40: 258-321, DOI: 10.1080/01652176.2020.1831708

Schetters, T.P., Kleuskens, J., Carcy, B., Gorenflot, A. and Vermeulen, A. 2007. Vaccination against large *Babesia* species from dogs. *Parassitologia.* 49 Suppl 1: 13-17. PMID: 17691601.

Singer, R.S., Finch, R., Wegener, H.C., Bywater, R., Walters, J. and Lipsitch. M. 2003. Antibiotic resistance: the interplay between antibiotic use in animals and human beings. *Lancet Infect. Dis.* 3: 47-51.

Singh, R., Shukla, D.C., Khanna, P., Singh, K. and Mehrotra, M. 1995. An outbreak of rabies in cattle and buffaloes in Uttar Pradesh. *Indian J. Anim. Sci.* 65: 166-168.

Soutter, F., Werling, D., Tomley, F.M., and Blake, D.P. (2020). Poultry Coccidiosis: Design and interpretation of vaccine studies. *Front. Vet. Sci.* 7: 101.https: // doi.org/10.3389/fvets.2020.00101

Srinivasan, V., Reddy, A., Rao, K. and Kihm, U. 2001.Serological response of bovines to combined vaccine containing foot and mouth disease virus, rabiesvirus, *Pasteurella multocida* and *Clostridium chauvoei* antigens. *Vet. arhiv.* 71: 37-45.

Thornton, D.H. and Muskett, J.C. 1975. Effect of infectious bronchitis vaccination on the performance of live Newcastle disease vaccine. *Vet. Rec.* 96: 467-468.

Tizard, I.R. 2020.Vaccination against coronaviruses in domestic animals. *Vaccine.* 38(33): 5123-5130. https: //doi.org/10.1016/j.vaccine. 2020.06.026

Tizard, I.R. 2021.Sheep and goat vaccines. In: Vaccines for veterinarians. 215-224. e1. DOI 10.1016/B978-0-323-68299-2.00026-5

Uchinuno, Y., Yamada, S., Fujikawa, H. and Koda, Y. 1973. Newcastle disease infectious bronchitis inactivated combined vaccine. *Jpn. J. Vet. Med. Res.* 26: 124-128.

Vitek, C.R. and Wharton, M. 2008. Diphtheria toxoid. In: Plotkin, S.A., Orenstein, W.A., Offit, P.A. (Eds.). Vaccines. Saunders/Elsevier, Philadelphia, PA. pp. 139-156.

Weston, W.M. and Klein, N.P. 2008. Kinrix: A new combination DTaPIPV vaccine for children aged 4-6 years. *Expert Rev. Vaccines.* 7: 1309-20.

Zanella, A. 1970. Researches on use of inactivated, oil emulsified vaccines in the control of the most important avian diseases. The IV World Veterinary Poultry Association Conference. pp.69-78.

Zimmer, C., Corum, J. and Wee, S. 2021.Coronavirus Vaccine Tracker. The New York Times. https: //www.nytimes.com/interactive/2020/science/coronavirus-vaccine-tracker.html Accessed on June 22, 2021.

Chapter 9

Zero-Valent-Iron Nanoparticles (ZVI-NPs) Utility in Environmental Remediation and Management

*Riya Panchal and Dharmender Kumar**

Department of Biotechnology,
Deenbandhu Chhotu Ram University of Science and Technology,
Murthal – 131 039 Sonepat, Haryana
**e-mail: dkbiology@gmail.com*

ABSTRACT

The heavy metals in waste water are treated by the use of a specific nano-zerovalent Iron (nZVI). With the help of nZVI, the contaminants and other unwanted substances present in the water and soil are completely removed. Their smaller size and the ability to remove non-specific degrading materials makes them suitable for decontamination processes and also enhances their utility in removal of contaminants from underground water. This review is an attempt for decontamination of the unwanted particles in the underground, soil surface for their environmental remediation. In addition to this, various metals like arsenic, lead etc. and other lethal drugs like nitrites are also studied with their effect on human and with the environment as well. Therefore, this chapter describes the current research conducted for the application of ZVI in environmental remediation and decontamination of harmful components of xenobiotics. The various applications have been discussed on the bases of studies conducted by different researchers to check the efficiency and applicability of nZVI in various fields. The emphases have also been made in terms of waste water implication and their degradation efficiency increased by deploying physical, chemical and biological methods.

Keywords: *Zerovalent Iron nanoparticles, Environmental remediation, Contamination, Decontamination, Heavy metals, Particle size.*

Introduction

Pollution has been a major issue for the world from the start and is mainly proven deadly for the environment (Briggs, 2003). The rapid increase in industrialisation is also causing degradation of our environment due to increase load of various pollutants in the environment. These contaminants get easily dissolved in water and are taken upby plants for their metabolic activity through irrigation and the plants are further consumed by animals (Kazninaetal., 2005). Later, it enters to food chain and reaches to various living organisms, humans and have deleterious effect on their health by affecting their metabolic activity (Briggs, 2003). The cast iron is more commonly used to remove the contaminants and other unwanted substances from the underground water surface by targeting the iron at the area of action and/or work specifically (Aycicek *et al.,* 2008). The chemical reaction generally occurs on the iron surface, rather than the bulk iron due to large surface area, mobility, high reactivity and its low cost. Many pollutants and contaminants are released into the water and lakes due to human activities and also more commonly due to anthropogenic activity. Through the drainage system and by chemical industries a lot of wastes has been released directly or indirectly into the water (Raymond and Felix, 2011). The various metals *viz.* cadmium (Cd),lead (Pb),mercury (Hg) and arsenic (As) are responsible for many respiratory diseases, neurodegenerative disorders, can lead to the development of cancer. Because of these harmful properties, these compounds can have adverse effect on the environment. Out of these metals the lead (Pb) is the most common toxic metal and is responsible for many diseases in human. This is produced by volcanic activities, human sources and geochemical weathering *etc.* This gets easily absorbed by the plants roots. This leads to toxicity to the living organism and effect to the environment (Hidalgo,2018).

Like lead the arsenic is a highly toxic metal and dangerous to human health. This metal is usually produced by industrial pollution and consequently its traces have been detected in soil, water and on ground surface. More precisely this has been found in underground wells where it forms a concoction of debasing materials in drinking water (Singh *et al.,* 2011). Another metal which also comes in this category is cadmium, which is more commonly found in soil and water systems. This directly affects human liver and kidney. It is given the tag of most toxic metal due to its properties of high toxicity and long half-life. Human and other animals can get exposed to cadmium through inhalation in the form of smoke of cigarettes and other industrial releases (Laspina *et al.,* 2005). Many studies have been conducted by researchers to eliminate these heavy metals from our surroundings by the help of various processes like- filtration, precipitation *etc.* But these processes are applied less because of their high cost and complex equipment (Hegedus *et al.,* 2001). For the environmental remediation, the ZVI alone failed to bring out the result in order to remove the arsenic and other metals, but the process involving the ferrous ions for co-precipitation of the arsenic compounds in the sand filter is a way to remove such contamination from the nature (Kowalski and Sogaard, 2014;

Ayicek *et al.*, 2008). In order to understand it in better way, the mode of action of ZVI need to be understood.

ZVI like many other variants of metals and forms of iron is the most widely used element in the world (Aycicek *et al.*, 2008). The most common use of the ZVI are in the remediation of toxic compounds from the environment. This serves as a purpose for the removal of the contaminants, various toxic drugs and unwanted substances (Raymond and Felix, 2011). When added, the water molecules form a barrier for the contamination and aid the removal of these substances. Many other processes are implemented for its proper usage and working. *In situ* and *ex situ* remediation methods are widely used (Gomes *et al.*, 2013). Similarly, the solid waste is recycled by the process of anaerobic digestion, which seems to be a great approach to combat such type of problems (Mackenzie and Georgie, 2019). ZVI is added in a proper condition and in a great flow of anaerobic digestion system. The other cases which involve the same procedure of anaerobic digestion such as manure, sludge and food sludge in the presence of ZVI has also been reviewed. The properties and applications of ZVI has been studied thoroughly which includes their particle size, pre-treatment method and amount of ZVI added. The dosage of ZVI matters in the working of digesters and environment in the tank, provided for the microbes, so that they can have a long term of stability and more volume of microbes are present, less presence of fatty acids and leads to the production of methane by the activities of microbes. Hence, for better results the quality of the digesters need to be maintained, the microbial working and anaerobic digestion mixed with the ZVI is to be kept in a suitable environment (Jiang *et al.*, 2018). The sources of pollutants from the water and the major sources of antibiotics and drugs are explained diagrammatically (Figures 9.1 and 9.2). There are various successful examples supporting the evidence of ZVI being helpful in removal of the contaminants from the environmental sources and these are described in the following section.

ZVI in powder form and macro-ZVI having amazing properties like oxidation, density, higher surface area makes it even more appealing to be used as a source for the removal of contaminants and pollutants from the soil and water via *in situ* conservation (Gomes *et al.*, 2013). This makes the NZVI more toxic to the pollutants so that they can be easily eliminated. The property of reactivity is achieved by NZVI through the oxidation process. This lead to completely oxidization of iron into ferrous oxide, which consequently provides electron, which removes harmful products formed during the Fenton reaction (Mareddy, 2017). In addition to this, nanoparticle also has low toxicity associated with it, which makes the soil stable (Gomes *et al.*, 2013). This remedial process has gained many appreciative comments and use for it effect on the environment. This is also proved by many researchers and scholars in their studies thoroughly with all the steps that it involves. Selection of the soil, then integration of NZVI in to the area where the actual contaminants exist (Mareddy, 2007; Lefevre *et al.*, 2016).

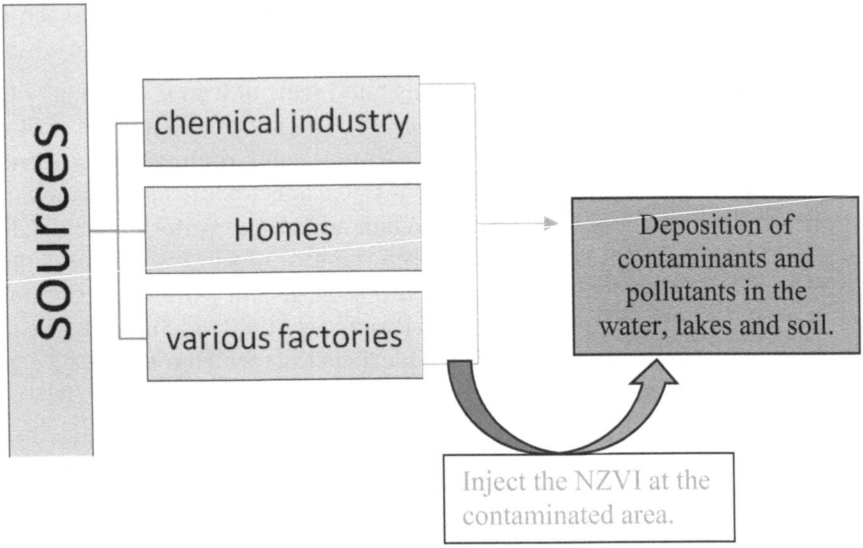

Figure 9.1: The Sources of Pollutants from Water.

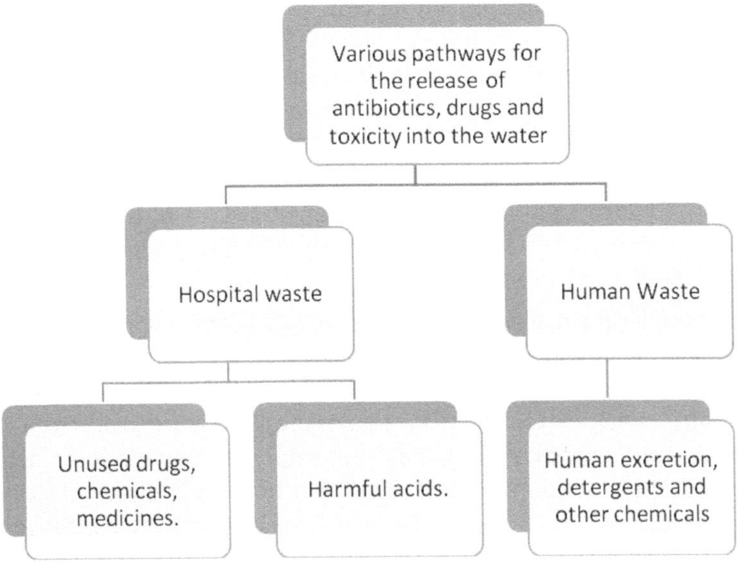

Figure 9.2: The Major Sources of Antibiotics and Drugs.

NZVI and the Effectiveness of Nanoemediation Process

Many countries where nanoremediation as a method to remove contaminants are much commonly used for *e.g.*, USA. Nanoremediation has been successful and effective in the case of pollution removal from the soil surface and under water treatment (Horvathova *et al.*, 2019). It is used in *in situ* conservation because of its promising properties like high reactivity and smooth mobility, it is easily incorporated in to the contaminants where they are situated and treated easily. By keeping in mind, the actual aim for the remediation purpose, the NZVI slurries should be where the contamination is present and injection should be provided at the actual area of action for the elimination of contaminants and the less formation of other side harmful products (Lefevre *et al.*, 2016). The nZVI modified with sulphide has shown the promising results for the removal and degradation of trichloroethylene (TCE) from the ground water (Liu *et al.*, 2017). Although the dechlorination of TCE by sulphide modified nZVI has less details (Liu *et al.*, 2017). TCE as the solvent is being used worldwide for the cleaning of metal greasing like activities. It has been proven to be a problematic solvent which causes respiratory problems in humans. It can be harmful for the human as it is classified as a carcinogen. Although, if we talk about its effect on humans and animals. It has not been found out completely and still under investigation, In some cases it has been found the main reason for the lung cancer (Mareddy, 2017).

Applications of n-ZVI in Various Remediation and Degradation Processes

The nanovalent iron has been recently discovered and researchers are keen for enhancing its properties for degradation field (Rajajayavel *et al.*, 2015; Wei and Zhang, 2003). The complete removal of selenium is not achieved like that, it has to be done in a way where the amount of selenium and its toxicity is reduced, so that it does not affect the quality and quantity of water (Mackulak *et al.*, 2015; Karn and Kuikan, 2019). Therefore, creating a healthy environment for humans and animals to live. Similarly removal of this heavy metal from the underground water is also important for the agricultural use. This also affects equally the soil surface and deteriorates its quality, making it futile for its future use. Use of pesticides and insecticides also make its quality poor and makes it difficult for the health of plants, humans and animals (Grigori and Raphael, 2013).

Removal of Selenium from Water

Many recent studies indicate that the nanoscale particle zero valent iron (nZVI) is very useful in the treatment of selenium from the water. Cilai *et al.* (2016) observed that, with the help of approx. 5g/L nZVI in the water within 3 min. 1.3 mM selenite was removed. In this process a method was used as described by (Cilai *et al.*, 2016; He *et al.*, 2007) Cs-STEM (Spherical aberration corrected scanning transmission electron microscopy) which includes X-ray energy dispersive spectroscopy (XEDS). Reaction mechanism of selenite was put up in the single core shell structured nanoparticle nZVI (Cilai *et al.*, 2016). By using this method, the translocation and transformation of Selenite was studied throughout the process to visualize its effect on the single nanoparticle corresponding to its surface area, diffusion and other properties. At a depth of 6nm from the surface the selenium layer first converted into Se (II) and Se (0) from Se (IV), was formed on the iron oxide Fe (0) interface. (Grigori and Raphael, 2013). The STEP mapping remains useful in finding out the problem, it shows that the acceleration of diffusion in selenite is there with the increased amount of nZVI for the elimination of selenium from the water completely (Ling *et al.*,2015). This study shows a better way for the starters to understand the importance and effectiveness of the ZVI, as told above. Although it is already known how much effective is ZVI in the removal of contaminants and pollutants as well as the toxic drugs from the water and soil. (Briggs *et al.*, 2003) The techniques involved in this study are also very productive. It explains thoroughly as how ZVI properly works in the field of contaminants present for promising results. The processes are needed to be done in a basic manner to achieve the results that are desired (Grigori and Raphael, 2013; He *et al.*, 2007).

Elimination of Dissolving Uranium with the Help of ZVI Nanoparticle

Uranium (U) is a threat to humans and many other animals and it should be degraded before releasing into environment. It is highly radio toxic and needs to adopt a geochemical method in which, the Uranium in the liquid form is inserted into the iron-based minerals (Jing *et al.*, Li., 2016). The iron-based mineral can fully eliminate the U from the water bodies and from the surface as well. Hence, ZVI is currently being studied and determined on to the removal of U from the contaminated water. Therefore, ZVI is to be tested for their efficient removal. This is to be done by considering the factors like; pH, osmotic pressure, density, oxygen and microorganisms (Gomes *et al.*, 2013). All of this is to be done by keeping in mind about the effects it will be having on the environment and the amount of damage it will be doing to all the marine plants and animals (Mareddy, 2017). In other case the immobilisation of Uranium in the iron-based minerals has been done and a batch system was prepared with all physical conditions like temperature and pH *etc.* It is important for the researchers to do the background check on the pH of water in the system, so the pH of the water was set in between 2-10, before the mixing of illite ZVI. The importance of removal of uranium from the water depending on the pH of water, which varies between 2.7 to 10 (Jing, *et al.*, 2016). It is concluded that the elimination of uranium was more successful in this modest amount of pH in the ground water. Determination of the left-over residue of uranium on the solid phase was also done. A leaching solution of HNO_3 was prepared and a test was run for 2 h and 0.07 per cent of uranium was leached on to the solid surface, with the considerable amount of pH 4.03 (Latifoglu and Gurol, 2003). After the whole process is done and completed the uranium can be discharged further, it can be left only as a residue in to the environment with higher concentration of acid only for short terms. If it is the opposite case, as in longer case it first needs to be monitored carefully before its discharged and even has to undergo some treatments before the complete removal. It is important and necessary to remove the uranium from the water as it can be harmful to the humans. The water which is consumed by the plants and animals and further on by creating a food chain, they are eaten by humans, creating enough diseases in them. (Pasinszki and Krebz, 2020).

Remediation Solution for Arsenic and Drug Polluted Soil

Studies show that nanoparticles with nGeothite when impaired in the soil can be helpful in remedy purpose. (Mackulak *et al.,* 2015; Wei and Zhang, 2003). The addition of ZVI nanoparticles in the soil does not increase the volume of iron in the soil. On the other hand, more usage of nGeothite can cause phytotoxicity. This particular study done by the authors show some exciting results. These two particles- nanoparticle and nGeothite, were specifically immobilised compared and visualised. (Gil -Diaz *et al.,* 2016). Further, the impact and parameters of nZVI and nGeothite were determined on the As samples and their effect on the pH of soil, phytotoxicity and iron availability (Gil-Diaz *et al.,* 2016). The other possibility to deteriorate the soil was not observed and nGeothite showed amazing results (Mackulak *et al.,* 2015). With less dose of 0.2 per cent it also decreases about 82 per cent. The As availability with no side effects on the soil. However, it was seen that with much usage of nGeothite, it increased the volume of phytotoxicity in the soil mainly due to its electrical conductivity. The increased volume of Fe was seen in the usage of both the nanoparticles which was inevitable. Thus, the study showed the effectiveness and strength of these nanoparticles to eliminate and immobilise from the soil which is very helpful in agricultural use (Gil-Diaz, *et al.,* 2016). The property of the soil such as pH, density, humidity and iron availability has been altered by the presence of these toxic elements (Jiang *et al.,* 2018). As described above, arsenic gets accumulated in to the soil surface and to be removed. ZVI has proven to be a solution for the remediation of hazardous compounds and other materials which makes use of water and soil. Arsenic is one of the most toxic heavy metal presents on the earth. Simultaneously, it has been shown by the authors that, ZVI has the ability to work (Jiang *et al.,* 2018). The As has been has been shown in the study that nGeothite as a compound alone but with ZVI it can be useful in the removal of the drugs and arsenic as well (Mackulak *et al.,* 2015).

Removal of Soil Pollutants and Contaminants from Various Metrics

The major origin of the contaminants, whether it be constituent or inorganic present in the water and its other parts is majorly liberated from the chemical (Mackenzie and Georgie, 2019). It is very important for these products to be removed from water as it can cause several diseases to human. Here, the nZVI is used due to their metallic iron core and thin surface oxide layer and core-shell structure. It played an important role in the liquid environment (Liu and Hong, 2017). But, in order for the removal of contamination from the water, the adsorption and absorption qualities are very helpful in the removal of these harmful substances (Pasinszki and Krebz, 2020). Iron, having the properties to remove the contaminated substances is sometimes used as a reductant in the process of filtration of wastewater (Jiang *et al.*, 2018). In order to achieve better results, the nZVI has to be directly transported to the area where contamination is present in more amount. It has been seen that the activities in soil and ground water are closely related, if the soil is contaminated then the water will get too, ultimately and *vice-versa* (Mackulak *et al.,* 2015). This particular application can be helped in two ways either by applying the reactive nanoparticles in the contaminants or by adding the reactive NPs in their areas. This area of remediation was well explained by Pasinszki and Krebz (2020). In addition to these the nZVI has applications in the betterment of environment.

ZVI has the property of removing the toxicity from the soil surface and ground water surface (Wei and Zhang, 2003). Many cases have been studied where nZVI helps in the purification of soil and water. In continuation, the effect of ferrous ions on *Artemiasalina* was studied with a varying concentration of 1, 10 and 100 mg/L. The process was assessed in time interval of 96 hrs. (Liu *et al.,* 2010) The chemical and biological means were also applied and it showed variation due to the properties of ZVI. In the time interval of 96 hrs, it was seen that also the iron acting upon *Artemia salina* increased the effect of iron at highest concentration of 100 mg/L (Rajendran *et al.,* 2019). Further, the studies of *Artemia salina* can be studied for various other applications in environmental water remediation and cleaning. This plant can be used as a biomarker in the aquatic system where this toxicity can be studied (Horvathova *et al.,* 2019).

The studies have shown the impact and effect of nano-nZVI on the environmental ground water (Maldotti *et al.,* 2000). In relation to this, many *in situ* remediation of water has been studied. Water is a major source of nutrition for humans, plants and animals. Contaminant's presence in water is not a new thing to exist. It is the major reason why many parts of our country still deprived of water. Therefore, this method is a very amazing approach in this field (Karn and Kuiken, 2009; Liu and Honng, 2017). But, the drawback still holds nZVI to achieve its work in *in situ* mode, because of its higher mobility and reactivity which should not be neglected, as the nZVI works when the coating is applied to its surface. Here the nano-nZVI is to be stabilised, which make its direct contact with the surface and thereby increase the nZVI reactivity (Lefevre *et al.,* 2016). A synthetic classic borohydride synthesis pathway has been formed which includes the stabilised nano-nZVI with cellulose nano crystal (CNC) rigid nano materials (He *et al.,* 2007). This material possesses two different classical groups (-OH and -COOH). The final steps were initiated by determining the nanoscale reinforced polymeric materials and analysed the effect of the nanoscale by adding methyl orange dye as its degrading potential (Lin *et al.,* 2010). A porous media was taken and the complete process of these polymeric substances were done in the glass beads and other transport column experiments. This was performed to check the movement of nanocomposites (Lin *et al.,* 2010; Mackenzi and Georgi, 2019).

Removal of Nitrate with the Help of Zero Valent Nanoparticle

The nanoparticles were modified in such a way that the nano particles iron core structure has changed (Win–Yingr *et al.*, 2005). With the help of material studio software, the process was analysed and studied in the presence of *Thiobacillus dinitrificans* and process was simulated for the removal of nitrate. Many equations were applied for the calculation of thermodynamic properties and molecular dynamics (MD). The results were obtained by analysing the software induced calculations by the help of radial distribution function, density potential energy and temperature graphs (Latifoglu and Gurol, 2003). It was studied in the case that the simulation, both in the presence of nano particle and *Thiobacillus dintrificans* has increased the removal of nitrate (Win–Yingr *et al.*, 2005). In another type of method, the removal of nitrate is mediated at high temperature. This was brought into action with batch and chemical columns. It was firmly believed that at high elevated temperature it would help in the nitrate removal by overcoming this barrier. After this, the assumptions and case study performed in the batch mode for all of its properties and pH were studied. It was proved that nitrates can be removed at high temperature by overcoming the energy barrier (Keum and Li, 2004). Microbial related removal of nitrate was also analysed in the presence of nZVI. This was very easy to access and to visualise the growth of bacteria, they were in the flasks in which the processes were performed in the lab condition. After 3 days, the nitrate to be removed was reduced kinetically and treated with ZVI alone without addition of inoculum. The reduction volume of nitrate in the presence of nZVI was neutral (Kuscu and Sponza, 2009). Nitrate removal in terms of nZVI - M was more efficient with respect to its pH, volume, mixing rate and dilution volume. The presence of several bacterial colonies in the flasks indicates the activity of ZVI-nano particle. Therefore, it showed the most acceptable way which increased volume of nZVI and removes bacteria more efficiently in the less volume. (Wei and Zhang, 2003). Different cases shown that the nitrate concentration is lethal to human health and marine animals. Studies also showed that for the major removal of this nitrate the process of denitrification of aqueous solution in the presence of nZVI as well as the already isolated *Paracoccus* sp. strain YF1 was taken into consideration (Wei and Zhang, 2003). The factors related to the nano particles were studied and was concluded that mild toxicity to the strain by the nano particle promoted the denitrification process by providing the electron sources in the aerobic conditions. (Liu *et al.*, 2017) (Neilson *et al.*, 1994). It was concluded that major factors to enhance the properties of nano particles are temperature and pH which helps in the removal of toxic nitrate from the water (Wei and Zhang, 2003). Furthermore, various study and developments are required of the nanoparticles as to determine. It was seen that it can be removed in the absence of it, when only the strain is present (Liu, *et al.*, 2017).

Use of Recycled Nano Particle ZVI for the Removal of Nitrobenzene from Water

Nitrobenzene just as the name suggests is a chemical substance which is very harmful and toxic (Haigler and Spain, 1991). It is used as the major production source of other chemicals like aniline, dyes, pharmaceuticals *etc.* They are mainly released into the water by various chemical and other industries and can be harmful to the environment (Kuscuand Sponza, 2009) Nitrobenzene, if released in to the water in raw form then it can be very poisonous to humans, animals and plants (Haigler and Spain, 1991). Nitrobenzene is a major concern due to its toxic properties to the living systems. Nano valent iron has been studied for the removal of nitrobenzene. All of the organic and inorganic contaminants related to nitrobenzene, are removed by this process, but on the other hand there is a thing that needs to be highlighted, which is that nitrobenzene is sometimes soluble in water surface and soil surface, which usually requires the solubilised form of ZVI (Zee *et al.,* 2001; Nielson *et al.,* 1994). These methods have been used from many years to remove the contaminants from water. This is because nitrobenzene doesn't contain any electron character, which makes it repel from the microorganisms, hence not making it filtered from the water as much easily. To make it more effective and useful the oxidation processes for treatment and remediation of waste water from the nitrobenzene has been employed, for example; oxidation process (Latifogluand Gurol, 2003) and by photo-catalysis methods (Maldotti *et al.,* 2000; Nomiya *et al.,* 2001). It is seen that ZVI has greater reduction potential in eliminating the nitro groups, thus leaving only the amino groups in the water which can be later easily degraded with simple processes (Keumet *et al.,* 2004). These amino groups being left in the water are not much difficult to destroy because of less toxicity and damage to marine animals and plants. Furthermore, the various other studies were conducted to compare the activity and efficiency of the nano-particle in the removal of nitrobenzene from water. The ZVI produced from the low-grade iron oxidation was compared with produced commercially. Both of the ZVI were seen and their activities were determined in the removal of nitrobenzene (Win -Ying *et al.,* 2005).

Remediation for the Contamination and Pollutants from Surface Groundwater and Soil Surface

Worldwide soil is being a major concern as a pollutant. Among all the types of pollutions we know, soil contains toxic pollutants which are harmful to every one of us (Mareddy, 2017). The conservation of soil has become a main priority for our future generation. Important processes that have been used these years for the maintenance of soil *in situ* and *ex situ* conservation. Both of these methods are great in their own ways (Wies *et al.,* 2004). *In situ* method is applied in a way where soil is majorly untouched, with the help of variant other methodologies like nano remediation, oxidation *etc.* (Rajajayavel and Ghoshal, 2015). This process requires only short amount of time to be completed and is nearly remained untouched which is not good and healthy for humans, plants and animals. Resulting the soil remained with no changes, as the characteristics of soils are still unaltered, density still the same (Latifoglu and Gurol, 2003). As a method involved in the remediation of water and soil, *in situ* is not compatible choice, as its time consuming, expensive and requires a lot of equipment management. This could prove to be a fatal attempt to improve the health of humans (Horvathova *et al.*, 2019). Large processes are carried out at the place where we are supposed to collect the contaminated soil (Gomes *et al.*, 2013). The soil taken or chosen for the treatment has to be digged from the desired place and has to be placed in the container and treated significantly. At the end of the whole process the soil treated has to be returned back to its original place without any furthermore implications. As the contaminated soil is treated, it is generally assumed that the contaminants present in the soils are highly toxic and cannot be completely neglected (Horvathova *et al.*,2019). The main priority for this method, involves certain volume of contamination has to be reduced so that it can be suitable for further agricultural use and also for the concerning health of the humans, plants and animals (Rajendran *et al.*, 2019). Many studies have indicated that ZVI compounds are used in the treatment of soil and water. One of the main processes which comes under this is the technology of nano particle nano remediation, which has proven to be the most effective in this case (Horvathova *et al.*, 2019).

Conclusions and Future Perspectives

The nZVI is known to be a well good absorbent and a source for the removal of organic and inorganic substances from the underground water and soil. It is a best way for the elimination of all the other toxic and unwanted contaminants present in the water and on the surface of the soil (Mackulak *et al.*, 2015). nZVI as the name suggests has been proven its worth in many studies. Its promising properties like higher mobility, higher adsorption activity, higher reactivity as well act as the main source of a better component for this purpose (Pasinszki and Krebz, 2020). For the last 100 years, the pollutant and degradation has been done to the environment due to the activities of chemical industries and also by the human activities. All of the things that profits human destroys environment in many ways like that. And in the midst of these problem, ZVI has come out to be the best and advantageous source for the good of everything. Iron being the main particle source for the removal of these contaminants and drugs from the water and soil has proven its worth even more in this area. (Mackenzie and Georgie, 2019). As compared to other particles present which also performs in the same way as NPs does it, it is much more practical to use nZVI. This is because many studies and cases as, itself has proven that how much ZVI is compatible in removal of these toxics. It injected into the main source of contamination and replace the toxicity with the iron. But as we talk about the iron, it is also not very much safe and healthy for the underground water, waste water and soil surface for much concentration of iron present in them (Mackulak *et al.*, 2015). The nZVI particles are of course known for their very much amazing properties as discussed above and to name some other of it like low equipment cost price, large specific surface area as well as low operational cost. But as we talk about the positive impact of its properties it has also very degrading effects also like the property of aggregation its impact of adsorption on the surface of water and soil. In today's period of time many researchers are still finding answers to these particular properties, so that they can also overcome this and can act it out properly. The nanoparticles are seeming be the promising source for the to remove the toxic materials from the water and soil surface. The future researches should see their widespread use and production. Water and soil remediation needed to find out the ways for the improvement in the economic efficiency of nonmaterial.

REFERENCES

Aycicek, M., Kaplan, O., Yaman, M. 2008. Effect cadmium on germination, seedling growth and metal contents of sunflower (*Helianthus annus* L.), *Asian J Chem.* PAGE NOAND DOI

Briggs, D. 2003. Environmental pollution and the global burden of disease. *Br. Med. Bull.* 68: 1-24. DOI: 10.1093/bmb/ldg019.

Cilai T, Yingping H, Zengqiang Z, Jianjun C, Hui Z, Yong H. 2016.Rapid removal of selenate in a zero-valent iron/Fe3O4/Fe2+ synergetic system, Applied Catalysis B: Environmental, DOI: 10.1016/j.apcatb.2015.11.045.

Gil-Díaz M, Diez-Pascual S, González A, Alonso J, Rodríguez-Valdés E, Gallego JR, Lobo MC. 2016. A nano remediation strategy for the recovery of an As-polluted soil. Chemosphere. DOI: 10.1016/j.chemosphere.

Gomes, H. I., Dias-Ferreira, C., and Ribeiro, A. B. (2013). Overview of in situ and ex situ remediation technologies for PCB-contaminated soils and sediments and obstacles for full-scale application. *The Science of the total environment, 445-446*, 237–260. https: //doi.org/10.1016/j.scitotenv.2012.11.098

Grigori, Z., Raphael, S., 2013. Selenium removal from water and its recovery using iron (Fe3+) oxide/hydroxide-based nanoparticles sol (NanoFe) as an adsorbent, Separation and Purification Technology. DOI: 10.1016/j.seppur.2012.10.037.

Haigler B., Spain J., 1991.Biotransformation of nitrobenzene by bacteria containing toluene degradative pathways. Appl Environ Microbiol. DOI: 10.1128/ AEM.57.11.3156-3162.1991.

He, F., Zhao, D., Liu, J. and Roberts, C.B. 2007. Stabilization of Fe–Pd Nanoparticles with Sodium Carboxymethyl Cellulose for enhanced transport and dechlorination of Trichloroethylene in soil and groundwater, *Ind Eng Chem Res.* DOI: 10.1021/ie0610896

Hegedüs, A., Erdei, S., Horváth, G. 2001. Comparative studies of H(2)O(2) detoxifying enzymes in green and greening barley seedlings under cadmium stress. *Plant Science* DOI: 10.1016/s0168-9452(01)00330-2.

Horváthová, H., Lászlová, K. and Dercová, K.2019. Bioremediation vs. Nanoremediation: Degradation of Polychlorinated Biphenyls (PCBS) Using Integrated Remediation Approaches. Using Integrated Remediation Approaches. *Water Air Soil*. Doi.10.1007/s11270-019-4259-xhttps: //doi.org/10.1016/S1381-1169(01)00258-8.

Jiang D, Zeng G, Huang D, Chen M, Zhang C, Huang C, Wan J. 2018. Remediation of contaminated soils by enhanced nanoscale zero valent iron. Environ Res. 2018 DOI: 10.1016/j.envres.2018.01.030.

Jing, C, Li Y., and Landsberger S. 2016.Review of soluble uranium removal by nanoscale zero valent iron. J Environ Radioact. DOI: 10.1016/j.jenvrad.2016.06.027.

Karn, K., Kuiken, T. 2009. Nanotechnology and *in Situ* Remediation: A Review of the Benefits and Potential Risks, Environmental health perspectives National Institute of environmental health sciences DOI: 10.1289/ehp.0900793

Kaznina, N.M., Laidinen, G.F., Titov AF., Talanov, A,V. 2005. Effect of lead on the photosynthetic apparatus ofannual grasses. *The Biological Bulletin* 32(2): 147-150

Keum, Y. S., and Li, Q. X. (2004). Reduction of nitroaromatic pesticides with zero-valent iron. *Chemosphere, 54*(3), 255–263. https: //doi.org/10.1016/j.chemosphere.2003.08.003

Kowalski, K. P., and Søgaard, E.G. 2014. Implementation of zero-valent iron (ZVI) into drinking water supply - role of the ZVI and biological processes. *Chemosphere, 117*, 108–114. DOI: 10.1016/j.chemosphere.2014.05.088

Kuscu, O. S., and Sponza, D. T. (2009). Effects of nitrobenzene concentration and hydraulic retention time on the treatment of nitrobenzene in sequential anaerobic baffled reactor (ABR)/continuously stirred tank reactor (CSTR) system. *Biores. Technol.*, 100: 2162–2170. https: //doi.org/10.1016/j.biortech.2008.10.049

Laspina,N.V., Groppa, M.D., Tomaro, M.L. 2005. Nitric oxide protects sunflower leavesagainst Cd-induced oxidative stress, *Plant Science*: An international journal of Experimental Plant Biology. DOI: 10.1016/j.plantsci.2005.02.007.

Latifoglu, A., and Gurol, M. D. (2003). The effect of humic acids on nitrobenzene oxidation by ozonation and O3/UV processes. *Water research, 37*(8), 1879–1889. https: //doi.org/10.1016/S0043-1354(02)00583-3

Lefevre, E.; Bossa, N.; Wiesner, M.R.; Gunsch, C.K. 2016. A review of the environmental implications of in situ remediation by nanoscale zero valent iron (nZVI): Behavior, transport and impacts on microbial communities. Sci Total Environ. doi: 10.1016/j.scitotenv.2016.02.003

Lin YH, Tseng HH, Wey MY, Lin MD. 2010.Characteristics of two types of stabilized nano zero-valent iron and transport in porous media. Sci Total Environ. DOI: 10.1016/j.scitotenv.2010.01.039

Liu, P., Honng, H. 2017. Magnetic Nanomaterials for Water Remediation. In: Eds.

Mackenzie, K., Georgi, A. 2019. NZVI Synthesis and Characterization. In: Phenrat T., Lowry G. (eds) Nanoscale Zerovalent Iron Particles for Environmental Restoration. Springer pp. 45-95 doi.10.1007/978-3-319-95340-3_2

Macku¾ak, T., Mosný, M., Grabic, R., Golovko, O., Koba, O., and Birošová, L. 2015. Fenton-like reaction: a possible way to efficiently remove illicit drugs and pharmaceuticals from wastewater. Environ ToxicolPharmacol. *39*, 483–488. https: //doi.org/10.1016/j.etap.2014.12.016

Maldotti, A., Andreotti, L., Molinari, A.,Tollari, S.,Penoni, A.,Cenini, S., 2000 Photochemical and photocatalytic reduction of nitrobenzene in the presence of cyclohexene,Journal of Photochemistry and Photobiology A: Chemistry,133, 2000, 129-133, DOI: 10.1016/S1010-6030(00)00212-4.

Mareddy, A.R. 2007.Environmental Impact Assessment, Butterworth-Heinemann eBook ISBN: 9780128112380

Nielsen H., Thomas, H.and Christensen. 1994.Variability of biological degradation of phenolic hydrocarbons in an aerobic aquifer determined by laboratory batch experiments, *J. Contaminant Hydrol.* DOI: 10.1016/0169-7722(94)90077-9.

Nomiya, K.,Hashino, K., Nemoto, Y., Watanabe, M., 2001. Oxidation of toluene and nitrobenzene with 30 per cent aqueous hydrogen peroxide catalyzed by vanadium(V)-substituted polyoxometalates. J Mol Catal A: Chemical, 176: 79-86.

Pasinszki, T., and Krebsz, M. (2020). Synthesis and Application of Zero-Valent Iron Nanoparticles in Water Treatment, Environmental Remediation, Catalysis, and Their Biological Effects. *Nanomaterials 10*(5), 917. https: //doi.org/10.3390/nano10050917

Rajajayavel, S. R., Ghoshal, S. 2015. Enhanced reductive dechlorination of trichloroethylene by sulfidated nanoscale zerovalent iron. *Water Research, 78,* 144–153. https: //doi.org/10.1016/j.watres.2015.04.009

Rajendran, S., Naushad, M., Raju, K. and Boukherroub, R., 2019 Emerging Nanostructured Materials for Energy and Environmental Science. Part of the Environmental Chemistry for a Sustainable World book series. DOI: 10.1007/978-3-030-04474-9

Raymond A,.W., FelixE., O., 2011. Heavy Metals in Contaminated Soils: A Review of Sources, Chemistry, Risks and Best Available Strategies for Remediation. DOI: 10.5402/2011/402647

Singh, R., Gautam, N., Mishra, A., Gupta, R. 2011. Heavy metals and living systems: An overview. *Indian J.Pharmacol.* DOI: 10.4103/0253-7613.81505.

Soto-H., 2018. Nanoscale Zero Valent Iron for Environmental Cadmium Metal Treatment. DOI: 10.5772/intechopen.72737 van-der Zee, F. P., Lettinga, G., and Field, J. A. (2001). Azo dye decolourisation by anaerobic granular sludge. *Chemosphere*, 44(5), 1169–1176. https: //doi.org/10.1016/s0045-6535(00)00270-8

Weis, J.S., Weis, P., 2004. Metal uptake, transport and release by wetland plants: implications for phytoremediation and restoration. *Environ.Int.* DOI: 10.1016/j.envint.2003.11.002.

Wei-X., Zhang. 2003. Nanoscale Iron Particles for Environmental Remediation: An Overview. *J Nanoparticle Res*, DOI: 10.1023/A: 1025520116015.

Wen-Ying X., Ting-Yao G., Jin-H., 2005. Reduction of nitrobenzene by the catalyzed Fe–Cu process, *J Hazard Mat.* DOI: 10.1016/j.jhazmat.2005.04.011.

Wiley-VCH Verlag GmbH and Co. KGaA, Weinheim, Germany, pp. 515–546. DOI: 10.1002/9783527803255.ch16

Chapter 10

Microbial Endophytes as Biofertilisers: Sustainable and Promising Remedy for Ailing Soils

*Kajal Chauhan[1], Disha Chahal[2] and Sulekha Chahal[1]**

[1]*Department of Biotechnology,*
Kurukshetra University, Kurukshetra – 136 119, Haryana
[2]*School of Sciences, Department of Geography,*
Indira Gandhi National Open University, New Delhi – 110 068
**e-mail: schahalbiotech@kuk.ac.in*

ABSTRACT

In the last century, intensive and indiscriminate usages of chemical fertilizers to increase yield and productivity of agricultural crops have been a common practice. Though chemical fertilizers definetely boost up crop production, but their long term usage toxicated the soil, strengthened pesticides, polluted air and water, released greenhouse gases, and thereby generated negative impact on human health and environment. Biofertilizers are sustainable and promising alternative for chemical fertilizers, which have the potential for enhancing yield to meet the demands of the growing human population at the same time generate conditioning effect on ailing soil. Endophytes (microbes that reside inside the host plant) are the naturally occurring agriculturally important microorganisms, which have cost efficient biofertilizing potential and are also environment friendly. They provide host with the variety of beneficial attributes for example, secrete plant growth regulators and have nitrogen fixing and phosphate solubilising abilities and much more. This review also sees the sights of endophytic distribution prototypes regarding their geographical locations, colonising behaviour and potentials for providing beneficial attributes to the host. Furthermore, explores the latest advancements of endophytic research including applications and limitations of using endophytes as biofertilisers.

Keywords: *Biofertilisers, Endophytes, Agriculture, Chemical fertilizers, Microorganisms*

Abbreviations

PGPRs: Plant Growth Promoting Rhizobias; GFP: Green Flourescent Protein; BNF: Biological Nitrogen Fixation; GAs: Gibberellic Acids; JA: Jasmonic Acids; ABA: Abcissic Acid; ACCD: 1-Aminocyclopropane-1-Carboxylic Acid Deaminase

Introduction

Agriculture is the art as well as science of cultivating plants and domesticating animals and its establishment is considered, a major turning point in human evolution and civilization *i.e.* transition from a nomadic hunter-gatherer lifestyle dependent on foraging and hunting to settled life. Agricultural practices allow humans to have surplus food throughout the year and relocation and migration for searching food was no longer required. As a result, humans were able to establish communities, villages, towns and civilizations. This shift in survival mode provided a relatively safer existence and in general more leisure time for analytical and creative pursuits resulting in complex language development, and the accelerated evolution of art, religion, and science. However, developments of agriculture also amplify population. This expanding population places a heavy burden on agricultural system. To produce higher yields of crops, agricultural revolution were required. The green revolution thereby was intended to overcome food shortages in India by increasing the yields of agricultural produce with the help of better irrigation systems, pesticides, fertilizers, agricultural machinery, *etc.* And, although yields increased substantially due to green revolution but reciprocally, it also gifted many unpleasant after-effects. The increased use of chemical fertilisers paved way for their bio-accumulation and biomagnifications in soil and environment (Mishra *et al.,* 2015; Bisen *et al.,* 2015; Keswani *et al.,* 2014). Their continuous and long term usage containment groundwater, decrease soil fertility, increased soil salinity and sodicity and leads to the biodiversity loss. The chemical fertiliser's incorporation in to food chains reflected through fatal health and diseases state of flora and fauna of respective regions. Under such circumstances, the biofertilisers can be a desired alternative for chemical pesticides which can increase food production on one side and pose no harmful effect to the environment and human beings on the other side. The other advantages of biofertilisers are that they are low in cost, can be used in the form of spray therefore easy to apply, fix atmospheric nitrogen in the form usable for the host, can convert insoluble phosphate in the form suitable for the plant, produce phytohormones to stimulate plant growth hence provide nutrition to the soil and balance the fertility status of soil.

Microbial Endophytes

Endophytes are the microorganisms which lives inside the vascular plants (Fadiji and Babalola, 2020a; Brader *et al.,* 2017). Their intimate contact with the host is the attention seeker point which makes them suitable for acting as biofertilisers. Due to this intimacy, the endophytes are protected from the biotic and abiotic factors of the environment (Hallmann *et al.,* 1997; Rajkumar *et al.,* 2009). This leads to their unrestricted growth in host plant. This is a competitive advantage of endophytes over the PGPRs. The re-introduction of endophytes does not cause any change in the indigenous bacterial populations of the host whereas PGPRs causes change or shift in the microbial populations (Conn *et al.,* 2004). So, endophyte serves as a better option.

Endophytes are the microorganisms (bacteria and fungi) which colonise the host intracellularly and intercellularly asymptomatically (Wilson, 1995). Plant restricts the growth of endophytes but the endophytes acquire the capability to find a new way for their adaptation in the restricted conditions (Dudeja *et al.,* 2012).

Distinct endophytes may be located in varying host plant parts, for example, roots, stem and leaves (Fürnkranz *et al.,* 2012). In the course of association with the host plant, endophytes evolve novel ways to establish this association (Goyal *et al.,* 2016). The most favourable place of location of endophytes is roots followed by stem and least in the leaves.

Geographical locations and season also affect the pattern of distribution of endophytes. Some reports suggested that a seasonal fluctuation do affect the growth and presence of the endophytes in host plants (Gao *et al.,* 2005; Forchetti *et al.,* 2010) therefore, different types of endophytic species are likely to inhabit plant in different seasons. Soil type and rainfall distribution are major factors which determines the diversity of endophytes in plants (Hardoim *et al.,* 2012). The *Bacillus, Pseudomonas* and *Azospirillum* species which generally present in soil, usually found to be associated with plants as endophytes. In the study conducted by Duhan *et al.* (2020) on medicinal plant *Tinospora cordifolia*, the samples were collected during summer and winter seasons for one year of Kurukshetra region. The Kurukshetra district lies between 29°- 3401500 and 30°_1501500 North latitude and 76° to 1000 and 77°_17.500 East longitude. This region has a sub-tropical continental climate. Environmental conditions of this location also affect the distribution of endophytes. The rainfall distribution annually is comparatively satisfactory *i.e.* 290.5 mm. Saraswati, Markanda and Ghaggar are the central rivers of this region and the soil is commonly alluvial loam and clay. It is one of the affluent districts from an agricultural point of view. They have identified species from various genera such as *Bacillus, Aneurinibacillus* and *Pseudomonas*. Some different species were also reported from the similar plant of different geographical locations. The type of species screened can also be influenced by the culture medium used for isolation of bacterial endophytes from the plant tissue.

Endophytes can furthermore be classified on the basis of the method they use while living inside the host as follows-obligate endophytes, facultative endophytes and passive endophytes. Facultative endophytes are the one most preferable in plant growth promoting activities. This group resides in the soil freely and can approach the host roots in slight opportunity. Obligate endophytes are those which are transferred through seeds. They do not reproduce outside the host tissue. Passive endophytes are the one which colonise the host due to the effect of some events like wounds in root hairs instead of original colonisation (Hardoim *et al.,* 2015).

Endophytes colonise the host plant primarily through roots and can distribute systemically in the whole plant via vascular tissues. It is a site from where they launch into their proliferation. In an experiment, the distribution of endophytes was visualised by labelling the root and shoot tisuses with GFP – tagging methodology. As a result it was concluded that endophytes introduced in stem moved to the roots suggesting the uninterrupted distribution of endophytes in root microbiome (Johnston-Monje and Raizada, 2011).

Isolation and Identification of Endophytes

After a plant is selected, plant material can be collected and placed in sealed plastic bags to store at 4°C until isolation procedures begin. In the laboratory, plant materials are thoroughly surface sterilized to eliminate surface-contaminating microbes. Then, outer tissues are removed from the samples and the inner tissues are excised and placed on appropriate media. The endophytes are encouraged to sporulate on specific plant materials. After several days of incubation, hyphal tips of the fungi are removed and transferred to potato dextrose agar. Bacterial forms also emerge from the plant tissues and transferred on nutrient agar media. Eventually, when an endophyte is acquired in pure culture by subsequent subculturing, it can be identified via standard morphological and molecular biological techniques and methods. Further, isolated microbe can be tested for desired characteristics/potential and on confirmation protocol for mass multiplication can be standardized. Consequently, further their formulation for using them as biofertizer can be finalized.

Applications of Endophytes as Biofertilisers

See Figure 10.1

Nitrogen Fixation

Artificial nitrogen fertilisers available in market are expensive in manufacturing and are also one of the main causes of soil and groundwater pollution with nitrate contamination (Spaink *et al.,* 2012). Nitrogen can be fixed by the endophytes without formation of nodule like structure within the host plant (Ladha and Reddy, 2003). There are some endophytic bacteria that naturally have BNF genes in their genome which convert nitrogen into plant usable forms such as nitrate and ammonia (Bhattacharjee *et al.,* 2008).

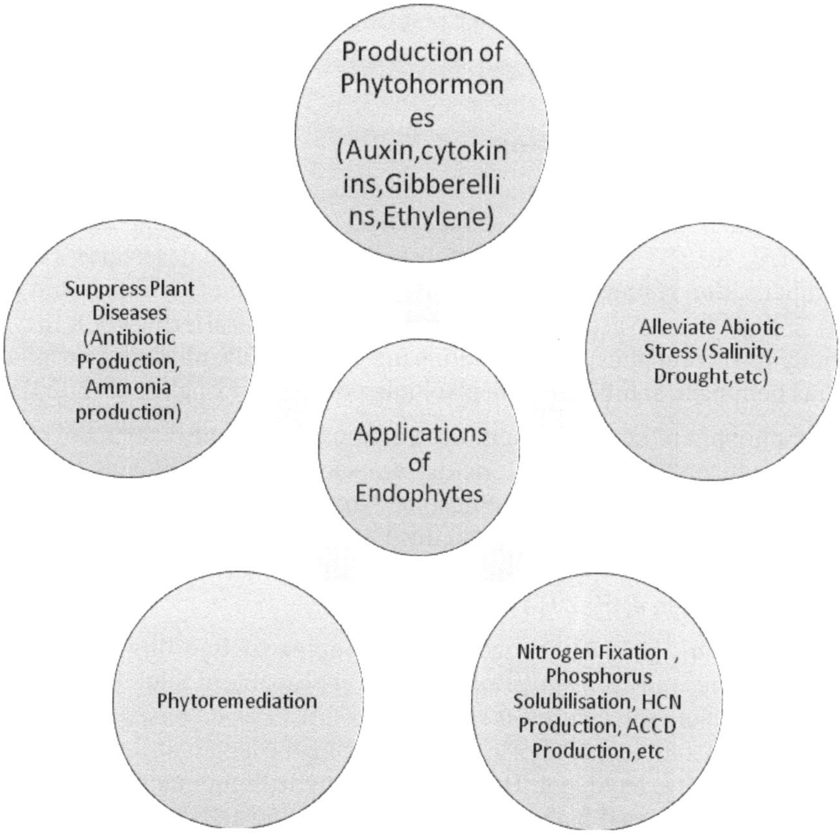

Figure 10.1: Showing Various Applications of Endophytes in Various Fields.

Triplett (1996) conducted an experiment on *Saccharum officinarum* L. The plants were cultivated without using artificial nitrogen fertilisers but were inoculated by diazotrophic endophytes like *Herbaspirillum seropedicae* and *Acetobacter diazotrophicus.* As a result, host plants were able to complete its nitrogenous needs by using solely atmospheric nitrogen. Andrade *et al.* (2014) suggested that out of 20 isolates from banana roots (Musa L) which had the prospective of growing on nitrogen free media, there were 4 strains belonging to *Bacillus spp* (Firmicutes) showed the capability of fixing atmospheric nitrogen. This ability was checked by kjeldahl and acetylene reduction assay methods. And also, the same strains exhibited other plant growth promoting abilities like phosphorus solubilisation and IAA production. The nitrogen fixing bacteria namely, *Herbaspirillum* spp., *Enterobacter cloacae, Burkholderia* spp., *Azospirillum* spp., *Klebsiella oxytoca, Pantoea* sp and *Klebsiella pneumonia* were isolated from the roots, leaves and stems of sugarcane plant by Govindarajan *et al.,* 2007,2008 and Mendes *et al.,* 2007. A nitrogen-fixing bacterium, *Klebsiella variicola* was isolated from the ROC22 sugarcane cultivar roots which has the ability of fixing atmospheric nitrogen in correlation with sugarcane plants under aseptic conditions, promoting plant uptake of N, K, and P and GT21 cultivar growth under greenhouse conditions (Wei *et al.,* 2014).

Phosphorus Solubilisation

Plants use phosphate in monobasic and dibasic form from the soil, but most of the phosphate in the soil is in unusable form *i.e.,* immobilised and insoluble form (Gouda *et al.,* 2018). That is why, artificial or chemical phosphate fertilisers are used for compensation of phosphate deficiency in the host plant (Rodriguez and Fraga, 1999). Endophytes can act as an alternative to these chemical fertilisers as they have an ability to secrete phosphorus solubilising enzymes like phytase, phosphatase, *etc.* thus helping in solubilising the insoluble phosphate (Singh *et al.,* 2020).

For phosphate solubilisation, acetic, fumaric, malonic, tartaric, glycolic, citric, and gluconic acid (organic acids) are secreted by endophytes as a main mechanism (Heijden *et al.,* 2008). Phosphate uptake was promoted in the cassava plant roots (*Manihot esculenta* Crantz) due to the salicylic acid and benzene acetic acids secretion by endophyte *Pantoea dispersa* for more effective phosphorus solubilisation (Chen *et al.,* 2014).

Verma *et al.* (2015) recorded two species of Bacillus strain *i.e.,* B. *amyloliquefasciens* and B. *megaterium*, which were capable of solubilising insoluble zinc and potassium salts. Kurek *et al.* (2013) isolated *Pseudomonas letiola* capable of mobilizing insolublephosphorus and enhancing the length of shoot and growth of the apple tree variety Ligol. The three endophytic Pseudomonas strains L111, L228, and L321 showed high phosphorus solubilising activity isolated from the leaves of *Miscanthus giganteus* (Oteino *et al.,* 2015).

The faster growth and the better performance of the host plant named *Phyllanthus amarus* due to the better uptake of phosphorus, promotion of germination, and stimulation of the immune system was observed to be exhibited by the two phosphate-solubilizing and salt-tolerant bacteria (*Bacillus* sp. and *Acinetobacter* sp.) by enhancing the biosynthesis of antioxidative enzymes, phenolic compounds and the radical scavenging system (Joe *et al.,* 2016).

Phytohormones Production

Endophytes have the potential to generate plant hormones for example - auxins, cytokinins, gibberellins, ethylene, *etc.* An endophyte namely, *Burkholderia vietnamiensis* was isolated from wild cotton wood (*Populus trichocarpa*) which was capable of secreting IAA leading to plant growth promotion. As a result, the inoculated plants were having more nitrogen content and dry weight biomass than the control ones (Xin *et al.,* 2009).

Porostereum spadiceum AGH786 is an endophytic fungus which secreted phytohormones in host plant (soyabean seedling) when inoculated under NaCl stress. Gibberellic Acids, Jasmonic Acids, Abcissic acid and Isoflavones were secreted by this fungal endophyte but GAs were more in secretion when compared with control (Hamayun *et al.*, 2017).

Endophytic bacteria such as *Micromonosporaspp., Rhizobium spp., Microbacterium trichothecenolyticum, Brevibacillus choshinensis and Endobacter medicaginis* have been studied for their presence in the root nodules of host plant (Igiehon and Babalola, 2018).

The endophytic fungi isolated from *Asclepias sinaica* and recognized as *Alternaria alternata* and *Penicillium chrysogenum* (Fouda *et al.*, 2015). The results showed that the isolates were secreting ammonia and IAA leading to improved root elongation and growth. According to the studies by Abdallah *et al.* (2016) an endophyte *Alcaligenes faecalis* isolated from *Withania somnifera* was capable of producing IAA and enhance phosphate solubilisation. The main function of IAA is cell elongation and differentiation (Asgher *et al.*, 2015). Common endophytic IAA producers belong to the bacterial genera *Azospirillum, Azotobacter, Alcaligenes, Herbaspirillum, Enterobacter, Pseudomonas, Klebsiella, Rhizobium, Burkholderia, Pantoea, Bacillus, Acetobacter* and *Rhodococcus* (Rashid *et al.*, 2012; Duca *et al.*, 2014; Apine and Jadhav, 2011; Costacurta and Vanderleyden, 1995).

Ethylene, a stress hormone is involved in various stress responses. Bacterial endophytes can possess 1-ACCD which formulates nitrogen sources α-ketobutyrate and ammonia from ACC which is an ethylene precursor (Sun *et al.*, 2009). For example, *Pseudomonas brassicacearum* SVB6R1 secrete ACCD in sorghum plants thereby enhancing salt stress tolerance (Gamalero *et al.*, 2020). Moreover, *Paenibacillus polymyxa* from the bulbs of *Lilium lancifolium* promotes the growth of two *Lilium* varieties through secretion of ACCD, among other effects (Khan *et al.*, 2020).

An endophyte, *Azospirillum lipoferum* was introduced in maize plant treated with gibberellins inhibitor synthesis and subjected to stress (drought) conditions. After inoculation, the host plant was performing well as compared to the control one under stress conditions because of the bacterial gibberellins (Cohen *et al.*, 2009).

Siderophore Production

The essential co-factor for number of enzymes accounting for the biochemical process of fixing nitrogen in nodules is iron but iron in the soil is accessible in the unusable form to the host plant. Soil iron is available in the insoluble ferric form of oxides, carbonates, phosphatase and hydroxides (Ma *et al.*, 2016). During iron deficiency, endophytes can produce siderophores which can bind divalent metal ions like Ferrous ion, ferric ion, *etc.* (Saha *et al.*, 2016). Sideropores help in prevention of uptake of undesirable metals like Zn, Cd, Cr, Pb, and Al, *etc.* through plant roots by scavenging these undesired metals present in rhizosphere thereby reducing toxicity for the host (Neubauer *et al.*, 2000). In the condition of iron limitation, an experiment showed the growth promotion of common bean by siderophores secreted by two endophytic *Bacillus* spp. strain (Sabate *et al.*, 2018).

The citrus endophyte *methylobacterium mesophilicum* released siderophores which when applied to *Xylella fastidiosa*, it promoted the chlorophyll content and growth of the plant (Lacava *et al.*, 2008).

Stress Tolerance

Both biotic and abiotic stresses are responsible for 30-50 per cent agriculture loss all over the world (Kumar and Verma, 2018). The major abiotic stresses include salinity, temperature, drought, trace metals, flooding, *etc.* So we need to have environment– friendly method to overcome these problems. Endophytes serve a very good option so that host plant could be stronger and healthier having a good immune system. Endophytic bacteria help plants to overcome drought stress via the production of volatile compounds, abscissic acid, ACC-deaminase, and IAA (Ullah *et al.*, 2019). Afridi *et al.* (2019) reported the improved growth and stress tolerance of two wheat varieties inoculated with ACC-producing bacterial endophytes.

Plants under stress produce exceeding amount of ethylene which is synthesised from ACC (Khan and Khan, 2014; Tiwari *et al.*, 2018). ACC-deaminase produced by the endophyte can alleviate the impact of elevated ethylene concentrations on stressful plants through the hydrolysis of ACC into α-ketobutyrate and ammonia. The ammonia and energy released from this process could be used by host for growth (Glick, 2014; Ali *et al.*, 2014).

Challenges and Limitations in the Use of Endophytes

Most endophytes can be isolated and screened by using customary microbiology techniques but some are hard to culture. So, un-culturable endophytes offer major challenge during structure and diversity measurement and identification process of endophytic community. Berg *et al.* (2014) suggested the use of fluorescence confocal microscopy and *in situ* hybridation techniques to explore endophytes. The other modern biotechnological tools such as a polymerase chain reaction, DNA fingerprinting and cloning techniques can be helpful to detect uncultivable species. The use of 16S and 18 S rRNA gene sequencing is a reliable and convenient way to survey bacterial and fungal diversity and to resolve the phylogenetic relationships among genera or species staring from lower levels (Nair and Padmavathy, 2014). The ITS sequences and large subunit analysis were exploited as tools to identify non-sporulating fungi and to confirm the identified fungi from Basidiomycetes. Furthermore, the bioinformatics tools such as genomic, proteomics and metagenomics can help in identification and to predict the functions of culturable and nonculturable microbial endophytes present in the parent samples (Brader *et al.*, 2017). So, the collective usage of conventional culture-dependent methodology along with modern culture-independent advances can help to discover abundant endophytic species which are difficult to notice using cultivable routine methodology.

Moreover, there is need of elaborate studies to screen competent endophytes that have extensive application as biofertilizers and biopesticides for specific agricultural crops. This is because numerous challenges are faced when isolated endophytes are applied to field crops. The most noticeable limitation is the poor plant–microbial endophytic interaction due to reduced rhizospheric competence in the presence of prevalent preexisting microbes (Schulz *et al.*, 2002). The endophytic flora also gets influenced by change in the season, climate, rainfall and soil physical, chemical and biological properties.

Apart from evaluating the potential of endophytes as biofertilizers other parameters such as marketing, optimum formulation, production methods and mode of application are some other restrictive factors for using these microbes for farming practices. One more fear is that endophytes may act as opportunistic pathogens and their subsequent application can cause mild to severe diseases in plants, animals or humans which may also create epidemic and pandemic scenario. So, judicious formulation, application- mode and methodology should first be completely workout before their commercial release for application as biofertilizers to agricultural crops.

Conclusion and Future Outlook

In the present scenario, there is an emergent need for new agrochemicals that are highly effective, less expensive, possess low toxicity, and have minor environmental impact. Endophytes pose all these attributes *i.e.* plant growth promoting prospective for maintaining vibrant plant physiology, restoring nutrition availability, eco-friendly nature and sustainable way to enhance yield of crop plants. Considering the significance of endophytes, more studies should be done to explore the way the endophytes react with the plant host in order to ascertain the best mode to make them effective for continuous crop production. Besides, most endophytes were screened through culture-dependent methods; there is still a need to investigate culture-independent techniques such as genomics and metagenomics, in order to explore more novel functions and species. Also, the mechanisms of action and interactions between endophytes and host plant and with other microbes in the tissue of plants are not fully understood. Factors influencing patterns of distribution are not fully understood as endophytic species differ from one plant to the other. Perhaps, elaborate research using metagenomics and bioinformatics tools could help us to have insight in to the mechanisms involved in plant–microbe interactions, evolutionary relationship and establish genomic determinants of endophyte lifestyle and diversity.

REFERENCES

Abdallah, R. A. B., Mejdoub-Trabelsi, B., Nefzi, A., Jabnoun-Khiareddine, H., and Daami-Remadi, M. (2016). Isolation of endophytic bacteria from Withania somnifera and assessment of their ability to suppress Fusarium wilt disease in tomato and to promote plant growth. *J Plant Pathol Microbiol*, *7*(5).

Afridi, M. S., Mahmood, T., Salam, A., Mukhtar, T., Mehmood, S., Ali, J., and Chaudhary, H. J. (2019). Induction of tolerance to salinity in wheat genotypes by plant growth promoting endophytes: Involvement of ACC deaminase and antioxidant enzymes. *Plant Physiology and Biochemistry, 139*, 569-577.

Ali, S., Charles, T. C., and Glick, B. R. (2014). Amelioration of high salinity stress damage by plant growth-promoting bacterial endophytes that contain ACC deaminase. *Plant Physiology and Biochemistry, 80*, 160-167.

Andrade, L. F., de Souza, G. L. O. D., Nietsche, S., Xavier, A. A., Costa, M. R., Cardoso, A. M. S. and Pereira, D. F. G. S. (2014). Analysis of the abilities of endophytic bacteria associated with banana tree roots to promote plant growth. *Journal of Microbiology, 52*(1), 27-34.

Apine, O. A., and Jadhav, J. P. (2011). Optimization of medium for indole 3 acetic acid production using Pantoea agglomerans strain PVM. *Journal of applied microbiology, 110*(5), 1235-1244.

Asgher, M., Khan, M. I. R., Anjum, N. A., and Khan, N. A. (2015). Minimising toxicity of cadmium in plants—role of plant growth regulators. *Protoplasma, 252*(2), 399-413.

Bhattacharjee, R. B., Singh, A., and Mukhopadhyay, S. N. (2008). Use of nitrogen-fixing bacteria as biofertiliser for non-legumes: prospects and challenges. *Applied microbiology and biotechnology, 80*(2), 199-209.

Berg, G., Mahnert, A., and Moissl-Eichinger, C. (2014). Beneficial effects of plant-associated microbes on indoor microbiomes and human health. *Front. Microbiol.* 5: 15. doi: 10.3389/fmicb.2014.00015

Bisen, K., Keswani, C., Mishra, S., Saxena, A., Rakshit, A., and Singh, H. B. (2015). Unrealized potential of seed biopriming for versatile agriculture. In *Nutrient use efficiency: from basics to advances* (pp. 193-206). Springer, New Delhi.

Brader, G., Compant, S., Vescio, K., Mitter, B., Trognitz, F., Ma, L. J., and Sessitsch, A. (2017). Ecology and genomic insights into plant-pathogenic and plant-nonpathogenic endophytes. *Annual Review of Phytopathology, 55*, 61-83.

Chen, Y., Fan, J. B., Du, L., Xu, H., Zhang, Q. H., and He, Y. Q. (2014). The application of phosphate solubilizing endophyte Pantoea dispersa triggers the microbial community in red acidic soil. *Applied soil ecology, 84*, 235-244.

Cohen, A. C., Travaglia, C. N., Bottini, R., and Piccoli, P. N. (2009). Participation of abscisic acid and gibberellins produced by endophytic Azospirillum in the alleviation of drought effects in maize. *Botany, 87*(5), 455-462.

Conn, V. M., and Franco, C. M. (2004). Analysis of the endophytic actinobacterial population in the roots of wheat (Triticum aestivum L.) by terminal restriction fragment length polymorphism and sequencing of 16S rRNA clones. *Applied and environmental microbiology, 70*(3), 1787-1794.

Costacurta, A., and Vanderleyden, J. (1995). Synthesis of phytohormones by plant-associated bacteria. *Critical reviews in microbiology, 21*(1), 1-18.

Duca, D., Lorv, J., Patten, C. L., Rose, D., and Glick, B. R. (2014). Indole-3-acetic acid in plant–microbe interactions. *Antonie Van Leeuwenhoek, 106*(1), 85-125.

Dudeja, S. S., Giri, R., Saini, R., Suneja Madan, P., and Kothe, E. (2012). Interaction of endophytic microbes with legumes. *Journal of Basic Microbiology, 52*(3), 248-260.

Duhan P., Bansal, P., and Rani S. (2020). Isolation, identification and characterization of endophytic bacteria from medicinal plant *Tinospora Cordifolia.* South African Journal of Botany,134: 43-49. https: //doi.org/10.1016/j.sajb.2020.01.047.

Fadiji, A. E., and Babalola, O. O. (2020). Elucidating mechanisms of endophytes used in plant protection and other bioactivities with multifunctional prospects. *Frontiers in Bioengineering and Biotechnology, 8*, 467.

Forchetti, G., Masciarelli, O., Izaguirre, M.J., Alemano, S., Alvarez, D., Abdala, G.(2010). Endophytic bacteria improve seedling growth of sunflower under water stress, produce salicylic acid, and inhibit growth of pathogenic fungi. Current Microbiology 61, 485–493.

Fouda, A. H., Hassan, S. E. D., Eid, A. M., and Ewais, E. E. D. (2015). Biotechnological applications of fungal endophytes associated with medicinal plant Asclepias sinaica (Bioss.). *Annals of Agricultural Sciences, 60*(1), 95-104.

Fürnkranz, M., Lukesch, B., Müller, H., Huss, H., Grube, M., and Berg, G. (2012). Microbial diversity inside pumpkins: microhabitat-specific communities display a high antagonistic potential against phytopathogens. *Microbial ecology, 63*(2), 418-428.

Gamalero, E., Favale, N., Bona, E., Novello, G., Cesaro, P., Massa, N., and Lingua, G. (2020). Screening of bacterial endophytes able to promote plant growth and increase salinity tolerance. *Applied Sciences, 10*(17), 5767.

Gao, X.X., Zhou, H., Xu, D.Y., Yu, C.H., Chen, Y.Q., Qu, L.H.(2005). High diversity of endophytic fungi from the pharmaceutical plant Heterosmilax japonica Kunth revealed by cultivation-independent approach. FEMS Microbiology Letters 249, 255–266.

Glick, B. R. (2014). Bacteria with ACC deaminase can promote plant growth and help to feed the world. *Microbiological research, 169*(1), 30-39.

Gouda, S., Kerry, R. G., Das, G., Paramithiotis, S., Shin, H. S., and Patra, J. K. (2018). Revitalization of plant growth promoting rhizobacteria for sustainable development in agriculture. *Microbiological research, 206*, 131-140.

Govindarajan, M., Balandreau, J., Kwon, S. W., Weon, H. Y., and Lakshminarasimhan, C. (2008). Effects of the inoculation of Burkholderia vietnamensis and related endophytic diazotrophic bacteria on grain yield of rice. *Microbial Ecology, 55*(1), 21-37.

Govindarajan, M., Kwon, S. W., and Weon, H. Y. (2007). Isolation, molecular characterization and growth-promoting activities of endophytic sugarcane diazotroph Klebsiella sp. GR9. *World Journal of Microbiology and Biotechnology, 23*(7), 997-1006.

Goyal, S., Ramawat, K. G., and Mérillon, J. M. (2016). Different shades of fungal metabolites: an overview. *Fungal metabolites, 1*, 29.

Hallmann, J., Quadt-Hallmann, A., Mahaffee, W. F., and Kloepper, J. W. (1997). Bacterial endophytes in agricultural crops. *Canadian journal of microbiology, 43*(10), 895-914.

Hamayun, M., Hussain, A., Khan, S. A., Kim, H. Y., Khan, A. L., Waqas, M.,. and Lee, I. J. (2017). Gibberellins producing endophytic fungus Porostereum spadiceum AGH786 rescues growth of salt affected soybean. *Frontiers in microbiology, 8*, 686.

Hardoim, P.R., Hardoim, C.C., van Overbeek, L.S., van Elsas, J.D.(2012). Dynamics of seedborne rice endophytes on early plant growth stages. PLoS ONE 7, e30438.

Hardoim, P. R., Van Overbeek, L. S., Berg, G., Pirttilä, A. M., Compant, S., Campisano, A., and Sessitsch, A. (2015). The hidden world within plants: ecological and evolutionary considerations for defining functioning of microbial endophytes. *Microbiology and molecular biology reviews, 79*(3), 293-320.

Igiehon, N. O., and Babalola, O. O. (2018). Rhizosphere microbiome modulators: contributions of nitrogen fixing bacteria towards sustainable agriculture. *International journal of environmental research and public health, 15*(4), 574.

Joe, M. M., Devaraj, S., Benson, A., and Sa, T. (2016). Isolation of phosphate solubilizing endophytic bacteria from Phyllanthus amarus Schum and Thonn: Evaluation of plant growth promotion and antioxidant activity under salt stress. *Journal of applied research on medicinal and aromatic plants, 3*(2), 71-77.

Johnston-Monje, D., and Raizada, M. N. (2011). Conservation and diversity of seed associated endophytes in Zea across boundaries of evolution, ethnography and ecology. *Plos one, 6*(6), e20396.

Keswani, C., Mishra, S., Sarma, B. K., Singh, S. P., and Singh, H. B. (2014). Unraveling the efficient applications of secondary metabolites of various Trichoderma spp. *Applied microbiology and biotechnology, 98*(2), 533-544.

Khan, M. S., Gao, J., Chen, X., Zhang, M., Yang, F., Du, Y; and Zhang, X. (2020). Isolation and characterization of plant growth-promoting endophytic bacteria Paenibacillus polymyxa SK1 from Lilium lancifolium. *BioMed research international, 2020*.

Khan, N. A., and Khan, M. I. R. (2014). The ethylene: from senescence hormone to key player in plant metabolism. *J Plant Biochem Physiol, 2*, e124.

Kumar, A., and Verma, J. P. (2018). Does plant-microbe interaction confer stress tolerance in plants: a review. *Microbiological research, 207*, 41-52.

Kurek, E., Ozimek, E., Sobiczewski, P., S³omka, A., and Jaroszuk-Œcise³, J. (2013). Effect of Pseudomonas luteola on mobilization of phosphorus and growth of young apple trees (Ligol)—Pot experiment. *Scientia horticulturae, 164*, 270-276.

Lacava, P. T., Silva-Stenico, M. E., Araújo, W. L., Simionato, A. V. C., Carrilho, E., Tsai, S. M., and Azevedo, J. L. (2008). Detection of siderophores in endophytic bacteria Methylobacterium spp. associated with Xylella fastidiosa subsp. pauca. *Pesquisa Agropecuária Brasileira, 43*(4), 521-528.

Ladha, J. K., and Reddy, P. M. (2003). Nitrogen fixation in rice systems: state of knowledge and future prospects. *Plant and soil, 252*(1), 151-167.

Ma, Y., Rajkumar, M., Zhang, C., and Freitas, H. (2016). Beneficial role of bacterial endophytes in heavy metal phytoremediation. *Journal of environmental management, 174*, 14-25.

Mendes, R., Pizzirani-Kleiner, A. A., Araujo, W. L., and Raaijmakers, J. M. (2007). Diversity of cultivated endophytic bacteria from sugarcane: genetic and biochemical characterization of Burkholderia cepacia complex isolates. *Applied and environmental microbiology, 73*(22), 7259-7267.

Mishra, S., Singh, A., Keswani, C., Saxena, A., Sarma, B. K., and Singh, H. B. (2015). Harnessing plant-microbe interactions for enhanced protection against phytopathogens. *Plant Microbes Symbiosis: Applied Facets*, 111-125.

Nair D. N., Padmavathy S. (2014). Impact of endophytic microorganisms on plants, environment and humans. *Sci. World J.* 2014 250693 10.1155/2014/250693

Neubauer, U., Furrer, G., Kayser, A., and Schulin, R. (2000). Siderophores, NTA, and citrate: potential soil amendments to enhance heavy metal mobility in phytoremediation. *International journal of phytoremediation, 2*(4), 353-368.

Otieno, N., Lally, R. D., Kiwanuka, S., Lloyd, A., Ryan, D., Germaine, K. J., and Dowling, D. N. (2015). Plant growth promotion induced by phosphate solubilizing endophytic Pseudomonas isolates. *Frontiers in microbiology, 6*, 745.

Rajkumar, M., Ae, N., and Freitas, H. (2009). Endophytic bacteria and their potential to enhance heavy metal phytoextraction. *Chemosphere, 77*(2), 153-160.

Rashid, S., Charles, T. C., and Glick, B. R. (2012). Isolation and characterization of new plant growth-promoting bacterial endophytes. *Applied soil ecology, 61*, 217-224.

Rodríguez, H., and Fraga, R. (1999). Phosphate solubilizing bacteria and their role in plant growth promotion. Department of Microbiology, Cuban Research Institute on Sugarcane By-Products (ICIDCA), Havana, Cuba. *Havana. Cuba.*

Sabaté, D. C., Brandan, C. P., Petroselli, G., Erra-Balsells, R., and Audisio, M. C. (2018). Biocontrol of Sclerotinia sclerotiorum (Lib.) de Bary on common bean by native lipopeptide-producer Bacillus strains. *Microbiological research*, 211, 21-30.

Saha, M., Sarkar, S., Sarkar, B., Sharma, B. K., Bhattacharjee, S., and Tribedi, P. (2016). Microbial siderophores and their potential applications: a review. *Environmental Science and Pollution Research, 23*(5), 3984-3999.

Schulz, B., Boyle, C., Draeger, S., Römmert, A. K., and Krohn, K. (2002). Endophytic fungi: a source of novel biologically active secondary metabolites. *Mycological research, 106*(9), 996-1004.

Singh, B., Boukhris, I., Kumar, V., Yadav, A. N., Farhat-Khemakhem, A., Kumar, A., and Alghamdi, O A. (2020). Contribution of microbial phytases to the improvement of plant growth and nutrition: a review. *Pedosphere, 30*(3), 295-313.

Spaink, H. P., Kondorosi, A., and Hooykaas, P. J. (Eds.). (2012). *The Rhizobiaceae: molecular biology of model plant-associated bacteria*. Springer Science and Business Media.

Sun, Y., Cheng, Z., and Glick, B. R. (2009). The presence of a 1-aminocyclopropane-1-carboxylate (ACC) deaminase deletion mutation alters the physiology of the endophytic plant growth-promoting bacterium Burkholderia phytofirmans PsJN. *FEMS Microbiology Letters, 296*(1), 131-136.

Tiwari, G., Duraivadivel, P., Sharma, S., and Hariprasad, P. (2018). 1-Aminocyclopropane-1-carboxylic acid deaminase producing beneficial rhizobacteria ameliorate the biomass characters of Panicum maximum Jacq. by mitigating drought and salt stress. *Scientific reports, 8*(1), 1-12.

Triplett, E. W. (1996). Diazotrophic endophytes: progress and prospects for nitrogen fixation in monocots. *Plant and Soil, 186*(1), 29-38.

Ullah, A., Nisar, M., Ali, H., Hazrat, A., Hayat, K., Keerio, A. A., and Yang, X. (2019). Drought tolerance improvement in plants: an endophytic bacterial approach. *Applied microbiology and biotechnology, 103*(18), 7385-7397.

Van Der Heijden, M. G., Bardgett, R. D., and Van Straalen, N. M. (2008). The unseen majority: soil microbes as drivers of plant diversity and productivity in terrestrial ecosystems. *Ecology letters, 11*(3), 296-310.

Verma, P., Yadav, A. N., Khannam, K. S., Panjiar, N., Kumar, S., Saxena, A. K., and Suman, A. (2015). Assessment of genetic diversity and plant growth promoting attributes of psychrotolerant bacteria allied with wheat (Triticum aestivum) from the northern hills zone of India. *Annals of microbiology, 65*(4), 1885-1899.

Wei, C. Y., Lin, L., Luo, L. J., Xing, Y. X., Hu, C. J., Yang, L. T. and An, Q. (2014). Endophytic nitrogen-fixing Klebsiella variicola strain DX120E promotes sugarcane growth. *Biology and fertility of soils, 50*(4), 657-666.

Wilson, D. (1995). Endophyte: the evolution of a term, and clarification of its use and definition. *Oikos*, 274-276.

Xin, G., Zhang, G., Kang, J. W., Staley, J. T., and Doty, S. L. (2009). A diazotrophic, indole-3-acetic acid-producing endophyte from wild cottonwood. *Biology and Fertility of Soils, 45*(6), 669-674.

Chapter 11

Isopteran Termite Gut Microbes: An Overview of Significant Interactions

Nidhi Kakkar, Bhanupriya and Sanjeev Kumar Gupta*

Department of Zoology, Institute of Integrated and Honors Studies (IIHS), Kurukshetra University, Kurukshetra – 136 119, Haryana
**e-mail: nidhikakkar12@yahoo.com*

ABSTRACT

Xylophages termites (Isoptera) have great diversity in their habitat, feeding habit and gut microbes. These termites have ability to digest cellulosic, hemicellulosic and lignocellulosic biomass with the help of fungus, protists, bacteria etc. These microbes show various interactions with termites such as pathogenicity, obligate mutualism etc. Fungus live inside termites predigests the wood and makes them edible to termites. These termites are not capable of producing their own cellulases. The gut microbes make availability of cellulases to digest wood or cellulosic material. A great variety of bacteria exists in the termite gut. These bacteria can be useful in the breakdown of complex material into simpler one and these reduced forms can be used in different industries. The enzymes of these microbes are very useful for human welfare: textile industry, biofuel production, bioethanol production etc.

Keywords: *Xylophages, Isoptera, Termites, Cellulase, Biofuel, Bioethanol.*

Introduction

Various insects groups have symbiotic association with fungi, bacteria, protists etc. (Scharf and Tartar, 2008; Bignell, 2011; Nobre *et al.*, 2011; Lu *et al.*, 2016).The unique interaction between termite and gut microbes make termite dependency for carbon, nitrogen and energy nutrition. They cannot survive longer without these microbes (Hungoh *et al.,* 2003). The termite gut provides a distinct ecological environment to potentially cellulolytic, hemi-cellulolytic and lignocellulosic degrading bacteria. These bacteria can maintain nutrient and microbial balance in nature by producing or secreting extracellular hydrolytic enzymes (Varma *et al.,* 1994).

Phylogenetic division shows two lineages: lower termite and higher termite. These two are mainly different in the presence of symbiotic and diversified population of prokaryotes. Protozoans of lower termites are responsible for the breakdown of lignocellulosic material. This signifies their obligate mutualism between protists and termites (Ohkuma, 2008). Protists found in the lower termite gut but not of higher termite gut (Ohkuma, 2003; Telmadarrehei *et al.,* 2020) and harbor only prokaryotes in their highly structured guts, both higher and lower termites can produce their own cellulolytic enzymes (Miyata *et al.,* 2007; Köhler *et al.,* 2012; Tokuda *et al.,* 2012).

The degradation efficiency of higher termites have been exceeding over millions of years that of many other known natural system (Xie *et al.,* 2013), that makes them desirable models for bioprospecting with emphasis on industrial conversion of lignocellulose and the production of biofuels and other commodity biochemical (Marynowska *et al.,* 2020). The antagonistic property of *Bacillus* of termite gut and other invertebrates can be used to control some plant pathogens such as fungus and nematodes (König, 2006). The mutualistic behavior of these Bacilli can enhance the plant growth by stimulation plant hosts and microbial symbionts (Mc. Spadden Gardener, 2004).

Lower and higher termites bothshow association with gut fungus (Cornelius *et al.,* 2002; Aanen *et al.,* 2007). Various types of interactions like: Symbiotic, pathogenicity *etc.* (Rosengaus *et al.,* 2010) used by termites as defensive strategies against parasite and pathogens (Mueller *et al.,* 2005). These strategies may be antimicrobial substances in defensive glandular secretion, feces and body exudates of worker termites (Rosengaus *et al.,* 2004, 2010).

Interaction between Termites and Microbes

Subterranean termites are dependent on a beneficial symbiosis with a dense and diverse flora of microorganisms in the hindguts of the workers to digest lignocellulosic compounds and to acquire supplemental nutrition (Breznak, 2000; Brune, 2006). All the three domain of life harbors the termite gut: Eukarya (Protozoa, yeast, fungi), the Archaea and eubacteria (Breznak, 2000; König *et al.,* 2002, 2006).

Termites and Fungus

The fungus- growing termites (Termitidae: Macrotermitidae) are a well-known example of fungus-growing insects. These termites establish fungus gardens in their nests and consume both the aggregates of the fungal mycelia *i.e.*, the fungus nodules, and the gardens themselves (Moriya *et al.*, 2005). These termites have evolved next level of their social life style that they can grow their own fungal food (Aanen and Boomsma, 2006).

Fungus- growing termites (Termitidae: Macrotermitidae) in the old World (Aanen and Eggleton. 2005) play major roles in decomposition of plant material and nutrients cycling (Corrêa *et al.*, 2010; Jouquet *et al.*, 2011; Ohkuma, 2003; Da Costa *et al.*, 2019). They release mineral nutrients for plants reuse (Issoufou *et al.*, 2019).

Cellulosic and hemicellulosic components of wood can be depolymerized with the help of fungal products. Wood- eating termites prefer fungal infected or decayed wood. These fungus predigest the wood and make them edible for termites. The most preferable brown rot fungus of Basidiomycetes help them to predigest the wood (Becker, 1976; Amburgey and Smythe, 1977; Cornelius *et al.*, 2002).

Contrary to this, certain fungal products from decayed wood also act as repellents and cause enormous death in termites (Roessler, 1932). Fungus can also influence termite survival, caste development and suppress pathogenic effects (Zobri and Grace 1990). The wood inhabiting fungi are potential facilitators of the foraging behavior and survivorship of termites (Viana-Junior *et al.*, 2018).

Termites and Flagellates

The historic literature showed the presence and role of the flagellated protists in the termite gut and their impact on termite social behavior. These protists die prior to the molt of their termite hosts (reviewed by Nalepa, 2017) and must be gained by feeding on the hindgut fluids of a donor nest mate a behavior called proctodeal trophallaxis (McMahan, 1969). Lower termites show collaborative relationship with flagellates to digest lignocellulose (Slaytor, 1992).

The gut protozoan have ability to digest the cellulose from the wood particles and convert it to acetate and other products. Termites absorb and oxidize the acetate released by these flagellates. The cellulose digestion depends on the mutualistic protozoan as the host can not synthesize cellulase (Upadhyay, 2011).

Termite's exhibit behavioral adaptions for facultative transfer of microbial cultures from adult to young (Reinhard and Kaib, 2001). Termites' show stomodeal and proctodeal feeding that ensures transfer of a fresh infection of flagellates from one termite to another (Raina *et al.*, 2008).

Termites and Bacteria

Higher termites depend mainly on the bacterial communities for digestion of food because protists are harbored in lower termites only (Russmassler *et al.*, 2015). Termites gut contains approximately 15 phyla of bacteria which are more than bacterial diversity found in human and animal intestine (Ley *et al.*, 2006).

Several strains of cellulolytic bacteria were isolated from the guts of worker and soldier termites: *Odontotermes parvidens* (Kakkar *et al.*, 2015), *O. obesus* (Kakkar *et al.,* 2016), *Coptotermes heimi* (Kakkar *et al.,* 2018) and cellulolytic and amylolytic bacteria both can be isolated from termite worker and soldiers gut (Kakkar *et al.,* 2017). The bacterial community composition varies in termite species. Environment also affects which gut bacteria dominant the colony at a given points of time (Otani *et al.,* 2016). The bacterial isolates obtained from termite worker and soldier have more number of cultural isolates (Kakkar *et al.,* 2017).

Termites and Bioservices

Termites provide ecosystem functioning in such a way to recycle the nutrient and organic matter and redistribution, improve soil fertility, generation and regulation of soil biodiversity and ecosystem restoration (Bignell and Eggleton, 2000, Sugimoto *et al.,* 2000; Jouquet *et al.,* 2011). They are major contributors to the ecological stability of their habitats. By preparing different substrates, like wood on leaf litter, into forms easily accessed by microorganisms, termites play a major role as ecosystem conditions (Bignell and Eggleton, 2000). The degradation and recycling of lignocellulose in natural ecosystems is performed by bacteria, fungi and animals (Cragg *et al.,* 2015).

These termites form value- added products such as arabinose, glucose, galactose, mannose and xylose by producing xylanases which hydrolyse hemicellulose (Kumar *et al.,* 2009). Lignocellulose in one of the most abundant renewable types of biomass available on earth and cellulose based biofuels are considered to be a sustainable alternative to reduce our dependence on fossil fuels (Ragauskas *et al.,* 2006; Yang and Wyman, 2008). Xylan is the major abundant hemicellulosic component useful in biofuel industries (Liu *et al.,* 2003). Xylanases is one of the most important enzymesfor industrial and biotechnological applications: bioconversion of lignocellulose to fermentable sugars, production of cellophane and rayon, textile industry, production of chemicals such as cellulose ethers and esters, wine clarification, animal feeds, improvement of bread quality and also in pulping and bleaching process (Ghoshal *et al.,* 2015). The enzymes of gut microbes useful in bioethanol production (Sun and Scharf, 2010) also in bioremediation, because of their ability to degrade toxic substances (Ke *et al.,* 2012). The termite symbionts can improve soil fertility by nitrogen fixation (Hussender, 2010; Du *et al.,* 2012; Thong-On *et al.,* 2012). Wood eating termites are smallest and most efficient natural bioreactor with diverse enzyme machinery.

REFERENCES

Aanen, D.K. and Eggleton, P. 2005. Fungus-growing termites originated in African rain forest. *Curr. Boil.* 15: 851-855.

Aanen, D.K. and Boomsma, J.J. 2006. Social-insect fungus farming. *Curr. Biol.* 16: 1014-1016.

Aanen, D.K., Ros, V.I.D., de Fine Licht, H.H., Mitchell, J., De Beer, Z.W., Slippers, B., Rouland- Lefevre, C. and Boomsma, J.J. 2007. Patterns of interaction specificity of fungus- growing termites and *Termitomyces* symbionts in South Africa. *BMC Evol. Bio.* 7: 115-126.

Amburgey, T.L. and Smythe, R.V. 1977. Factors influencing the production of termite trail following an arrest ant stimuli by isolates of *Gloeophyllum trabeum*. *Sociobiology*. 3: 13-25.

Becker, G. 1976. Termites and fungi. *Mat. Org. Beiheft.* 3: 465-478

Bignell, D.E. and Eggleton, P. 2000. Termites in ecosystems. In: Abe, T., Bignell, D.E. and Higashi, M. (Eds.). Termites: evolution, sociality, symbioses, ecology, Kluwer Academic Publishers, Dordrecht, The Netherlands, pp. 363-387.

Bignell, D.E. 2011. Morphology, physiology biochemistry and functional design of the termite gut. In: Bignell, D. E., Roisin, Y. and Lo, N. (Eds.). Biology of termites: A modern synthesis, Springer- Verlag, Berlin, pp. 375-412

Breznak, J.A. 2000. Ecology of prokaryotic microbes in the guts of wood and litter-feeding termites. In: Abe, T., Bignell, D.E. and Higashi, M. (Eds.). Termites: Evolution, Sociality, Symbioses, Ecology, Kluwer Academic Publisher, Dordrecht, The Netherlands, pp. 209-213.

Brune, A. 2006. Symbiotic associations between termites and prokaryotes. *Prokaryotes*. 1: 439-474.

Chaurasia, B., Pandey, A., Palni, L.M.S., Trivedi, P., Kumar, B. and Colvin, N. 2005. Diffusible and volatile compounds produced by an antagonistic *Bacillus subtilis* strain cause structural deformations in pathogenic fungi in vitro. *Microbiol. Res.* 160: 75-81.

Cornelius, M.L., Daigle, D.J., Connick, W.J., Tellez, M., Williams, K.S. and Lovisa, M.P. 2002. Interactions between Formosan subterranean termites (Isoptera: Rhinotermitidae) and wood decaying fungi. In *Proceedings of* the 4th International Conference on Urban Pests, Charleston, SC, USA. pp. 7-10.

Corrêa, M.M., Silva, P.S., Wirth, R., Tabarelli, M. and Leal, I.R. 2010. How leaf-cutting ants impact forests: Drastic nest effects on light environment and plant assemblages. *Oecologia*. 162: 103-115.

Cragg, S.M., Beckham, G.T., Bruce, N.C., Bugg, T.D., Distel, D.L., Dupree, P. 2015. Lignocellulosic degradation mechanisms across the Tree of Life. *Curr. Opin. Chem. Biol.* 29: 108-119.

Da Costa, R.R., Hu, H., Li, H. and Poulsen, M. 2019. Symbiotic plant biomass decomposition in fungus-growing termites. *Insects*. 10: 87.

Du, X., Li, X., Wang, Y., Peng, J., Hong, H. and Yang, H. 2012. Phylogenetic diversity of nitrogen fixation genes in the intestinal tract of *Reticulitermes chinesis* Snyder. *Curr. Microbiol.* 65: 547-551.

Ghoshal, G., Banerjee, U.C., Shivhare, U.S. 2015. Utilization of agrowaste and Xylanase production in solid state fermentation. *J. Biochem. Technol.* 6(3): 1013-1024.

Husseneder, C. 2010. Symbiosis in subterranean termites: a review of insights from molecular studies. *Environ. Entomol.* 39: 378-388.

Issoufou, A.A., Soumana, I., Maman, G., Konate, S., Mahamane, A. 2019. Effects of termites growth on litter decomposition: a modeling approach. *International Journal of recycling of organic Waste in Agriculture* 8: 415-421.

Jouquet, P., Traoré, S., Choosai, C., Hartmann, C. and Bignell, D. 2011. Influence of termites on ecosystem functioning. Ecosystem services provided by termites. *Eur. J. Soil Biol.*47: 215-222.

Kakkar, N., Gupta, S.K. and Saharan, B.S. 2015. Studies on Cellulolytic Activity and Structure of Symbiotic Bacterial Community in *Odontotermes parvidens* Guts. *Int. J. Curr. Microbiol. App. Sci.* 4: 310-315.

Kakkar, N., Gupta, S.K. and Saharan, B.S. 2016. Comparison of different stains in determination of extracellular cellulase activity of *Odontotermes obesus* gut bacteria. *Int. J. Entomol. Res.* 1: 54-56.

Kakkar, N., Gupta, S.K. and Rohitsaini, 2017. Isolation of cellulolytic and amylolytic bacteria of worker and soldier termite gut (Isoptera).*IRJNAS.* 4: 190-200.

Kakkar, N., Gupta, S.K., Bura, A. and Gandhi, A. 2018. Diversity of cellulose hydrolyzing bacteria from the gut of C*optotermes heimi* (Rhinotermitidae). *Asian. J. Biol. Sci.* 7: 28-32.

Ke, J., Singh, D. and Chen, S. 2012. Metabolism of polycyclic aromatic hydrocarbons by the wood-feeding termite *Coptotermes formosanus* (Shiraki). *J. Agr. Food Chem.* 60: 1788-1797.

Köhler, T., Dietrich, C., Scheffrahn, R.H. and Brune, A. 2012. High-resolution analysis of gut environment and bacterial microbiota reveals functional compartmentation of the gut in wood-feeding higher termites (*Nasutitermes spp.*). *Appl. Environ. Microbiol.* 78: 4691-4701.

König, H. 2006. *Bacillus* species in the intestine of termites and other soil invertebrates. *J. Appl. Microbiol.* 101: 620-627.

König, H., Fröhlich, J., Berchtold, M. and Wenzel, M. 2002. Diversity and microhabitats of the hindgut flora of termites. *Recent Res. Devel. Microbiol.* 6: 125-156.

König, H., Fröhlich, J. and Hertel, H. 2006. Diversity and lignocellulolytic activities of cultured microorganisms. In: Konig, H. and Varma, A. (Eds.). Intestinal Microorganisms of termites and other invertebrates, *Springer, Berlin, Heidelberg*, pp. 271-301.

Kumar, S., Singh, S.P., Mishra, I.M., Adhikari, D.K. 2009. Recent advances in production of Bioethanol from lignocellulosic biomass. *Chem. Eng. Technol.* 32: 517-526.

Liu, C.J., Suzuki, T., Hirata, S., Kawai, K., 2003. The processing of high molecular weight xylanase (XynE,110KDa) from *Aeromonas caviae* ME-I to 60 KDa Xylanase (XynE60) in *Escherichia coli* and purification and characterization of XynE 60. *J. Biosci. Bioeng.* 95: 95-101.

Ley, R.E., Peterson, D.A. and Gordon, J.I. 2006. Ecological and evolutionary forces shaping microbial diversity in the human intestine. *Cell.* 124: 837-848.

Marynowska, M., Goux, X., Sillam-Dussès, D., Rouland-Lefèvre, C., Halder, R., Wilmes, P., Gawron, P., Roisin, Y.,Defosse, P. and Calusinska, M. 2020. Compositional and functional characterization of biomass-degrading microbial communities in guts of plant fibre-and soil-feeding higher termites. *Microbiome.* 8: 1-18.

McMahan, E. A. 1969. Feeding relationships and radioisotope techniques. In: Krishna, K. and Weesner, F. M. (Eds.). Biology of termites, Academic Press, New York, pp. 387-406

McSpadden Gardener, B.B. 2004. Ecology of *Bacillus* and *Paenibacillus* spp. in Agricultural Systems. *Phytopathology*. 94: 1252–1258.

Miyata, R., Noda, N., Tamaki, H., Kinjyo, K., Aoyagi, H., Uchiyama, H. and Tanaka, H. 2007. Influence of feed components on symbiotic bacterial community structure in the gut of the wood-feeding higher termite *Nasutitermes takasagoensis. Biosci. Biotechnol. Biochem.* 71: 1244-1251.

Moriya. S., Inoue, T., Ohkuma, M., Yaovapa, T., Johjima, T., Suwanarit, P., Sangwanit, U., Vongkaluang, C., Noparatnaraporn, N. and Kudo, T. 2005. Fungal Community Analysis of Fungus gardens in termite nests. *Microbes. Environ.* 20: 243-252.

Mueller, U.G., Gerardo, N.M., Aanen, D.K., Six, D.L. and Schultz, T.R. 2005. The evolution of agriculture in insects. *Annu. Rev. Ecol. Evol. Syst.* 36: 563-595.

Nalepa, C.A. 2017. What kills the hindgut flagellates of lower termites during the host molting cycle?. *Microorganisms.* 5: 82.

Nobre, T., Rouland-Lefèvre, C. and Aanen, D. K. 2011. Comparative biology of fungus cultivation in termites and ants. In: Bignell, D.E., Roisin, Y. and Lo, N. (Eds.). Biology of termites: A modern synthesis, Springer- Verlag, Berlin, pp. 193-210.

Ohkuma, M. 2003. Termite symbiotic systems: efficient bio-recycling of lignocellulose. *Appl. Microbiol. Biotechnol.* 61: 1-9.

Ohkuma, M. 2008. Symbioses of flagellates and prokaryotes in the gut of lower termites. *Trends Microbiol.* 16: 345-352

Otani, S., Hansen, L.H., Sorensen, S.J. and Poulsen, M. 2016. Bacterial communities in termite fungus combs are comprised of consistent gut deposits and contributions from the environment. *Microb. Ecol.* 71: 207-220.

Ragauskas, A.J., Williams, C.K., Davison, B.H., Britovsek, G., Cairney, J., Eckert, C.A., Frederick, W.J., Hallett, J.P., Leak, D.J., Liotta, C.L., Mielenz, J.R., Murphy, R., Templer, R. and Tschaplinski, T. 2006. The path forward for biofuels and biomaterials. *Science.* 311: 484-489.

Raina, A., Park, Y.I. and Gelman, D. 2008. Molting in workers of the Formosan subterranean termite *Coptotermes formosanus. J. Insect. Physiol.* 54: 155-161.

Reinhard, J. and Kaib, M. 2001. Food exploitation in termites: indication for a general feeding-stimulating signal in labial gland secretion of Isoptera. *J. Chem. Ecol.* 27: 189-201.

Roessler, E. S. 1932. A preliminary study of the nitrogen needs of growing *Termopsis.* Pnbl Zool. University of California press, 36: 357

Rosengaus, R.B., Traniello, J.F.A., Lefebvre, M.L. and Maxmen, A.B. 2004. Fungi static activity of the sternal gland secretion of the damp wood termite *Zootermopsis angusticollis. Insectes Soc.* 51: 259-264.

Rosengaus, R.B., Traniello, J.F.A. and Bulmer, M.S. 2010. Ecology, behavior and evolution of disease resistance in termites. In: Bignell, D.E., Roisin, Y. and Lo, N. (Eds.). Biology of termites: a modern synthesis, Springer- Verlag Berlin, pp. 165-191.

Rossmassler, K., Dietrich, C., Thompson, C., Mikaelyan, A., Nonoh, J.O., Scheffrahn, R.H., Sillam-Dusses, D. and Brune, A. 2015. Metagenomics analysis of the microbiota in the highly compartmented hindguts of six wood-or soil-feeding higher termites. *Microbiome*. 3: 56.

Scharf, M.E. and Tartar, A. 2008. Termite digestomes as sources for novel lignocellulases. *Biofuel. Bioprod. Bior.* 2: 540-552.

Slaytor, M. 1992. Cellulose digestion in termites and cockroaches: what role do symbionts play?. *Comp. Biochem. Physiol. B, Biochem.* 103: 775-784.

Sugimoto, A., Bignell, D.E. and Macdonald, J.A. 2000. Global impact of termites on the carbon cycle and atmospheric trace gases. In: Abe, T., Bignell, D.E. and Higashi, M. (Eds.). Termites: Evolution, Sociality, Symbioses, Ecology, Kluwer Academic Publisher, Dordrecht, The Netherlands, pp. 409-435.

Sun, J.Z. and Scharf, M.E. 2010. Exploring and integrating cellulolytic systems of insects to advance biofuel technology. *Insects Sci.* 17: 163-165.

Swain, M.R., Ray, R.C. and Nautiyal, C.S. 2008. Biocontrol efficacy of *Bacillus subtilis* strains isolated from cow dung against postharvest yam (Dioscorea rotundata L.) pathogens. *Curr. Microbiol.* 57: 407-411.

Telmadarrehei, T., Tang, J.D., Raji, O., Rezazadeh, A., Narayanan, L., Shmulsky, R. and Jeremic, D. 2020. A study of the gut bacterial community of *Reticulitermes virginicus* exposed to chitosan treatment. *Insects.* 11: 681.

Thong-On, A., Suzuki, K., Noda, S., Inoue, J.I., Kajiwara, S. and Ohkuma, M. 2012. Isolation and characterization of anaerobic bacteria for symbiotic recycling of uric acid nitrogen in the gut of various termites. *Microbes. Environ.* 27: 186-192.

Tokuda, G., Watanabe, H., Hojo, M., Fujita, A., Makiya, H., Miyagi, M., Arakawa, G. and Arioka, M. 2012. Cellulolytic environment in the mid-gut of the wood-feeding higher termites *Nasutitermes takasagoensis. J. Insect Physiol.* 58: 147-154.

Upadhyay, R.K. 2011. Symbiotic and non-symbiotic micro flora of termite gut: a unique nonhuman agricultural system that can recycle photo-synthetically fixed carbon and nutrients. *J. Pharm. Res.* 4: 1161-1166.

Varma, A., Kolli, B.K., Paul, J., Saxena, S. and König, H. 1994. Lignocellulose degradation by microorganisms from termite hills and termite guts: a survey on the present state of art. *FEMS Microbiol. Rev.* 15: 9-28.

Viana- Junior, A.B., Cortes, M.O., Cornelissen, T.G. and De Siqueira Neves, F. 2018. Interactions between wood- inhabiting fungi and termites: a meta- analytical review. *Arthropod- Plant Interactions.* 12: 229-235.

Xie, L., Liu, N. and Huang, Y. 2013. Lignocellulose degradation in termite symbiotic systems, in biological conversion of biomass for fuels and chemicals: explorations from natural utilization system. Royal society of chemistry.

Yang, B. and Wyman, C. E. 2008. Pretreatment: the key to unlocking low-cost cellulosic ethanol. *Biofuel. Bioprod. Biorefin.* 2: 26-40.

Zoberi, M.H. and Grace, J.K. 1990. Fungi Associated with the subterranean Termite *Reticulitermes flavipes* in Ontario. *Mycologia* 82(3): 289-294.

Chapter 12

Biological Control of Agricultural Pests: A Step Towards Greener Remedies

*Lubhan Cherwoo, Ishika Gupta, Geetika Sharma, Nainpriya Babbar, Hema Setia and Ranjana Bhatia**

Department of Biotechnology,
University Institute of Engineering and Technology,
Panjab University, Chandigarh – 160 014
**e-mail: ranjanabhatia20@gmail.com*

ABSTRACT

Intensive infestation by pests and other crop diseases greatly affects agricultural productivity every year. Farmers general use chemical pesticides to treat or prevent pest infection, but excessive use of pesticides and other chemical products on crops has a negative impact on human health and the ecosystem. Use of biocontrol agents supersedes all the other control methods to control crop diseases because they do not include the use of any chemicals or machinery, do not leave any residues and are economically feasible. Overall, they have little to no negative impact on the environment. Biocontrol is the reduction of agricultural pests and diseases by using various strategies like importation, augmentation, and conservation, these techniques employ cyanobacteria, nematodes, lytic fungi, allelochemicals and biopesticides for pests like molluscs, pathogens, parasites and insects. Pest specific management strategies are often employed. Specific production protocols are followed for developing the biocontrol products and have resulted in the availability of various products in the market. Lastly, the benefits of using biocontrol strategies outweigh the limitations and thus biocontrol methods still remain a far better alternative to any other control methods. Exploiting accurate strategies for production and use of biocontrol agents will open up a range of new applications in crop management.

Keywords: *Biological control, Crop management, Pest, Plant pathogen, Biocontrol agent, Pesticides.*

Introduction

Agriculture is an important sector in the Indian economy, accounting for 18 per cent of the country's GDP. But this backbone of the country's economy is greatly affected by pests and crop diseases every year (around 30-35 per cent loss in crop yield). Most of these pests are insects, mites, nematodes, and gastropod molluscs. They reduce productivity in various ways by attacking the flowers, fruit, leaves and branches. Lepidopterous fruit borers are the most important pests, affecting production. Other important species include various leaf- and flower-eating caterpillars and beetles, bark borers, scales, leaf mites, fruit-sucking bugs, fruit-piercing moths, and fruit flies. The pests are also classified based on their impacts; stand reducers (damping-off pathogens), photosynthetic rate reducers (fungi, bacteria, viruses), leaf senescence accelerators (pathogens), light stealers (weeds, some pathogens), assimilate sappers (nematodes, pathogens, sucking arthropods) to name a few [1].

Therefore, to protect and control the crop damage countries have adapted various methods: physical (cultivation, mechanical weeding), biological (cultivar choice, crop rotation, antagonists, and predators) and chemical measures (pesticides) [2], amongst which the use of chemical pesticides has been reported widely.

Recent studies on crop management have revealed that the use of pesticides and other chemical products on crops has a negative impact on human health and the ecosystem [3]. Build-up of pesticide resistance has become very common. This has unfortunately led to an increase in the amount and frequency of pesticides applied. This, in turn, leads to crop losses, lowers crop yields; there is a surge in production costs, and a significant increase in health risks of people and wildlife. Other effects include toxicity of pesticides towards non-target organisms and severe impact on soil health and water quality [3, 4]. Table 12.1 summarizes some of the pesticides and their ill effects on human health. Hence, there is an intense need for an alternative to chemical control techniques.

Therefore, the risks associated with the extensive use of chemical pesticides have shifted the focus to the development and application of biocontrol strategies to control crop diseases. Biocontrol can be defined as the use of living organisms and their metabolites to suppress or eradicate the pest population or making it less damaging [5]. In simple words, it is the reduction of agricultural pests and diseases by using their natural enemies. Biocontrol agents are perfect alternatives to chemical pesticides because unlike agrochemicals they do not negatively impact the environment by utilizing chemicals, machinery or leaving residues. They supply long-term control without any additional costs or intervention. They are economically feasible and sustainable.

Table 12.1: Common Chemical-Based Pesticides and their Health Impacts

Pesticides	*Health Impacts*
Organophosphates	Affects the function of cholinesterase enzymes, disrupts normal cellular metabolism of proteins, carbohydrates and fats
Carbamate	Neuro behavioural effects, risk for dementia, and non-Hodgkin's lymphoma
Organochlorine	Endocrine disorders, effects on embryonic development, haematological and hepatic alterations
Pyrethroids and pyrethrins	Endocrine-disrupting activity, affects reproductive parameters in experimental animals including reproductive behaviour
Triazines	Endocrine-disrupting effects and reproductive toxicity

There are three main techniques used for conventional biocontrol practices. These include: importation or classical biocontrol, augmentation, and conservation control. Each of these techniques can be used either alone or in association with other techniques in a biological control program. Importation or classical biocontrol is a strategy used to 'import' natural enemies when exotic pests are introduced (by accident or intentionally) into regions (*e.g.*, countries) where they are not native. The conservation approach is about conserving and protecting the natural enemies and involves carefully choosing and applying pesticides in order to pose a minimal impact on natural enemies and also providing habitat and resources for natural enemies. Augmentation involves the manipulation of natural enemies to increase their effectiveness and their release in measured quantities to stamp out the pests. Augmentation, however, does not furnish permanent suppression of pests like other approaches [6].

Biological Control Agents

Biological control of pests can be achieved through the application of different agents mentioned below:

Cyanobacteria

The blue-green algae or the photosynthetic prokaryote is often employed as fertilizer due to the presence of biologically active and biocidal substances (hapalindoles), antibiotics and fungicides. It has shown promising effects on multiple species of pests and insects, such as common agricultural pest, *Helicoverpa armigera* [7].

Allelochemicals

Allelochemicals are a wide range of secondary metabolites produced by certain crops and inhibit the growth of other plant species (Figure 12.1). The use of these natural substances is preferred over traditional chemicals due to their biodegradation properties and ease of handling. For this purpose, plant extracts of different plants such as neem, tobacco, ginger are prepared and used. For example, neem is one of the most widely used allelopathic plants which release allelochemicals such as azadirachtin, salannin and nimbin which are inhibitory to *Cicadellidae* (*Jacobiasca lybica*) and whitefly (*Bemisia tabaci*). Allelochemicals are also proven to promote plant growth through nutrient regulation [8].

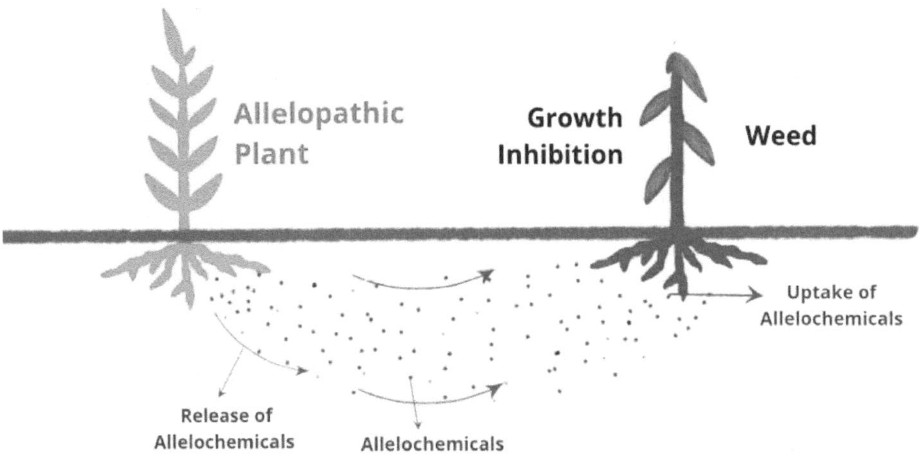

Figure 12.1: Mechanism of Allelopathy.

Natural/Biopesticides

Different microorganisms have been identified and used as very efficient biological agents. These include *Bacillus thuringiensis*, Nuclear Polyhedrosis Virus, *Trichogramma* spp., *Pseudomonas* spp. *Verticillium lecanii* and many more which have been used in combination or independently for different crops [9]. Table 12.2 enlists various biological agents used for different crops.

Pest Specific Management Strategies

Different kinds of pests may require different strategies for their management and control. This section describes the various biological management strategies that are employed for pests like molluscs, pathogens, parasites and insects.

Table 12.2: Biological Agents Beneficial on different Crops

Crop	*Pest*	*Biological Agents*	*References*
Cotton	*Pectinophora gossypiella*	*Trichogramma* spp., *Bacillus cereus*	[10], [11]
	Acariformes	*Chrysoperla* spp., *Verticillium* spp., *Beauveria* spp., *Coccinellidae*	[9], [12]
Rice	*Scirpophaga incertulas, Pycnarmon cribrata*	*B. thuringiensis, Trichogramma* spp.	[9]
Tomato and capsicum	Heliothis	*B. thuringiensis,* Nuclear Polyhedrosis Virus, *Trichogramma* spp., Baculovirus	[9], [13]
Brinjal and okra	*Leucinodes orbonalis*	*B. thuringiensis,* Nuclear Polyhedrosis Virus, *Trichogramma* spp.	[9]

Pathogens Management

Plant pathogens are disease causing organisms in agriculture, some close relatives of these plant pathogens cause disease in humans and animals also. These pathogens include *Indian cassava mosaic virus, Cotton leaf curl virus, Sunflower necrosis virus, Potato virus Y* and more as discussed in Table 12.3.

Table 12.3: Various Plant Pathogens and their Host Crops

Disease	*Disease Type*	*Crop*	*Pathogen*	*Pathogen Family*	*Pathogen Group*
Mosaic	Sterility mosaic	Pigeonpea	Pigeonpea sterility mosaic virus	*Phenuiviridae*	Virus
	Yellow mosaic	Mungbean	Mungbean yellow mosaic virus	*Geminiviridae*	
	Mosaic	Cassava	Indian cassava mosaic virus		
Leaf curl	Yellow leaf curl	Tomato	Tomato leaf curl virus		
	Leaf Curl	Cotton	Cotton leaf curl virus		
Necrosis	Bud necrosis	Groundnut	Groundnut bud necrosis virus	*Tospoviridae*	
	Leaf and stem necrosis	Sunflower	Sunflower necrosis virus	*Bromobiridae*	
Blight	Leaf blight	Wheat	*Alternaria triticina*	*Pleosporaceae*	Fungi
	Sheath blight	Rice	*Rhizoctonia solani*	*Ceratobasidiaceae*	

Disease	*Disease Type*	*Crop*	*Pathogen*	*Pathogen Family*	*Pathogen Group*
	Early blight	Tomato, potato	*Alternaria solani*	*Pleosporaceae*	
	Late blight	Potato	*Phytophthora infestans*	*Peronospora-ceae*	
Wilt	Fusarium wilt	Tomato	*Fusarium oxysporum f.* sp. *lycopersici*	*Nectriaceae*	
	Fungal wilt	Potato	*Sclerotium rolfsii*	*Atheliaceae*	
Rots	Black rot	Brassicas	*Xanthomonas campestris* pv. *Campestris*	*Botryosphaeria-ceae*	
	Mucor rot	Apples, pears	*Mucor piriformis*	*Mucoraceae*	
	Sphaeropsis rot	Apples	*Sphaeropsis pyriputrescens*	*Botryosphaeria-ceae*	
Aster yellows		Onion, wheat, barley, lettuce, celery, carrot	*Phytoplasmas*	*Acholeplasma-taceae*	Bacteria
Wilt	Bacterial wilt	Cucumber, muskmelon, squash, pumpkin, gourds	*Erwinia tracheiphila, Curtobacterium flaccumfaciens* pv *flaccumfaciens*	*Erwiniaceae, Microbacteria-ceae*	
Blight	Fire blight	Roses, pears, apples, almond, cherry, plum, quince, medlar, apricot, crabapple	*Erwinia amylovora*	*Erwiniaceae*	
	Rice bacterial blight	Rice	*Xanthomonas oryzae* pv *oryzae*	*Xanthomonada-ceae*	

Biocontrol

Management of plant pathogens is usually done with bacterial or fungal strains isolated from the endosphere and rhizosphere. Viruses can also be used as biocontrol agents and interest is developing in using bacteriophages for bacterial pathogens. Biological control, though not a permanent cure, disease suppression is comparable to chemical control. Genomic and genetic modifications have been made over the years to introduce natural protection against diseases like those done in cotton, corn, soybean, potatoes, apples [14, 15]. Plant pathogens are generally hard to manage once infection occurs, but certain interspecies antagonisms lead to biological control of plant pathogens.

a) Hyper-parasitism: Here the pathogen is directly affected by a hyperparasite in varying degrees of effectiveness. Hyperparasites can be subclassified into hypoviruses, obligate bacterial pathogens and facultative parasites like *Pasteuria penetrans* which parasitizes root-knot nematodes of genus *Trichoderma*, mycorrhizal and endophytic fungi are effective biocontrol agents for this parasite [14, 19].

b) Predation: Microbial predation is more widely used. It offers certain special features like non-specificity for pathogens and acts in a variety of modes of action, which can be classified into direct (inducing resistance or priming plants) or indirect (nutrient competition) methods. For example, nemato-pathogenic fungus *Arthrobotrys robusta* was used against *Cooperia punctate* larvae (L3) and showed promising results [16]. Similarly, nematode trapping fungi have been extensively tested against parasitic nematodes, for example, activity of *Arthrobotrys oligospora var. oligospora* and *Arthrobotrys cladodes var. macroides* against *Haemonchus contortus H. contortus* 3rd stage larvae (L3) [17, 18].

c) Lytic enzymes: Secretion and excretion of different factors that interfere with growth and development of pathogens is another mode to control their population. Pathogen exoskeletons are made of polymers such as chitin, cellulose and hemicellulose which can be digested or hydrolysed by different microorganisms producing a variety of lytic enzymes. For example, biocontrol of *Sclerotium rolfsii* by *Serratia marcescens* as a chitin degrader is possible because of the production of chitinase enzymes. Lytic enzymes are abundantly produced by microorganisms such as *Trichoderma, Bacillus, Lysobacter* and *Myxobacteria* and are very effective against fungal plant pathogens [14, 15, 16, 19].

d) Induction of host resistance: Plants react to stimuli by different microorganisms in the endosphere and rhizosphere. Induced host resistance works by inducing or enhancing crops' natural resistance against plant pathogens. This can happen through multiple pathways like induced systemic resistance or systemic acquired resistance which are further discussed in Table 12.4 and Figure 12.2 [14, 16, 20].

Competition Management

Competition between different species allows the stronger competitor to survive through natural selection. Bearing a negative impact on the crop, competition should be eliminated so that the productivity of the crop and economic losses do not incur.

Table 12.4: A Comparison of Induced Systemic Resistance (ISR) and Systemic Acquired Resistance (SAR)

Differential Factors	*Induced Systemic Resistance*	*Systemic Acquired Resistance*
Definition	Resistance activated by colonization of beneficial microorganisms in plant roots.	Resistance activated in healthy systemic tissues of plants with local infection.
Induction methodology	Induced by non-pathogenic bacteria in the rhizosphere.	Self-induced by signalling molecules.
Salicylic acid dependence	Non-dependent	Dependent
Signalling molecules	Jasmonic acid and ethylene signaling	Salicylic acid signalling

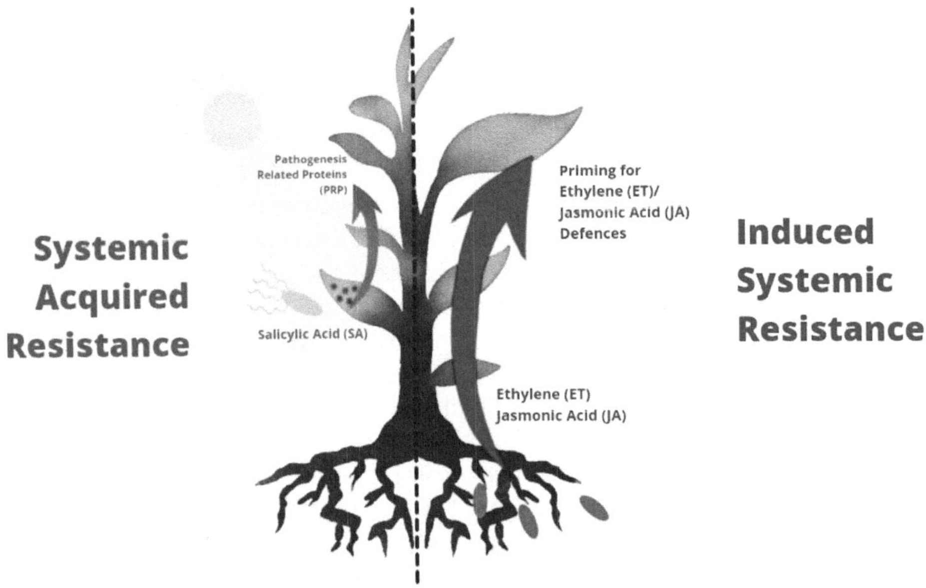

Figure 12.2: Mechanism of Systemic Acquired Resistance and Induced Systemic Resistance.

In an ecosystem, it refers to the unfavourable competition between crop and the weed. Weeds are one of the major competitions that result in higher losses and poor crop productivity [21]. The crop and weed competition can be direct (where both compete for light, nutrients, space and moisture) or indirect (through allelopathic weeds that release allelochemicals causing inhibition in the growth of the crop).

Biocontrol

a) *Natural Toxins*

Microorganisms are known to produce high levels of phytotoxins, which are considered safe from an environmental perspective. *Streptomyces hygroscopicus*, a soil bacterium, releases L-phosphinothricin (L-phosphinothricin and D-phosphinothricin racemise to produce glufosinate) that inhibits the glutamine synthetase causing a blockage in the photorespiratory pathway and accumulation of ammonia. This causes photodynamic damage due to the inhibition of carbon fixation. *S. hygroscopicus* is also known to produce a compound called tripeptide bialaphos, which acts as a protoxin towards the target weed [22].

Genetic engineering is a way to introduce selectivity by producing crop transgenics. For example, incorporation of *bar* genes (responsible for acylation of phosphinothricin, rendering its phytotoxic nature inactive), produced by *S. hygroscopicus*, can act as a selectable marker in the crop of interest. Application of a phosphinothricin-based herbicide would enable the selection only of the crop of interest, hence, eliminating the weeds. The use of competitive crop genotypes has also been considered as an alternative to the existing biological controls. These often cause the crop to grow rapidly with a higher germination rate, rapid leaf expansion and canopy development along with an extensive root system, to overcome the competition with weeds [23].

Another microbial toxin, called tentoxin, produced by fungus *Alternaria alternata*, is highly phytotoxic and selective towards multiple categories of weeds and kills them by inducing chlorosis. Similarly, there exist many other microbial toxins (*e.g.* AAL-toxin) which are highly specific to several weeds; however, they may be toxic towards mammals as well [23].

b) *Allelochemicals*

The use of allelopathic plants is one of the most effective and common methods to control weeds (Table 12.5). Due to the diversity observed in their structures, they display different modes of action such as interference with cell division, hormone biosynthesis, and mineral uptake and transport [24]. A commonly known allelopathic plant is sorghum (incorporated in soil) that produces allelochemicals such as hydrophobic p-benzoquinone, phenols, and acyanogenic glycoside. Rye (used as a cover crop), is another widely available and known allelopathic plant that expresses around 16 allelochemicals [21].

Table 12.5: Allelopathic Activity of Crops against Certain Weed Species

Allelopathic Source	*Weed Species*	*Weed Reduction (Per cent dry weight)*	*References*
Sorghum	*Phalaris minor*, *Chenopodium album*	48.0 - 56.0	[25]
Rye	*Portulaca oleracea*, *Amaranthus* spp.	5.0 - 95.0	[26]
Sunflower	*Rumex dentatus* L., *Chenopodium album* L.	57.0 - 67.0	[27]

Molluscs Management

Molluscs are soft bodied invertebrates. Some examples of pest molluscs include *Deroceras reticulatum, Achatina fulica, Arion vulgaris, Deroceras laeve, Arion hortensis, Bradybaena ravida.* Molluscs not only cause damage to crops on ground but also below the ground. Most arable crops that are affected tend to develop a thin crop stand and there is germination failure due to grain hollowing under extreme conditions [30]. Terrestrial slugs and snails are serious pests in a lot of plantations around the world (Table 12.6), due to their high reproductive potential and high spread rate in areas where they are newly introduced in absence of native natural control. These molluscs cause damage to raw succulent vegetables, seedlings, roots, seeds, tuber crops. In India, common molluscs like *Achatina fulica, Helix aspersa, Laevicaulisalte* cause damage to crops like portulaca, yam, potatoes, cereal, carrots, maize, balsam, rice, dahlia, marigold [28, 29].

Table 12.6: Various Mollusc Pests, their Distribution and Affected Plantations

Molluscs	*Molluscs Family*	*Affected Area*	*Affected Crop*
Achatina fulica	*Achatinidae*	India, Brazil, Africa, South America	Vegetable crops, staple yam, rice paddies, bananas, coconuts
Isidorella newcombi Bradybaena similaris, Acusta tourannensis, Bradybaena ravida	*Planorbidae, Bradybaenidae, Lymnaeidae*	Southeast Asia, Palearctic, and Nearctic Regions	Cereals, cotton, citrus plants, apple rotations, rice, grapevines
Limacus maximus,Limacus flavus	*Limacidae*	North Africa, China, India, Central and Southern Europe	Cultivated plants, mushrooms
Milax gagates	*Milacidae*	Britain, Argentina	Carrot, potato and tomato, soybean, sunflower, oilseed rape
Sarasinula plebeia	*Veronicellidae*	Central America	Bean, ornamental and horticultural crops
Cantareus aspersus, Theba pisana,Cornu aspersum	*Helicidae*	California, Western Europe, Morocco	Citrus plantations, cereal crops
Deroceras reticulatum	*Deroceras*	Europe, North Africa	Sugar beet, potato, barley, maize

Biocontrol

Mollusc management was done using chemical methods, which were only partially effective and caused unwanted damage to flora and fauna. Biological control, however, essentially includes permanent removal of molluscs from the target system [30, 31, 32, 33]. Snails and slugs can be managed using natural enemies like viruses, fungi, bacteria, parasites, predators, nematodes.

a) Bacteria: Bacterial genera like *Escherichia, Bacillus, Alcaligens* have shown molluscicidal properties. The combination of *Bacillus thuringiensis* and parasitic nematode *Rhabditis* sp. has shown effective control of terrestrial snails and slugs like *Eobania vermiculata.* The bacterium, *Moraxella osloensis,* has naturally been found to be symbiotically associated with the nematode *Phasmarhabditis hermaphrodita* leading to the production of endotoxins which are virulent to infectious slug species. *Paenibacillus glabratella* infects snails causing high mortality making it a good biocontrol agent against molluscs [34, 35, 36].
b) Nematodes: The parasitic nematode *Phasmarhabditis hermaphrodita* is one of the most widely established and commercially available biocontrol agents for mollusc pests in Europe, sold under the product name Nemaslug. It is effectively used for almost nine pest slug species.
c) Fungi: The fungi *Verticillium chlamydosporium* attacks mainly eggs of *Deroceras reticulatum.* Though as promising as fungus seem they are less likely to show any significant effect on mollusc population in an agricultural setting [38].
d) Parasites and Predators: Protozoa, lung worm, glow worm, flatworm, carabid beetles and larvae of lampyrid beetles or *Sciomyzidae* are the major parasitic predators. Black carp fish has commercially been the most viable biocontrol agent for snails [36, 37].

Biocontrol Products and their Production

Development and commercialization of biocontrol products is a multi-step process involving the following steps [39]:

a) Isolation of microorganisms from the natural system.
b) Evaluation of bio-agent.
c) Testing of the most effective isolate under field conditions.
d) Production
e) Formulation
f) Delivery
g) Compatibility
h) Registration and release

Robust choice criteria are followed for selecting biocontrol agents of specific pests and diseases. Selection relies on the target infective agent, crop and the cropping system. There is intensive screening of natural soils for the presence or absence of isolates with required traits and it frequently involves the selection of healthy plants from fields plagued with the target infective agent. Microorganisms are recovered from leaves, roots or rhizosphere, on the belief that they possess properties antagonistic to the infective agent [40]. These isolates are then screened for their activity against the infective agent in laboratory and greenhouse conditions. Potential isolates are evaluated for their efficiency in field conditions. Isolates exhibiting robust management and potency are further formulated for production and packaging [39]. A formulation is a mixture of active ingredients such as fungal spores with inert material like diluents and surfactants to modify the physical characteristics to a more effective form. A final formulation must be:

a) Easy to use.
b) Stable over a wide range (-5°C to 35°C) of temperature.
c) Have a durable shelf life (min. of 2 years) at normal temperature [40].

The most crucial obstacle within the application of bioagents is the loss of viability of the biological agents over time. The lifespan of the biocontrol product depends on storage temperature and the carriers utilized in the formulation of the biocontrol product. Some of the internationally and nationally used bioproducts for pests are described in Table 12.7 [40].

Table 12.7: List of Commercially available Biocontrol Products

Products		***Biocontrol Agents***	***Target Organisms/ Diseases***	***Crops***
Inter-national	Regalia (Marrone Bio Innovations)	*Reynoutria sachalinensis*	Alternaria fruit rot, Anthracnose, Blackspot, Botrytis, Crown rot, Verticillium wilt, *Pythium, Phytophthora*	Fruits, nuts, leafy vegetables
	Serenade (Bayer Cropscience Ltd.)	*B. subtilis*	Botrytis, mildew, Sclerotinia, *Xanthomonas, Erwinia*	Fruits, vegetables
	Actinovate (Mycorrhizal Applications)	*Streptomyces lydicus* WYEC 108	Mildew, botrytis, leaf blight, bacterial spot, *Fusarium, Pythium, Phytophthora, Rhizoctonia, Sclerotinia, Verticillium*	Ornamental crops
Indian	NIPROT (Pest Control India Pvt. Ltd.)	*Trichoderma viride/ Trichoderma harzianum*	Soil-borne fungal diseases, root rhizome foot, collar rot, root knot nematode	Food crops, cereals, vegetables
	Biolep (Biotech International Ltd.)	*B. thuringiensis subsp. Kurstaki*	*Lepidoptera*	Cotton, cereals, pulses, oil crops, fruit trees

Products		**Biocontrol Agents**	**Target Organisms/ Diseases**	**Crops**
	Biomonas (Biotech International Ltd.)	*Pseudomonas fluorescens*	Stem rots caused by *Sclerotinia* and *Rhizoctonia*, blights/ leaf spots caused by *Alternaria, Myrothecium, Ramularia, Xanthomonas, Erwinia*	Cotton, cereals, pulses, vegetables, oilseeds, fruit plants
	Daman WP (International Panacea Ltd.)	*Beauveria bassiana*	Lepidopterous caterpillars (*Helicoverpa, Spodoptera*), borers, hairy caterpillars, white fly aphids, scale insects, locust, Colorado beetles	Cereals, pulses, cotton, cauliflower, plantation crops

Registration of Biocontrol Agents

Commercialization of any biological agent requires its prior registration. Two necessary factors within the registration of bioagents are toxicity and environmental fate. Data needed for the registration of any biopesticide include:

a) Systemic name and common name of the biocontrol agent.
b) Natural incidence.
c) Morphological description of the bioagent.
d) Details of the producing method.
e) Impact on Human health.
f) Environmental toxicity.
g) Residual analysis

Merits and Demerits of Biocontrol in Agriculture

Biocontrol is essentially a non-toxic and eco-friendly approach with the benefit of being more cost effective and sustainable as compared to chemical or mechanical methods. It is the only method with high target pest specificity, high reproducibility and greater spread potential than any other method. Most of the biocontrol products and formulations are easy to handle, manufacture and non-residual. By acting as control agent biocontrol agents also enhance the natural microflora in the rhizosphere and endosphere. They are also a potentially permanent solution with less or negligible long-term maintenance. Combination of bio-pesticides and bio-fertilizers with biocontrol agents can further enhance the control strategy.

However, some limitations of this technology include their sometimes unpredictable impact on target weed. Also, their application may not be structured for small scale applications, sometimes leading to non- elimination of the weed but modulation of the weed to admissible levels especially under situations when the biocontrol agent depends on the weed for its nutrition. Though being less expensive in the long run it requires highly skilled application expertise and supervision during initial use. Some biocontrol agents can induce unwanted strains and novel mutations in native plantations. Furthermore, the level of control is not always high as compared to chemical control. Biocontrol methods are not mixable with chemical control agents and any interaction could lead to worsened infestations. But as we look upon the greater picture the benefits outweigh the limitations and thus biocontrol methods still remain a far better alternative to any other control methods.

Conclusion and Future Prospects

The recent growth of biopesticides over the chemically or synthetically derived approaches that possess high environmental repercussions is quite promising. Biopesticides prevent the build-up of pesticide resistance in the pests are non-pathogenic and result in increased crop yields. Biocontrol agents hold multiple applications in agriculture to manage pests, competition (weed and algae) and molluscs. Some methods such as allelopathy, use of lytic enzymes, microbial phytotoxins, nematodes, parasites and predators hold promising results when used against pests or crop competition.

Policies for the promotion of biocontrol agents among farmers should be created to reduce chemical control dependency. Future research should be focused on the biochemical and genetic analysis of crops to examine for natural allelopathic molecules. Further, genetic engineering can help to enhance the efficacy of multiple microbial biocontrol agents. With the advent of custom-made formulations to combat the enemies or competitors of a specific crop, regulatory and quality requirements must be maintained to ensure the availability of biocontrol agents at an affordable cost in developing countries. Overall, the application of biocontrol strategies would help achieve the aim of sustainable agriculture, food security, environmental protection and conservation of natural resources.

REFERENCES

1. Liliane, T. N., and Charles, M. S. (2020). Factors affecting yield of crops. *Agronomy-Climate Change and Food Security*, 9.
2. Hagstrum, D. W., and Flinn, P. W. (2018). Integrated pest management. In *Integrated management of insects in stored products* (pp. 399-407). CRC Press.
3. Aktar, M. W., Sengupta, D., and Chowdhury, A. (2009). Impact of pesticides use in agriculture: their benefits and hazards. *Interdisciplinary toxicology*, *2*(1), 1.
4. Nicolopoulou-Stamati, P., Maipas, S., Kotampasi, C., Stamatis, P., and Hens, L. (2016). Chemical pesticides and human health: the urgent need for a new concept in agriculture. *Frontiers in public health*, *4*, 148.

5. Mousa, W. K., and Raizada, M. N. (2016). Natural disease control in cereal grains.
6. Landis, D. A., and Orr, D. B. (1996). Biological control: approaches and applications.
7. Singh, N. K., Dhar, D. W., and Tabassum, R. (2016). Role of cyanobacteria in crop protection. *Proceedings of the National Academy of Sciences, India Section B: Biological Sciences, 86*(1), 1-8.
8. Farooq, M., Bajwa, A. A., Cheema, S. A., and Cheema, Z. A. (2013). Application of allelopathy in crop production. *International Journal of Agriculture and Biology, 15*(6), 1367-1378.
9. Mohan, S., Devasenapathy, D., Venilla, C., and Gill, M. S. (2010). Pest and disease management in organic ecosystem. *IPM Booklet, Tamil Nadu University, India.*
10. Mahfouz, S. A., and Abou-el-Ela, A. A. (2011). Biological control of pink bollworm Pectinophora gossypiella (Saunders) by Bacillus cereus MA7. *Journal of Microbial and Biochemical Technology, 3*(2), 030-032.
11. Naik, V., Nagrare, V. S., Subbireddy, K. B., Kumbhare, S., Wawdhane, P., and Prabhulinga, T. (2019). Management of cotton pink boll worm pectinophora gossypiella (saunders) with Trichogramma bactrae and T. brasiliensis. *Indian Journal of Entomology, 81*(4), 744-748.
12. Biddinger, D. J., Weber, D. C., and Hull, L. A. (2009). Coccinellidae as predators of mites: Stethorini in biological control. *Biological Control, 51*(2), 268-283.
13. Hoover, K., Schultz, C. M., Lane, S. S., Bonning, B. C., Duffey, S. S., McCutchen, B. F., and Hammock, B. D. (1995). Reduction in damage to cotton plants by a recombinant baculovirus that knocks moribund larvae of Heliothis virescens off the plant. *Biological Control, 5*(3), 419-426.
14. Pal, K. K., and Gardener, B. M. (2006). Biological control of plant pathogens.
15. Nega, A. (2014). Review on concepts in biological control of plant pathogens. *Journal of Biology, Agriculture and Healthcare, 4*(27), 33-54.
16. Araujo, J. V., Gomes, A. P. S., and Guimarães, M. P. (1998). Biological control of bovine gastrointestinal nematode parasites in Southeastern Brazil the nematode-trapping fungus Arthrobotrys robusta. *Revista Brasileira de Parasitologia Veterinária, 7*, 117-122.
17. Ranjbar-Bahadori, S., Razzaghi-Abyaneh, M., Bayat, M., Eslami, A., Pirali, K., Shams-Ghahfarokhi, M., and Lotfollahzadeh, S. (2010). Studies on the effect of temperature, incubation time and in vivo gut passage on survival and nematophagous activity Arthrobotrys oligospora var. oligospora and A. cladodes var. macroides. *Global Veterinaria, 4*(2), 112-117.
18. Duval, D., Galinier, R., Mouahid, G., Toulza, E., Allienne, J. F., Portela, J., and Gourbal, B. (2015). A novel bacterial pathogen of Biomphalaria glabrata: a potential weapon for schistosomiasis control?. *PLoS neglected tropical diseases, 9*(2), e0003489.

19. Tripathi, A., Sharma, N., and Tripathi, N. (2010). Biological Control of Plant Diseases: An overview and the Trichoder masystem as Bio-control Agents. *Management of fungal plant pathogens*, 121.

20. Chisholm, S. T., Coaker, G., Day, B., and Staskawicz, B. J. (2006). Host-microbe interactions: shaping the evolution of the plant immune response. *Cell, 124*(4), 803-814.

21. Jabran, K., Mahajan, G., Sardana, V., and Chauhan, B. S. (2015). Allelopathy for weed control in agricultural systems. *Crop protection, 72*, 57-65.

22. Duke, S. O., Cantrell, C. L., Meepagala, K. M., Wedge, D. E., Tabanca, N., and Schrader, K. K. (2010). Natural toxins for use in pest management. *Toxins, 2*(8), 1943-1962.

23. Lamichhane, J. R., Devos, Y., Beckie, H. J., Owen, M. D., Tillie, P., Messéan, A., and Kudsk, P. (2017). Integrated weed management systems with herbicide-tolerant crops in the European Union: lessons learnt from home and abroad. *Critical reviews in biotechnology, 37*(4), 459-475.

24. Cheema, Z. A., and Khaliq, A. (2000). Use of sorghum allelopathic properties to control weeds in irrigated wheat in a semi arid region of Punjab. *Agriculture, Ecosystems and Environment, 79*(2-3), 105-112.

25. Reberg-Horton, S. C., Burton, J. D., Danehower, D. A., Ma, G., Monks, D. W., Murphy, J. P.,. and Creamer, N. G. (2005). Changes over time in the allelochemical content of ten cultivars of rye (Secale cereale L.). *Journal of chemical ecology, 31*(1), 179-193.

26. T Anjum, T., and Bajwa, R. (2008). Screening of sunflower varieties for their herbicidal potential against common weeds of wheat. *Journal of Sustainable Agriculture, 32*(2), 213-229.

27. Schwarz, D., and Krienitz, L. (2005). Do algae cause growth-promoting effects on vegetables grown hydroponically. *Fertigation: optimizing the utilization of water and nutrients. International Potash Institute, Beijing*, 161-170.

28. Routray, S., and Dey, D. (2016). Snails and slugs as crop pests. *Rashtriya Krishi, 11*(1), 40-41.

29. Das, P. P. G., Badal Bhattacharyya, S., Bhagawati, E. B. D., Manpoong, N. S., and Bhairavi, K. S. (2020). Slug: An emerging menace in agriculture: A.

30. Howlett, S. A. (2012). Terrestrial slug problems: classical biological control and beyond. *CAB Rev, 7*(51), 1-10.

31. Port, G. R., Glen, D. M., and Symondson, W. O. C. (2000). Success in biological control of terrestrial molluscs. In *Biological control: measures of success* (pp. 133-157). Springer, Dordrecht.

32. Zala, M. B., Sipai, S. A., Bharpoda, T. M., and Patel, B. N. (2018). Molluscan pests and their management: A review. *AGRES-An International EJ, 7*(2), 126-132.

33. Kumar, P. (2020). A Review—On Molluscs as an Agricultural Pest and Their Control. *International Journal of Food Science and Agriculture, 4*(4), 383-389.

34. PEAReE, J. D., and RODGERS, P. B. (1995). Monoxenic culture of the slug parasite Phasmarhabditis hermaphrodita (Nematoda: Rhabditidae) with different bacteria in liquid and solid phase. *Fundam. Appl. Nematol, 18*(2), 159-166.

35. Tan, L., and Grewal, P. S. (2001). Pathogenicity of Moraxella osloensis, a bacterium associated with the nematode Phasmarhabditis hermaphrodita, to the slug Deroceras reticulatum. *Applied and environmental microbiology, 67*(11), 5010-5016.

36. Tan, L., and Grewal, P. S. (2002). Endotoxin activity of Moraxella osloensis against the grey garden slug, Deroceras reticulatum. *Applied and environmental microbiology, 68*(8), 3943-3947.

37. Sharma, A., Diwevidi, V. D., Singh, S., Pawar, K. K., Jerman, M., Singh, L. B.,. and Srivastawav, D. (2013). Biological control and its important in agriculture. *International Journal of Biotechnology and Bioengineering Research, 4*(3), 175-180.

38. Barua, A., Williams, C. D., and Ross, J. L. (2021). A Literature Review of Biological and Bio-Rational Control Strategies for Slugs: Current Research and Future Prospects. *Insects, 12*(6), 541.

39. Teixido, N., De Cal, A. L., Usall, J., Guijarro, B., Larena, I., Torres, R.,. and Köhl, J. (2015, June). Biocomes: new biological products for sustainable farming and forestry. In *III International Symposium on Postharvest Pathology: Using Science to Increase Food Availability 1144* (pp. 469-472).

40. Junaid, J. M., Dar, N. A., Bhat, T. A., Bhat, A. H., and Bhat, M. A. (2013). Commercial biocontrol agents and their mechanism of action in the management of plant pathogens. *International Journal of Modern Plant and Animal Sciences, 1*(2), 39-57.

Chapter 13

Soil Composition and Microbial Ecosystem

Manoj Kumar[1], *Mohit Kumar*[1], *Sudhir Kumar Kataria*[2], *Jitender Sharma*[1] *and Sunita Dalal*[1]*

[1]*Department of Biotechnology, Kurukshetra University, Kurukshetra – 136 119, Haryana*

[2]*Department of Zoology, Maharishi Dayanand University Rohtak – 124 001, Haryana*

**email: sdalal@kuk.ac.in*

ABSTRACT

Soil is a fundamental source of a wide variety of ecosystem and environmental services characterized as the commodities and biological system works that give advantage to human populations. It acts as a designing medium, recycling framework of nutrients and natural waste, controller of water quality, rearrangement platform for air and medium needed for plants to grow and develop. It is a territory for many organisms including invertebrates (like earthworms, woodlice, nematodes etc.), bacteria, fungi, archaebacteria, algae etc. In this review, we have discussed about soil, its type and composition, microorganisms present in soil and their role in fertility as well as the extracellular enzymes that are secreted by these microorganisms needed for various nutrient recycling and fertilization of soil.

Keywords: *Composition of soil, Soil microorganisms, Extracellular enzymes, Soil organic matter, Role of microorganisms.*

Introduction

Soil is mainly considered as a combination of organic matter, minerals, fluids, gases and living matter which mutually sustain life. The body of earth comprising soil is called Pedosphere. Pedolith was an ancient Greek word usually referred to soil, ground or earth that means to grind rocks in the elemental stones. Soil Science has two fundamental branches *i.e.* Edaphology included that how soils effect the living things of soils and Pedology referred as the creation, (morphology) depiction and classifying the soils into their ordinary atmosphere (Amundson, 2009). Soil comprises of a concrete phase of organic substance and mineralsalongwith a porous part that is filled with gases and water. Soil also take an interest in the arrangement of regulation services (environment guideline by controlling fluxing of greenhouse gas and C sequestration; flood manage, detoxifying the waste, security of plants from pests) by affecting natural organic substance and the broad-ranging consequences for soil morphological properties. The littlest Protists and Nematodes of the microfauna below 200 micrometer size on average reside in the H_2O-filled porosity. Enchytraeidae, Microarthropods and the numerous gatherings of the mesofauna (0.2–2 milimeter) reside all around in gas occupied soil porosity. Soil was considered as the principle supplier of ecosystem services. Agricultural and woodland formation, assurance against flooding and disintegration (erosion), water loading, obsession of air C and N are among thc frequently referred to environmental services that soil provided us. Soils are fundamental source of broad variety of environment services characterized as the goods and biological system works that give advantage for human populations (Bedatou *et al.*,2009). Every gram of soil can hold upto ten billion microorganisms of perhaps a abundance of various species (Torsvik and Ovreas, 2002). Soils give a wide scope of environment benefits that address human issues. These included procurement of services, the arrangement of food, crude materials, fiber, wood and actual help to framework; controlling services, filtering of impurities and supplements, being flood moderation, restoring of carbon and ozone harming substance regulation, recycling of wastes, detoxification and guideline regulating pests and population of disease and social aids, being recreation, heritage esteems,aesthetics and social identity (Kopittke *et al.*, 2019).

Soil acts as an essential carbon pool, (Amelung *et al.*, 2020) and is potentially the most receptive to human interruption and environmental transformation (Davidson and Janssens, 2006). Soil acts like a designing medium,recycling framework for supplements and natural waste, modifier of air arrangement, habitat for soil organisms,controller for quality of water along with a platform needed for plant growth and development (Dominati *et al.*, 2010). It contains the greater part of genetic diversity found on Earth. One gm soil can have a large number of creatures, belonging to thousands of genus, generally microorganisms and many more still not explored. Soil has an average prokaryotic concentration of approximately 10^8 organism entities for every gram (Dykhuizen, 1998).

Properties of soils, for example, crop production, structure, mass thickness, texture (sand, silt and clay), porosity, consistency, temperature, shading and resistivity impacted the environment services. Soil structure known as peds (more usually soil aggregates) are made up of soil isolates when carbonates, iron oxides,clay,silica along with humus, cover the molecules causing them to follow in bigger steady secondary form. Porosity of soil comprises of empty porous fraction in soil amount that is engaged by water and gas (Six *et al.*, 2000).

For agriculture purpose, soil is needed for support, development and growth of plants by acting like an anchor and essential supplement base to crops. Soil assets are basic to climate along with food and fiber making, delivering 98.8 per cent consumption of food to people (AlShrouf, 2017). Plants take water as well as minerals from soil as per a few cycles engaged in nutrition of plants. It assimilates water and make it available for later, also flood and drought regulation being one of the most significant biological system services of soil. As water permeates through soil, it gets refreshed (Pan and Zhao, 2007). It is the territory for billions of organisms and significant fraction of known and unexplored diversity of organisms is present in the soil, as microbes, algae, fungi, archaea, invertebrates (millipedes,earthworms, centipedes, woodlice,slugs, parasites,snails, mites, enchytraeids, springtails, nematodes, protists)and other organisms living over the ground have some portion (plants) or consume a fraction of their life span on earth (insect) beneath earth (Kopittke *et al.*, 2019). Over the ground and underground biodiversities are firmly interconnected (Melero *et al.*, 2007) making soil assurance of fundamental significance. The organic segment of soil is a critical sink ofcarbon as almost 57 per cent of biotic substance is considered to be carbon. Indeed, lichens, cyanobacteria and mosses found even in desert form organic coverings of soil that catch and seize a lot of carbon during photosynthesis (Van Cuyk *et al.*, 2001).

Types of Soil

On the basis of soil texture, porosity and other physical parameters, it is categorized to 4 types *i.e.* sand, clay, silt, and loam.

Due to weathering of rocks, fine or littlest particles produced during this process forms sandy soil. This is poorest kind of soil to be used for farming and growing plants because they have exceptionally less healthy benefit and helpless limit to hold water. The range of larger molecule size of sand is 2 – 4.75milimeter, Medium range from 0.425 – 2 milimeter and littlest particle size range is 0.075 – 0.425 milimeter (Bekchanova *et al.*, 2021).

Clay is essentially comprised of the littlest molecules of soil, that are thickly loaded by too little/no airspace and water is adequately held by them. Clay soil isn't reasonable for agriculture because entry of moisture and air is more difficult into this soil. The molecules of the soil are firmly arranged along one another having atleast/no airspace. Its molecule size is 0.002 milimeter (Firoozi and Baghini, 2016).

Silt is having more modest molecules compared to sand and is composed of rocks and many other mineral compound, that are much bigger than clayand smaller than sand. Siltis smoother and having fine quality of soil with better water holding capacity compared to sand. Silt can be easily transported through waves and mostly present close to lake,stream, river and other water bodies. Silt is much fertile as compared to sand soil and clay soil. Consequently, it is utilized for farming to increase fertility of soil. The molecule size range of silt is 0.002 - 0.06 milimeter.

Loam soil is 4th type of soil. This type is a mixture of silt, clayand sand to achieve the aim that the valuable features can be incorporated from each type. For instance, it holds moisture along with supplements. This makes it more reasonable in cultivation. It is additionally referred as agriculture soil because it has harmony with all the 3 kinds of soil and also contain humus. Beside these features, higher level of calcium and pH is present due to inorganic origin. It is appropriate for agriculture because root of plants get an adequate measure of supplements and water required for their development,growth and improvement. This soil is having practically the same amounts of silt and sand with somewhat low amount of clay. 40 per cent of each *i.e.* silt and sand and 20 per cent of clay makes a decent proportion for loam soil.

Alluvial soil is finely grained fertile soil deposited and stored by water streaming over flood fields or in river beds. It is most appropriate soil for farming because the proportion of silt, clay and sand is with the end goal that the water, air and supplement retention limit of this soil become high.

Composition of Soil

Soil is essentially made out of approx. 50 per cent voids or pores (that are filled with gas and water)and 50 per cent solids (45 per cent mineral and 5 per cent organic matter). Fraction of natural substance and mineral may be considered as steady (temporarily), while the gas content and soil water is viewed as exceptional factor with high variability yet this variation is also adjusted by them (for example elevation in amount of one factor get adjusted by decrease in another). The void space takes into consideration the invasion and flowing of water and wind, the two that are basic for existence of life inside soil. Tough packaging is typical issue within soil particles that decreases the space, keeping away water and air from arriving at root of plants and fauna in soil. After a particular fraction of time, consistent/uniform soil develops a profile that comprises at least 2 layers, alluded as soil horizons. Horizons contrast in at least one property, for example, in their thickness, porosity, consistency, texture, structure, temperature, color and reactivity. The biological effects on soil properties are strong mostly close to surface, but geochemical impacts on soil features rise with deepness (Zanella *et al.*, 2019). Developed soil profiles ordinarily incorporate three fundamental master horizons from surface to depth: A (living segments), B (parent mineral parts + living segments), and C (geochemical segments) (Zanella *et al.*, 2019).

The texture of soil is controlled by general extents of every molecule of clay, silt and sand which make soils. The association of each mineral molecule with water, gases and organic substances through abiotic and biotic ways make those molecules to remain as one for forming soil aggregations. Water is crutial agent for soil formation because of its participation in disintegration, dissolution,transport, precipitation and accumulation of materials which made soil (FAO report, 2021). The combination of water and broke down or suspended materials (many dissolved organic and mineral substances) that consume the soil void space is known as soil solution (Valentin *et al.,* 1999).

Soils are the sources of nutrients for plants, the majority of which are held by the molecules of organic matter (colloids) and clay. The supplements or nutrients might be adsorbed on elemental surfaces of clay, bound inside the clay minerals (consumed), or bound inside organic mixtures as a feature of the living organisms or dead soil matter. These bound supplements associate with soil water to support the soil solution entities (weaken changes in the soil solution) as soil gets wet or dry out, as plants take up nutrients, as salts are drained, or as acids or soluble bases are added. Plant supplement accessibility is influenced by soil pH, which is a proportion of the hydrogen atom action in the soil solution. Soil pH is a function of many soil framing factors, and is for the most part lower (more corrosive) where weathering is further advanced (Miller, 2016).

Some nitrogen originates from rain as diluted HNO_3 and NH_3, (Goulding *et al.,* 1998) yet the majority of the nitrogen is accessible in soils because of nitrogen fixation by microbes. Once in the soil-plant framework, most of the nutrients are recycled through living organisms, plant and microbial parts (soil organic matter), mineral-associated forms and the soil solution. Both soil organic substances and living soil organisms (microorganisms, plant roots and animals) are of crutial significance to this recycling, and subsequently to soil fertility and soil development (Pukalchik *et al.,* 2018). Microbial soil enzymes may deliver nutrients from organic substances or minerals to get used by plants and different microorganisms, incorporate them into cells of living organisms, or cause their removal from the soil by volatilisation (lost to the environment as gases) or Leaching.

Soil molecules can be characterized by their composition (mineralogy) along with their size. The molecular size distributed in a soil and its texture decides a large number of the properties of that soil, specifically water potential and hydraulic conductivity, (Saxton and Rawal, 2006) however the mineralogy of those molecules can firmly alter those properties. The mineralogy of the fine soil molecules, clay, is particularly significant.

a. Soil Organic Matter

Organic matter in soil is comprised of organic substances and includes animals, plants and microbial material, both dead and living. The formation, collection and degradation of organic matter are significantly reliant upon climate. Soil moisture, topography, and temperature are the central point influencing the accumulation of organic substances in soils. Organic matter will in general aggregate under wet or cold conditions where decomposer action is obstructed by low temperature (Wang *et al.*, 1996) or abundance moisture which brings about anaerobic conditions (Brodowski *et al.*, 2007). Conversely, high rainfall and high temperatures of tropical areas results in quick disintegration of organic matter and leaching of plant nutrients. Cellulose and hemicellulose go through quick disintegration by fungi and microbes, with a half-existence of 12–18 days in a temperate environment (Minayeva *et al.*, 2008). Brown colored rot fungi can decompose cellulose and hemicellulose, providing the lignin and phenolic compounds lastly. Starch, an energy storing framework for plants, goes through quick decay by microorganisms and fungi. Lignin in plant cell walls goes through extremely deliberate decay, fundamentally by white rot fungi and actinomycetes and its half-life under temperate environment is around half year (Minayeva *et al.*, 2008). These two components are additionally liable for add-on to the structure and measure of soil organic matter. A typical soil comprises of 22 per cent macrofauna, 8 per cent roots and largely distributed 70 per cent microorganisms. The living part of a section of soil may include 900 lb of earthworms, 1500 lb of bacteria, 2400 lb of fungi, 890 lb of arthropods and algae and 133 lb of protozoa (Parfitt *et al.*, 1995). A portion of the soil organic matter comprises of the microbial biomass and metabolites of bacteria, molds, and actinomycetes that functions to decay the dead organic matter (Hajnos *et al.*, 2003). The major part of organic matter is a complex array of little organic molecules, on the whole called humus or humic substances (Sparling, 1992).

Humus is organic matter that has been decayed by soil microflora and fauna to where it is resistant to additional breakdown. Humus generally comprises only 5 per cent of the soil or less by volume, however it's a fundamental source of nutrients and increase the significant textural characteristics essential for soil heealth and plant growth (Scheu, 2002). Humus additionally feed to arthropods, termites and earthworms which in turn improve the quality and fertility of soil (McDonald, 1994). Finally, produced humus is held in colloidal form in the soil solution and results in a weak acid that can degrade silicate minerals (Ponge, 2003). The cation and anion exchange capability of humus is extremely high and its dry weight is many times greater than the clay colloids. It is against the change in pH and soil moisture so, it also acts as a buffer (Pettit, 2004).

Humic acids and fulvic acids, which start as crude organic matter, are significant components of humus. After the death of plants, animals and microorganisms, microorganisms start to take food from the residues through secretion of extra cellular enzymes, finally completing the production of humus (Kappler and Brune, 2000). As the residues are decomposed, only aliphatic and aromatic hydrocarbons get assembled and stabilized by the hydrogen bonds, stay as complex assembly which is collectively called humus (Kappler and Brune, 2000). Humus is generally impure in the soil, as it always form complexes with metals and clays which are further responsible for the stability and soil structure (Pettit, 2004).

Mostly soil living matter including plants, animals, microbes, and fungi are subject to organic matter for nutrients as well as energy. Organic compounds in soil are degraded to different extents depending upon the soil moisture, temperature and air circulation. Fungi and bacteria feed on the crude organic matter, which are then taken up by protozoa, which are then fed upon by nematodes, annelids and arthropods that are able to decompose and transform crude or humified organic matter. This has been known as the soil food web, through which all organic matter is processed as in digestive system (Prescott, 2010).

b. Minerals (Elements)

Seventeen nutrients/supplements or elements are fundamental for plant development and propagation that are essentially present in the soil repository in effectively accessible form. They are {Carbon (C), Hydrogen (H), Oxygen (O), Nitrogen (N), Phosphorus (P), Potassium (K), Sulfur (S),Calcium (Ca), Magnesium (Mg), Iron (Fe), Boron (B), Manganese (Mn), Copper (Cu), Zinc (Zn), Molybdenum (Mo), Nickel (Ni) and Chlorine (Cl)}. Oxygen, carbon and hydrogen are exceptional as they are provided through nitrogen fixation (Mccahon *et al.*, 1997), the supplements are initially taken from the mineral segment of the soil (Osterholm *et al.*, 1983). A specific nutrient proportion of the soil solution is consequently obligatory for enhancing growth of the plants and when the values are different from nutrient proportion then it can be determined by composition of plant.

Plants can only take and consume nutrients when they are available in plant-accessible form. Mostly, supplements are taken as ions from (or along with) soil water. In spite of the fact that minerals are the beginning of most supplements, and most of the supplement components are held in crystalline/translucent structure inside essential and optional minerals, they disintegrate gradually to help quick plant development.

Clay molecules and soil organic matter give a more open repository of many plant supplements (for example K, Ca, Mg, P, Zn) as they have nutrients on their surfaces. As plants assimilate the supplements from the soil water, the surface-bound pool renewed the solvent pool. Another system of replenishment of nutrients in soluble pool is the disintegration of soil organic matter by microorganisms which is significant for supplying of plant-accessible Nutrients (N, S, P, and B) from soil (Van der Ploeg *et al.*, 1999).

Soil Microorganisms

The investigation of microorganisms in soil, their capacities, and what they mean for soil properties is known as soil microbiology. First microscopic organisms on earth accepted to came 2 to 4 billion years prior that could perform nitrogen fixation and subsequently delivered oxygen into the environment (Farquhar *et al.*, 2000). This prompted more amazing microorganisms, which influenced soil construction and fertility. Soil microorganisms can be categorized as bacteria, actinomycetes, fungi, virus, nematodes and protozoa. Each category have various unique qualities which determine the organism and various functions performed by them (Johns, 2017). Various groups of soil microorganisms are described briefly as follows:

Bacteria

Bacteria are usually unicellular and microscopic organisms. Around 10^8 to 10^9 microorganisms can be available in a teaspoon of wet, fertile soil (Tilak *et al.*, 2005). Up to 10^{10} bacterial cells possess every gram of soil near or in plant root, an area known as the rhizosphere. A group distinguished greater than 33,000 bacterial and archaeal species on sugar beet roots. They are mainly decomposers and make nutrients accessible that are not gained by various organisms. They play a significant part in different mineral cycles like Nitrogen cycle.

Examples of soil bacteria are *Agrobacterium, Arthrobacter, Bacillus, Alcaligens, Clostridium, Corynebacterium, Erwinia, Nitrosomonas, Nitrobacter, Pseudomonas, Rhizobium, Thiobacillus, etc.*

Actinomycetes and Fungi

Actinomycetes are soil microbes that look like as bacteria and fungi and regularly considered as a missing transformative connection between these two groups (Jog *et al.*, 2016). However, they are more like bacteria as compared to fungi. Their major roles are to provide a trademark smell to different soil and are sources of a few critical therapeutic medicines.

Fungi are unusual organisms containing strings called hyphae which further form mycelium which can be less than 0.8 mm to as long as several metres. They can be beneficial as well as harmful to other soil organisms as they can play anonymous role at the same time. They are beneficial as they can break complex nutrients and release them into the soil that is used by the other soil organisms (Ud din Khanday *et al.*, 2016). Also, fungi can make attachment with the plant roots (relationship called Mycorrhiza) which significantly increase the plant growth. It helps the plant by providing nutrients in reference to get carbohydrates from the plant. On the other hand, fungi can also behave as a parasite and get food by joining themselves to plants or different organisms for self-centered reasons.

Examples - The main groups of soil-borne root pathogenic parasites and oomycetes comprise of genera *Fusarium* (Michielse *et al.*, 2009), *Verticillium* (Klosterman *et al.*, 2009), *Rhizoctonia, Pythium, Phytophthora* and numerous others, of worldwide and neighborhood significance.

Algae

Algae are mainly found in soils where moisture and daylight are accessible. In soils, the number of algae found ranges from 100 to 10,000 in each gram of soil (Johns, 2017). They are ready for photosynthesis as they acquire carbon dioxide from environment and energy from daylight and synthesize their food (Shields and Durrell, 1964).

Examples – *Euglenophyta* (Euglenoids),*Chrysophyta* (Golden-brown algae and Diatoms), *Pyrrophyta* (Fire algae), *Chlorophyta* (Green algae), *Rhodophyta* (Red algae), *Paeophyta* (Brown algae), *Xanthophyta* (Yellow-green algae) *etc.*

Protozoa

These are single-celled organisms with no color. They are bigger than microscopic bacteria and differing in size from microns to a few millimeters (Stout, 1980). Their number in soil goes from 10^4 to 10^5 for every gram of soil and their number is high in surface soil. They can accommodate unfavorable conditions, as they are portrayed by an protective, non functional (dormant) stage in their life cycle (Johns, 2017).

Viruses

Viruses in soil are vital as they may impact the soil biology by capacity to move genes and basic reason of microbial death. Therefore, viruses are significant components in worldwide cycles, impacting the turnover and grouping of nutrients and gases. In spite of this significance, the virology in soil is subject to be understudied (Zhuang and Jin, 2003). To investigate the function of the viruses in soil quality and plant health, studies are conducted for virus diversity. It has been discovered that viruses are exceptionally higher in number in every area concentrated so far independent of the critical change in the bacterial number of similar conditions. A few analysts accept that researching this diversity of generally unexplored soil viruses can possibly change our understanding of the function of viruses in worldwide biological system measures and the development of microbial life itself (Johns, 2017).

Nematodes

These are worm like organisms commonly 50 micrometer in diameter and 1 mm long. They play a significant part in soil by working at a few levels of the food web in soil. Plants and algae (the first level) are common food for some nematodes while others are grazers that feed on microorganisms and fungi (second level), and some other feed on different nematodes (more elevated levels). Free-living nematodes can be bacterial consumers, Fungal consumers (feed by penetrating the cell walls of fungi and sucking out the interior substance) (Neher, 2001) or Predatory nematodes (eat a wide range of nematodes and protozoa). They eat more modest organic entities or held themselves to the outer skin of bigger nematodes, scratching away until the prey's inside body parts can be extracted. Nematodes are significant in mineralization or delivering supplements in plant-accessible forms. When microscopic organisms or parasites are eaten by the nematodes, release of ammonia take place as microbes and fungi contain substantially more nitrogen than the nematodes require. Nematodes may likewise be helpful markers of soil quality due to their gigantic diversity and their support in numerous capacities at various levels of the food web in soil (Johns, 2017).

Examples - *Longidorus, Meloidogyne* (root-knot nematodes), *Heterodera* (soybean cyst nematodes), *Xiphinema* (dagger nematodes), *Trichodorus, Nacobbus, Ditylenchus, Globodera* (potato cyst nematodes), *Pratylenchus* (lesion nematodes) and *Aphelenchoides* (foliar nematodes).

Role of Soil Microorganisms

Soil microorganisms have a fundamental role in decaying organic matter, recycling supplements and fertilization of soil. Soil microorganisms behave as components as well as producers both for organic carbon in soil. Abundance of soil organic carbon improves soil fertility as well as water-holding limit. Without the recycling of elements, it is impossible to continue life on earth. The responses associated with basic recycling are regularly chemical in nature, however biochemical responses, those are processed by organisms, also have a significant influence in the recycling of elements. Soil microorganisms have major significance in this cycle. Soil microbes are also significant for the advancement of healthy soil structure. They produce heaps of sticky substances (polysaccharides and mucilage) that assist to solidify soil aggregates. This cementation makes aggregates less prone to crumble when exposed to water. A few bacteria have the capability to colonize the minerals and help to alter weathering of rocks and the breaking of these minerals. More minerals in region can bring about a higher number of bacteria (Vieira *et al.,* 2020). Microscopic organisms are liable for the cycle of nitrogen fixation, which is the change of air N_2 into compounds (like alkali) such as NH_3 that can be utilized by plants. The most distinguished feature of bacteria is their biochemical flexibility (Falkowski *et al.,* 2008). Actinomycetes can create antibiotics that are utilized to treat bacterial infections. Fungi are significant in soil as food resource for various organisms like large organisms, pathogenic microbes, beneficial symbiotic associations with plants or other forms of organisms and soil health. Fungal constituents additionally balance out soil structure as these string like structures divide all through the soil, in nearly present molecules and aggregates same as hairnet. The fungi can be considered as the "strings" of the soil texture. It should be focused on that microorganisms for the most part apply little effect on changing the genuine physical structure of the soil that may be performed by bigger life forms algae are additionally capable for performing N_2 fixation.

Microorganisms are capable to make nutrients and minerals in accessible form to plants in soil, synthesize hormones that spike growth, stimulate the plant resistant framework and trigger or hose stress responses. There is a developing body of exploration that supports the theory that soil microorganisms, and fungi especially, can be used to release carbon out of the environment and sequester it in soil. Soil microorganisms may give a critical methods for decreasing air ozone depleting substances and help to restrict the effect of ozone depleting substance induced environmental change (Johns, 2017). Most soil enzymess are produced by bacteria, fungi and plant roots. The biochemical activity associated to them is a factor in both degradation and stabilisation of soil form. Activity of enzymes is higher in areas that are treated with manure as contrasted with inorganic fertilizers. The microflora of the rhizosphere may also alter action of catalysts.

Microorganisms and Mineral Cycles

Carbon Cycle

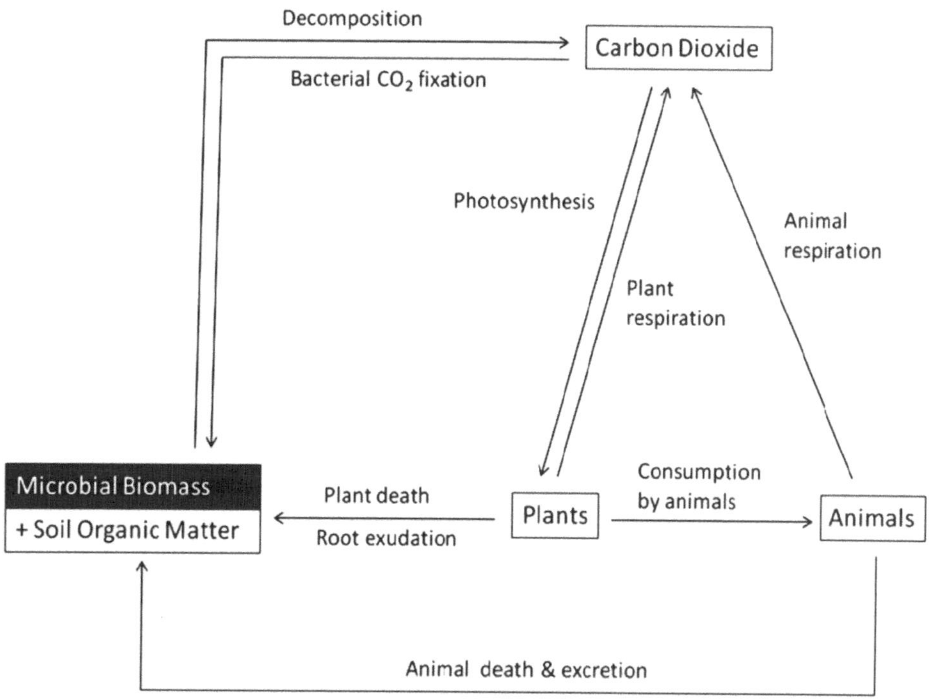

Carbon is the most important element in living beings. Main sources of carbon for the organic world are photosynthetic organisms like plants, algae and some bacteria that can fix atmospheric CO_2. Microbes decompose the organic material like cellulose, hemicellulose and lignin to return CO_2 into the atmosphere. Cellulose is the most abundant carbon compound found in plant that is decomposed by fungi (*Trichoderma, Aspergillus, Penicillium, Fusarium, Chaetomium etc.*) and bacteria (*Clostridium, Streptomyces, Bacillus, Pseudomonas, Nocardia etc.*). simple sugars llike pentose and hexose *etc.* can be polymerised to form hemicelluloses. Hemicellulose is present in the cell wall or intercellular space in the plant body. It is decomposed by fungi (*Chaetomium, Aspergillus, Penicillium, Trichoderma etc.*) and bacteria (*Bacillus, Pseudomonae, Cytophaga, Erwinia, Streptomyces*). Lignin is highly resistant to microbial degradation and the decomposition occurs at very slow rate by fungi (*Aspergillus* and *Polyporus*) and bacteria (*Streptomyces* and *Nocardia*) (Gougoulias *et al.*, 2014).

Nitrogen Cycle

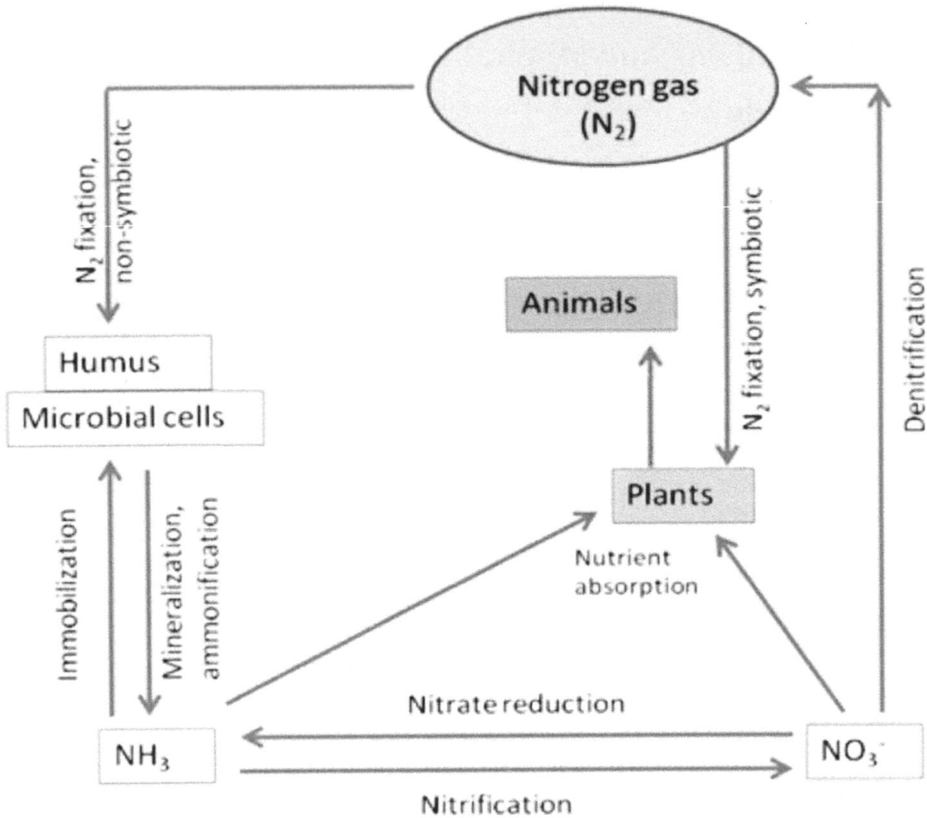

Nitrogen is present in highest concentration (78 per cent) in the atmosphere but in soil, it is found in very limited amount. Nitrogen is very essential nutrient substance for both plant and animals. Microbes help to increase the nitrogen content in the soil and play significant role to return nitrogen from the organism to the atmosphere (Rosswall, 1976). Nitrogen cycle is comprised of fixing atmospheric nitrogen by industrial, natural or biological means to ammonia and protein so that plants and animals can easily absorb it. It is a reduction process and carried out biologically by prokaryotes that are called diazotrophs. Nitrogen-fixing bacteria are of two types: the free-living (non-symbiotic) bacteria includes cyanobacteria (or blue green algae), *Anabaena* and *Nostoc* and symbiotic bacteria that include *Rhizobium*, attached to the leguminous plants. The symbiotic type invades the root hairs of plants, where they grow and encourage development of root nodules, enlargement of plant cells and make bacteria in close attachment. The bacteria change free N_2 to NH_3, which host plant utilizes for their growth. Further steps of nitrogen cycle include Ammonification, Nitrification and Denitrification. In ammonification, organic form of nitrogen (proteins, humus *etc.*) is converted to inorganic form like ammonia (NH_3). Microorganisms involved in the ammonification process are bacteria (*Pseudomonas, Bacillus, Clostridium Serratia*) and fungi (*Alternaria, Aspergillus, Mucor, Penicillium, Rhizopus*). Nitrification process is an oxidative process in which Ammonia (NH_4^+ form) is first converted to nitrite (NO_2^-) by a group of bacteria like (*Nitrosomonas, Nitrosococcus, Nitrosolobus, Nitrospira*) and then nitrite (NO_2^-) is further converted to nitrate (NO_3^-) by some soil bacteria like (*Nitrobacter, Nitrospira* and *Nitrococcus*). In denitrification process, conversion of nitrate (NO_3^-) to nitrogen or nitrous oxide (N_2, N_2O) occurs by a group of bacteria like (*Thiobacillus denitrificans, Micrococcus denitrificans, Pseudomonas* and *Bacillus*). These denitrifying bacteria are abundant in aerobic soil and their population is higher in rhizosphere soil (Soil around the plant root) (Mahmud *et al.*, 2020).

Phosphorus Cycle

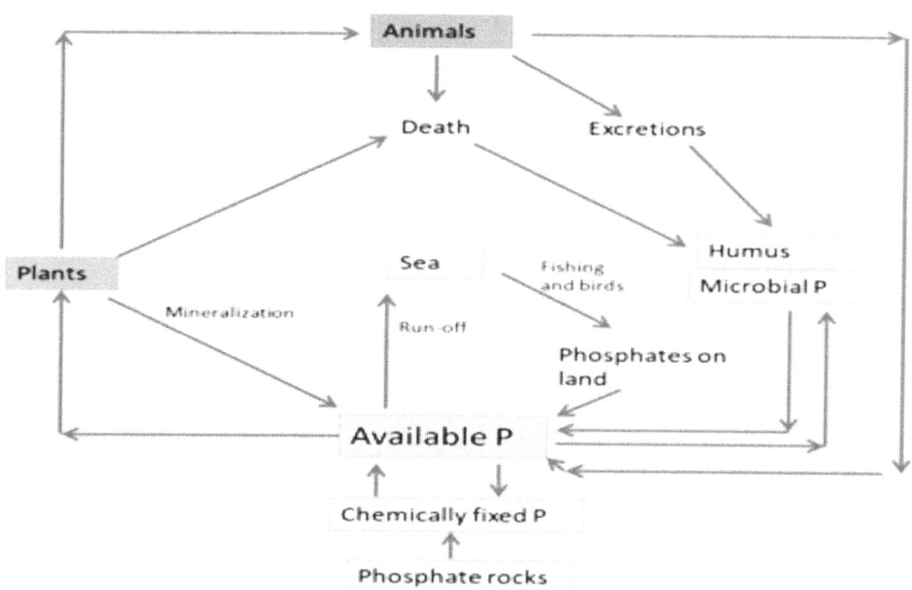

Phosphorus is an important constituent of protoplasm and required for metabolism of all living organisms. Main source of phosphorus is rocks. In soil, phosphorus is present in both inorganic and organic forms. The inorganic form of phosphorus is derived from rocks or through fertilizers and can be dissolved easily in water when found as phosphates of Na, K, Ca, Mg *etc.* The organic P containing compounds that are extracted from plants and microorganisms and are mainly having phospholipids, phosphorylated sugars, coenzymes, nucleic acid, phytin and other compounds. In soil, 15-85 per cent of total phosphorous is organic which is unavailable to plants unless it is microbiologically converted into inorganic form and is solubilized by many microbes into solution. Such microbes are abundant on root surface of plants. There are many species of bacteria (Species of *Psedomonas,*

Bacillus, Flavobacterium, Mycobacterium, Micrococcus) and fungi (Species of *Penicillium, Fusarium, Aspergillus etc.*) (Tian *et al.*, 2021).

Sulfur Cycle

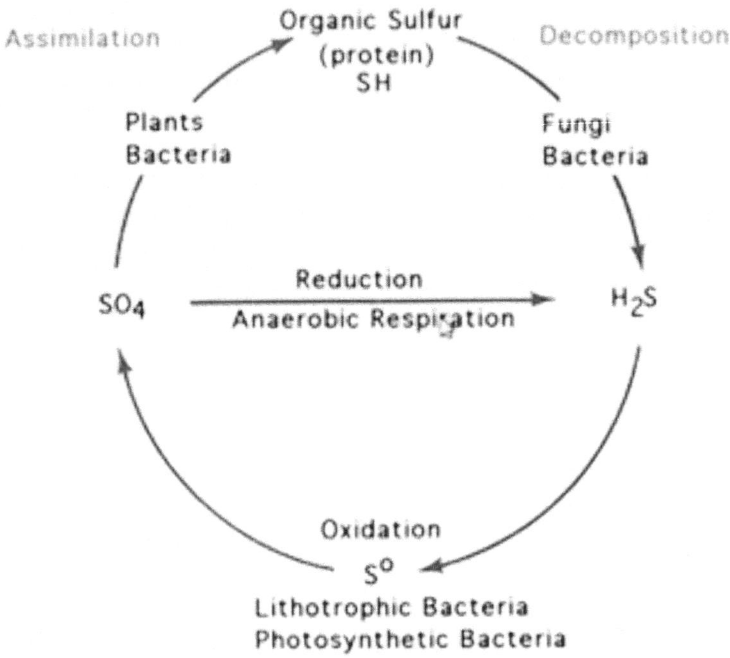

Sulfur is an essential element for all living beings. Sulphur found in soil is approx. 95 per cent organically bound mostly in two major forms; sulfate-esters and sulfonates. Many bacteria and fungi are capable of mineralizing Sulphur from sulfate-esters in soil. S- cycling in soil may include complex connections between numerous free living and symbiotic root linked microbial populations. In organisms and plants, Sulfur is present in organic form like amino acid (cysteine and methionine) or vitamins and inorganic form are also their like sulphates. In sulfur cycle some filamentous bacteria like (*Beggiatoa, Thiothrix and Thioloca*) and non-filamentous bacteria like (*Thiobacillus*) are involved while Some fungi (Aspergillus, Penicillium and Microsporium) and actinomycetes also help in oxidation (Gahan and Schmalenberge, 2014).

Enzymes

Enzymes act as biological catalysts commonly formed of proteins. Catalysts generally increases the rate of chemical reactions. The particles on which enzymes act are called substrates, and substrate is converted to the products into different molecules known as products. Enzymes on the basis of origin are of two types *i.e.* intracellular enzymes that are present in the host body and function inside them while other is extracellular enzymes which are secreted by the cell and function outside the cell (Baldrian, 2009).

In soil study, extracellular enzymes are of prime interest as most of the mineralization rear and tear processes of rocks require enzyme activity outside the microbial cells. Extracellular enzymes are produced by both prokaryotic and eukaryotic cells and are demonstrated to be a vital part of various biological activites. For many decades, Enzyme activity in soils had been studied and enzymes which are mostly related to the:

1. C cycle (glycosyl hydrolases, oxidases, and peroxidases),
2. N cycle (proteases, peptidases, urease, and chitinase),
3. P and S cycles (phosphatases and arylsulfatase) (Baldrian, 2009)

Soil Extracellular Enzymes

Enzymes in the soil system play important biochemical functions in the organic matter decomposition (Sinsabaugh *et al.,* 1991). These biological agents are important for catalyzing important reactions necessary for the vital processes of microorganisms in soils and for the stabilization of soil structure, the breakdown of organic wastes, formation of organic matter, and nutrient cycling. These enzymes in the soil are continuously being synthesized, accumulated, and/or decomposed and therefore they play a key role in nutrient cycling in ecosystems. Soil enzymes have a close relationship with the other biological properties of soil while they play an important role in mineralization process in the soil. Some of the major enzymes that perform extracellular activity in soil are as follows

Urease

Urease can be found in soils and microorganisms as intracellular structure and in humic substances and mostly combined with other soil colloids as extracellular structure. Urease in extracellular structure can protect its catalytic activity for a long time while free living extracellular urease can be rapidly breakdown by proteolytic enzymes in enzymes. Sources of urease enzyme in nature are plants, animals and microorganisms (Dotaniya *et al.,* 2019). It was reported that urease activity in nature ranged between 0.14 and 14.3 μmol NH_3-N g^{-1} h^{-1}. In another study, it was ranged between 24.12 and 39.03 mg N 100 g^{-1} dry soil. It was noted that urease activities in soils had an important relationship with soil properties like organic matter, texture, pH, cation exchange capacity (CEC) while urease activity was increased through the mixing of organic residues with soils. Urease is a hydrolytic enzyme that attacks the N and C bond (Peptide bond) in amide compounds such as urea and forms the alkaline end product ammonia and carbon-dioxide. it is one of the most important enzymes that play role in nitrogen cycle. Urea hydrolysis is a chemical reaction that occurs in soils and catalysed by urease. Urea hydrolysis is proven to be higher in alkaline soil (pH ~ 8.2 or higher) as it is suitable for urease to carry out the catalytic reaction. Various factor influencing enzyme activity (Cropping history, organic matter content, soil depth, management practices, heavy metals, temperature (60°C), pH (7.4 to 8.5) (Zantua *et al.*, 1977) *etc.*).

Urease is useful enzyme in soil as it helps to regulate nitrogen availability to plants (Antonious, 2003). Also, it can be used as a cytotoxic agent to kill the insects feeding on plant while also play role in germination of seed and increased plant growth.

Protease

Proteases in the soil play a crucial role in N mineralization. Microorganisms release extracellular protease into the soil. Source of Proteases enzymes are microorganisms such as *Aspergillus niger, Bacillus subtilis*, and plants. Its molecular size ranges from 15 to 30 kDa. Protease enzymes are involved in the initial hydrolysis of protein components of organic nitrogen to simple amino acids. Hydrolytic degradation of proteins is an important step in the nitrogen cycle (Rao *et al.*, 2014). Proteases in soils hydrolyze not only added proteins but also native soil added proteins to break down the protein into oligopeptides, amino acids and NH_4^+. Various factor influencing enzyme activity (Humic acid concentration, availability of C and N, temperature (40° to 60°C), pH (6 to 8.5), *etc.*).

Proteases are mostly used in leather, laundry, biocatalyst, and bioremediation procedures (Guoliang *et al.*, 2003).

Chitinase

Source of chitinase in soil are microorganisms and plants. It is 54 kDa in molecular size. It plays an important role in degradation and hydrolysis of chitin. In soil, chitinase enzyme play an important role in carbon and nitrogen cycle as the end product of its activity in the soil ecosystem is carbohydrates and inorganic nitrogen (Brurberg *et al.*, 2001). Various factor influencing enzyme activity(nitrogen availability, soil depth, atmospheric CO2 levels, temperature range (20 to 60 °C), pH range (5 to 5.5) *etc.*).

Chitinase is used mainly in protein and fertilizer engineering. It is used to production non-allergenic, nontoxic, biocompatible, and biodegradable materials. Chitinases can be used to convert chitin-containing biomass into useful (depolymerized) components. Chitinase can be usedto control plant pathogenicfungi and insects.

Fungal protoplasts have been used as a highly efficient experimental method to study the synthesis of cell wall, enzyme synthesis and secretion and strain enhancement for biotechnological applications (Dahiya *et al.*, 2005). Chitinase activity can also be used as an indicator of fungi activity in soil. It has a strong correlation between chitinase activity and fungal population in the soil. Therefore, chitinase activity seemstobe a suitable indicator for the active growthof fungi in the soil (Miller *et al.*, 1998).

Phosphatase

Phosphatases are enzymes that catalyze the hydrolysis of both esters and anhydrides of phosphoric acid. Soil phosphotriesterase activity has been determined by hydrolysis of tris-p- nitrophenyl phosphate. In P-cycle, both acidic and alkaline type of phosphatase activity are present. Source of phosphatase enzyme is plant and microorganisms. Its molecular weight is 58 kDa. The activities of phosphatases (activity of many hydrolytic enzymes) depend on several factors such as soil properties, interactions between soil organism, plant cover, management practices, pollution, crop species, and varieties, leachate inputs and in the presence of inhibitors and activators. Also, other various factor influencing enzyme activity {Organic matter content, pH (Acidic range- 4.7 to 6.5, alkaline range- 8 to 11), temperature (37°C)} (Rao *et al.,* 2014).

The ELISA test uses alkaline phosphatase-conjugated antibody as the detection system because it uses the substrate p-nitrophenyl phosphate and help in precipitation of heavy metals and enzyme immunoassays. Acid phosphotase is use in diagnostic use, biochemical markers *etc.*

Arylsulfatase

It play an important role in Hydrolysis of sulfate esters bonds. Its mainly participating in the sulfur (S) cycle, and releasing sulfate ions in the soil by the enzyme arylsulfatase. End product of this S-cycle is Sulfate (SO_4^{-2}). Sources of arylsulfatase are Microorganisms, plants, and animals. Its molecular weight is 53.2 kDa. Various factor influencing enzyme activity (Heavy metal pollution, pH (6), temperature range (30° to 70°C), composition of organic matter content and availability of organic sulfate esters in soil) (Rao *et al.,* 2014).

Thearylsulfutase of several sources has potential applications in many areas of industrial, research and agriculture. In industry, arylsulfatase derived from bacteria can be applied to desulfate the sulfated polysaccharides. This enzyme is also used in analytical endocrinology. Arylsulfatase A (or cerebroside-sulfatase) is an enzyme that degrades the sulfatides, namely cerebroside 3-sulfate into cerebroside and sulfate. Arylsulfatase B is in the group of arylsulfatase enzymes present in the lysosomes of the liver, pancreas, and animals kidney. The purpose of the enzyme is to hydrolyize sulfate in the body (Stressler *et al.,* 2016).

Peroxidase

Peroxidase plays a key role in the hydrolysis of lignin. The sources of peroxidase in nature are plants, animals and microorganisms. Its molecular weight is between 30 and 150 kDa. End product of peroxidase enzyme is C compounds (humic substances). In crop production peroxidases enzyme provide a positive role in the removal of toxicants and the decomposition of organic matter in soil. It also help in C-cycling. Various factor influencing enzyme activity (Soil pH (3.5 to 9), temperature range (20° to 40° C), precipitation and average annual temperature, SOM content, management measures, N enrichment, *etc.*) (Rao *et al.,* 2014).

Peroxidase can be used to treat of industrial waste waters. For example, phenols are important contaminants that can be removed by enzymatic polymerization using horseradish peroxidase. It also can be used to convert toxic materials into less harmful substances. Peroxidases are sometimes used as histological markers. Cytochrome c peroxidase is used as a soluble and easily purified model of cytochrome c oxidase. Extensive research has been conducted on the use of peroxidase in many manufacturing processes such as adhesives, computer chips, auto parts, barrels and tank linings. Other studies have shown that peroxidase can be used successfully used to polymerize anilines and phenols in an organic solvent matrix.

REFERENCES

AlShrouf, A. (2017). Hydroponics, aeroponic and aquaponic as compared with conventional farming. *American Scientific Research Journal for Engineering, Technology, and Sciences (ASRJETS)*, *27*(1), 247-255.

Amelung, W., Bossio, D., de Vries, W., Kogel-Knabner, I., Lehmann, J., Amundson, R., ... and Chabbi, A. (2020). Towards a global-scale soil climate mitigation strategy. *Nature Communications*, *11*(1), 1-10.

Amundson, R. (2009). Soil Preservation and the Future of Pedology La Conservation des Sols et le Futur de la Pédologie. *Anais XIV Simpósio Brasileiro de Sensoriamento Remoto. Natal: INPE*, 2801-2807.

Antonious, G. F. (2003). Impact of soil management and two botanical insecticides on urease and invertase activity. *Journal of Environmental Science and Health, Part B*, *38*(4), 479-488.

Baldrian, P. (2009). Microbial enzyme-catalyzed processes in soils and their analysis. *Plant, Soil and Environment*, *55*(9), 370-378.

Bedatou, E., Melchor, R. N., and Genise, J. F. (2009). Complex palaeosol ichnofabrics from Late Jurassic–Early Cretaceous volcaniclastic successions of Central Patagonia, Argentina. *Sedimentary Geology*, *218*(1-4), 74-102.

Bekchanova, M., Campion, L., Bruns, S., Kuppens, T., Jozefczak, M., Cuypers, A., and Malina, R. (2021). Biochar's effect on the ecosystem services provided by sandy-textured and contaminated sandy soils: a systematic review protocol. *Environmental Evidence*, *10*(1), 1-12.

Brodowski, S., Amelung, W., Haumaier, L., and Zech, W. (2007). Black carbon contribution to stable humus in German arable soils. *Geoderma*, *139*(1-2), 220-228.

Brurberg, M. B., Synstad, B., Klemsdal, S. S., van Aalten, D. M., Sundheim, L., and Eijsink, V. G. (2001). Chitinases from *Serratia marcescens*. *Recent Res Dev Microbiol*, *5*, 187-204.

Davidson, E. A., and Janssens, I. A. (2006). Temperature sensitivity of soil carbon decomposition and feedbacks to climate change. *Nature*, *440*(7081), 165-173.

Dominati, E., Patterson, M., and Mackay, A. (2010). A framework for classifying and quantifying the natural capital and ecosystem services of soils. *Ecological economics, 69*(9), 1858-1868.

Dotaniya, M. L., Aparna, K., Dotaniya, C. K., Singh, M., and Regar, K. L. (2019). Role of soil enzymes in sustainable crop production. In *Enzymes in Food Biotechnology* (pp. 569-589). Academic Press.

Dykhuizen, D. E. (1998). Santa Rosalia revisited: why are there so many species of bacteria?. *Antonie van Leeuwenhoek, 73*(1), 25-33.

Falkowski, P. G., Fenchel, T., and Delong, E. F. (2008). The microbial engines that drive Earth's biogeochemical cycles. *science, 320*(5879), 1034-1039.

Farquhar, J., Bao, H., and Thiemens, M. (2000). Atmospheric influence of Earth's earliest sulfur cycle. *Science, 289*(5480), 756-758.

Firoozi, A. A., Firoozi, A. A., and Baghini, M. S. (2016). A Review of Clayey Soils. *Asian Journal of Applied Sciences, 4*(6).

Gahan, J., and Schmalenberger, A. (2014). The role of bacteria and mycorrhiza in plant sulfur supply. *Frontiers in plant science, 5*, 723.

Gougoulias, C., Clark, J. M., and Shaw, L. J. (2014). The role of soil microbes in the global carbon cycle: tracking the below ground microbial processing of plant derived carbon for manipulating carbon dynamics in agricultural systems. *Journal of the Science of Food and Agriculture, 94*(12), 2362-2371.

Goulding, K. W., Bailey, N. J., Bradbury, N. J., Hargreaves, P., Howe, M., Murphy, D. V.,. and Willison, T. W. (1998). Nitrogen deposition and its contribution to nitrogen cycling and associated soil processes. *New Phytologist, 139*(1), 49-58.

Guoliang, J., Dong, Y., Yun, L., Guanghua, Z., Zhongjun, L., and Xinhua, Z. (2003). Studies on the effects of polyaspartate protease fertilizer enhancer in the absorptions of soil nutrition and the enzymatic activities of crops. *Journal of Ocean University of Qingdao, 2*(1), 62-64.

Hajnos, M., Jozefaciuk, G., Soko^{3}owska, Z., Greiffenhagen, A., and Wessolek, G. (2003). Water storage, surface, and structural properties of sandy forest humus horizons. *Journal of Plant Nutrition and Soil Science, 166*(5), 625-634.

Jog, R., Nareshkumar, G., and Rajkumar, S.(2016). Enhancing soil health and plantgrowth promotion by actinomycetes.In *Plant growth promoting actinobacteria*, 33-45.

Johns, C. (2017). Living soils: the role of microorganisms in soil health. *Fut Direct Intl*, 1-7.

Kappler, A., and Brune, A. (2000). Transformation and mineralization of synthetic 14C-labeled humic model compounds by soil-feeding termites. *Soil Biology and Biochemistry, 32*(8-9), 1281-1291.

Klosterman, S. J., Atallah, Z. K., Vallad, G. E., and Subbarao, K. V. (2009). Diversity, pathogenicity, and management of *Verticillium* species. *Annual review of phytopathology*, *47*, 39-62.

Kopittke, P. M., Menzies, N. W., Wang, P., McKenna, B. A., and Lombi, E. (2019). Soil and the intensification of agriculture for global food security. *Environment International*, *132*, 105078.

Li, L. C., Carroll, P. R., and Dahiya, R. (2005). Epigenetic changes in prostate cancer: implication for diagnosis and treatment. *Journal of the National Cancer Institute*, *97*(2), 103-115.

Mahmud, K., Makaju, S., Ibrahim, R., and Missaoui, A. (2020). Current progress in nitrogen fixing plants and microbiome research. *Plants*, *9*(1), 97.

McCahon, T. J., and Miller, K. B. (1997). Climatic significance of natric horizons in Permian (Asselian) palaeosols of north central Kansas, USA. *Sedimentology*, *44*(1), 113-125.

McDonald, P. (1994). Core monographs in soil science. *The Literature of Soil Science. Cornell University Press, Ithaca*, 161-256.

Melero, S., Madejón, E., Ruiz, J. C., and Herencia, J. F. (2007). Chemical and biochemical properties of a clay soil under dryland agriculture system as affected by organic fertilization. *European Journal of Agronomy*, *26*(3), 327-334.

Michielse, C. B., van Wijk, R., Reijnen, L., Cornelissen, B. J., and Rep, M. (2009). Insight into the molecular requirements for pathogenicity of *Fusarium oxysporum* f. sp. *lycopersici* through large-scale insertional mutagenesis. *Genome biology*, *10*(1), 1-18.

Miller, J. O. (2016). Soil pH affects nutrient availability. *Univ. Maryl. Ext.-Solut. Your Community*, 1-5.

Minayeva, T. Y., Trofimov, S. Y., Chichagova, O. A., Dorofeyeva, E. I., Sirin, A. A., Glushkov, I. V.,. and Kromer, B. (2008). Carbon accumulation in soils of forest and bog ecosystems of southern Valdai in the Holocene. *Biology Bulletin*, *35*(5), 524-532.

Neher, D. A. (2001). Role of nematodes in soil health and their use as indicators. *Journal of nematology*, *33*(4), 161.

Osterholm, M. T., Chin, T. D., Osborne, D. O., Dull, H. B., Dean, A. G., Fraser, D. W.,. and Hall, W. N. (1983). A 1957 outbreak of legionnaires'disease associated with a meat packing plant. *American journal of epidemiology*, *117*(1), 60-67.

Pan, X. Z., and Zhao, Q. G. (2007). Measurement of urbanization process and the paddy soil loss in Yixing city, China between 1949 and 2000. *Catena*, *69*(1), 65-73.

Parfitt, R. L., Giltrap, D. J., and Whitton, J. S. (1995). Contribution of organic matter and clay minerals to the cation exchange capacity of soils. *Communications in soil science and plant analysis*, *26*(9-10), 1343-1355.

Pettit, R. E. (2004). Organic matter, humus, humate, humic acid, fulvic acid and humin: their importance in soil fertility and plant health. *CTI Research, 10*, 1-7.

Ponge, J. F. (2003). Humus forms in terrestrial ecosystems: a framework to biodiversity. *Soil Biology and Biochemistry, 35*(7), 935-945.

Prescott, C. E. (2010). Litter decomposition: what controls it and how can we alter it to sequester more carbon in forest soils? *Biogeochemistry, 101*(1), 133-149.

Pukalchik, M., Panova, M., Karpukhin, M., Yakimenko, O., Kydralieva, K., and Terekhova, V. (2018). Using humic products as amendments to restore Zn and Pb polluted soil: a case study using rapid screening phytotest endpoint. *Journal of soils and sediments, 18*(3), 750-761.

Rao, M. A., Scelza, R., and Gianfreda, L. (2014). Soil enzymes. *Enzymes in agricultural science. Omics Group eBooks, USA. pp*, 10-24.

Rosswall, T. (1976). The internal nitrogen cycle between microorganisms, vegetation and soil. *Ecological Bulletins*, 157-1

Saxton, K. E., and Rawls, W. J. (2006). Soil water characteristic estimates by texture and organic matter for hydrologic solutions. *Soil science society of America Journal, 70*(5), 1569-1578.

Scheu, S. (2002). The soil food web: structure and perspectives. *European journal of soil biology, 38*(1), 11-20.

Shields, L. M., and Durrell, L. W. (1964). Algae in relation to soil fertility. *The Botanical Review, 30*(1), 92-128.

Sinsabaugh, R. L., Antibus, R. K., and Linkins, A. E. (1991). An enzymic approach to the analysis of microbial activity during plant litter decomposition. *Agriculture, Ecosystems and Environment, 34*(1-4), 43-54.

Six, J., Paustian, K., Elliott, E. T., and Combrink, C. (2000). Soil structure and organic matter I. Distribution of aggregate size classes and aggregate associated carbon. *Soil Science Society of America Journal, 64*(2), 681-689.

Soil and water. Food and Agriculture Organization of the United Nations, 2021.

Sparling, G. P. (1992). Ratio of microbial biomass carbon to soil organic carbon as a sensitive indicator of changes in soil organic matter. *Soil Research, 30*(2), 195-207.

Stout, J. D. (1980). The role of protozoa innutrient cycling and energy flow.In *Advances in microbial ecology*, 1-50.

Stressler, T., Seitl, I., Kuhn, A., and Fischer, L. (2016). Detection, production, and application of microbial arylsulfatases. *Applied microbiology and biotechnology, 100*(21), 9053-9067.

Tian, J., Ge, F., Zhang, D., Deng, S., and Liu, X. (2021). Roles of Phosphate Solubilizing Microorganisms from Managing Soil Phosphorus Deficiency to Mediating Biogeochemical P Cycle. *Biology*, 10, 158.

Tilak, K. V. B. R., Ranganayaki, N., Pal, K. K., De, R., Saxena, A. K., Nautiyal, C. S.,. and Johri, B. N. (2005). Diversity of plant growth and soil health supporting bacteria. *Current science*, 136-150.

Torsvik, V., and Ovreas, L. (2002). Microbial diversity and function in soil: from genes to ecosystems. *Current opinion in microbiology, 5*(3), 240-245.

Ud din Khanday, M., Bhat, R. A., Haq, S.,Dervash, M. A., Bhatti, A. A., Nissa, M., and Mir,M. R. (2016). Arbuscular mycorrhizal fungiboon for plant nutrition and soil health. In *Soil science: Agricultural and environmentalprospectives*, 317-332.

Valentin, C., d'Herbes, J. M., and Poesen, J. (1999). Soil and water components of banded vegetation patterns. *Catena, 37*(1-2), 1-24.

Van Cuyk, S., Siegrist, R., Logan, A., Masson, S., Fischer, E., and Figueroa, L. (2001). Hydraulic and purification behaviors and their interactions during wastewater treatment in soil infiltration systems. *Water Research, 35*(4), 953-964.

Van der Ploeg, R. R., Bohm, W., and Kirkham, M. B. (1999). On the origin of the theory of mineral nutrition of plants and the law of the minimum. *Soil Science Society of America Journal, 63*(5), 1055-1062.

Vieira, S., Sikorski, J., Gebala, A., Boeddinghaus, R. S., Marhan, S., Rennert, T. and Overmann, J. (2020). Bacterial colonization of minerals in grassland soils is selective and highly dynamic. *Environmental microbiology, 22*(3), 917-933.

Vogt, P. F., and Miller, M. J. (1998). Development and applications of amino acid-derived chiral acylnitroso hetero Diels-Alder reactions. *Tetrahedron, 54*(8), 1317-1348.

Wang, Y., Amundson, R., and Trumbore, S. (1996). Radiocarbon dating of soil organic matter. *Quaternary Research, 45*(3), 282-288.

Zanella, A., Katzensteiner, K., Ponge, J. F., Jabiol, B., Sartori, G., Kolb, E.,. and Hager, H. (2019). TerrHum: an iOS application for classifying terrestrial humipedons and some considerations about soil classification. *Soil Science Society of America Journal, 83*, S42-S48.

Zantua, M. I., Dumenil, L. C., and Bremner, J. M. (1977). Relationships between soil urease activity and other soil properties. *Soil Science Society of America Journal, 41*(2), 350-352

Zhuang, J., and Jin, Y. (2003). Virus retention and transport as influenced by different forms of soil organic matter. *Journal of environmental quality, 32*(3), 816-823.

Chapter 14

Microorganisms as Biocontrol Agents in Vegetable Crops

Siddhi Chugh[1], Jagdish Parshad[2], Dinesh Kumar[3] and Sarita Rani[4]*

[1]Department of Stem Cell and Tissue Engineering Punjab University, Sector- 14, Chandigarh,
[2]Department of Microbiology, [3]Department of Horticulture,
[4]Department of Agronomy,
CCS Haryana Agricultural University, Hisar – 125 001, Haryana
**e-mail: lect.jagdish@gmail.com*

ABSTRACT

Pathogens cause a major threat to stability of ecosystem especially in the case of the food production which indicates the need for cultivating novel methods manage the massive loss caused by these pathogens (Newell, et al., 2010; Oerke and Dehne 2004). There has been the usage of many ways to control them using plethora of chemical pesticides but it had been observed that these pesticides were related with ecological and health hazards and also have the risk to develop the better genetic version through mutation and causing more harm to the crops. These major issues force scientists to innovate some alternate methods for the sustainable management of the crops. Hence, to control crop diseases and produce pesticide-free produce throughout the world, there has been use of some natural microbes known as Biocontrol Agents (Hajek and Eilenberg 2018; Wilson and Backman 1999; van Lenteren, et al., 2018). These agents can act on pathogens via different mechanisms say; induction of plant resistance mechanisms, hyperparasitism, enzyme production, antibiosis, competition for essential nutrients and space (Junaid, et al., 2013; Kohl, et al., 2019). These agents not only control phyto pathogen but also enhance growth and stress tolerance in plants. These agents can be used as bio-pesticides or bio-insecticides. Recently, they can also be used for the control and management of postharvest diseases in vegetable crops.

During, recent decades the genetic engineered microbes are developed with improved biocontrol capability. Many commercially available agents such as Kodiak, Companion, GB34 and Serenade, have Bacillus as active constituent and are generally used for sustainable control of plant diseases with superior yield of different crops (Borriss 2015; Khasa 2017). Thus, usage of microorganisms such as fungi, yeasts, bacteria and viruses have the huge potential to work as biocontrol agents to reinstate conventional ways.

Introduction

Vegetables and fruits are cheap source of essential nutrients and play a major role in development of our body. Due to their incredible nature of providing an individual with the high nutritional value such as minerals and vitamins, which in turn keeps the body healthy and prevent from various diseases such as cardiovascular, eye and cancer (Khattak and Rahman 2017), hence, the quality of fruits and vegetables is the major concern of the scientists but unfortunately, the recent data shows that the quality of fruits and vegetables is gradually decreasing from 35 per cent to 55 per cent throughout the passage of primary production and the end consumer. The major reason for loss in production is due to damage, decay or spoilage and is estimated to be 30 per cent. Large quantity of fruits and vegetable spoilage occurs during storage and transportation from field to grocer shops and homes (Dhall 2016; Lemaire and Limbourg 2019; Porat, *et al.,* 2018)

Since ages, the decay of crops has been observed which is due to the proliferation of pathogens on the crops that causes various types of plant diseases such as downy mildews, sclerotinia rots and many more (Dordas 2008; Koike, *et al.,* 2007). Hence, plant diseases are major concern for the security and quality of food that need to be achieved. The wide range of crops is affected by the pathogens' attack. There were many successful attempts made in past to overcome this problem such as use of resistant plant varieties, crop rotation and many more but all were in vain. Although the use of chemical pesticide is much more efficient than any insecticides but their usage also have certain drawbacks, such as the excessive use pose the major threat to health and environment (Damalas, *et al.,* 2011; Gentz, *et al.,* 2010; Mahmood, *et al.,* 2016). To overcome this, scientists have come with the better option *i.e.,* the use of natural antagonistic activity of microbes as Biocontrol Agents (BCA) which are environment friendly and in some cases is the only option available. BCA not only prevents the disease but also promote plant growth by releasing certain growth hormones. Furthermore, they increase stress tolerance and induce disease resistance, the products of BCA are used as plant strengtheners, biofertilizers and bio-pesticides (Jensen, *et al.,* 2016; Nega and Healthcare 2014).

Biocontrol is a process to control the growth of an insect or pest by using biological agents or living organisms. The term biological control was first coined by Harry Smith of the university of California in 1939, who defined it as "the suppression of insect populations by the actions of their native or induced enemies" (Gnanamanickam, *et al.,* 2002; Smith 1939). According to the US National Research Council, BCA are natural or GMO's diminishes the harmful effects of plant pathogens by action of their genes or gene products (Cook 1993; Sundh and Goettel 2013). The major principle involved in the biological control is the natural suppression of population of harmful plant pathogen growing on crops. Biological control does not completely eliminate the pest; it maintains the balance between the pest and their natural enemy at lesser density in the agro ecosystem (Schmitt and Seddon 2005).

Use of bio control agents has many advantages over the other conventional methods:

1. Reduce the use of chemical fungicides and their undesirable effects on crop.
2. Play a key role in integrated management of diseases.
3. Safe for the users, environment and the farming community.
4. Provide natural long-term immunity to crops and soil.
5. Biological control is less costly and cheaper than any other methods.
6. Biocontrol agents give protection to the crop throughout the crop period.
7. They do not cause toxicity to the plants.
8. Their multiplication is very easy in the soil and left with no residual problem.
9. They improve the crop yield.
10. Easy to handle their application to the target pathogens.
11. They can be used along with bio-fertilizers.
12. Their manufacturing process is easy (Junaid, *et al.,* 2013; Powell and Jutsum 1993; Singh, *et al.,* 2020).

This chapter would be dealing with the various mechanisms that are been used by the microbes such as bacteria, yeast and fungi to protect the crops against the phytopathogens and the ways by which it promotes the plant growth.

Role of Biocontrol Agents in Management of Phytopathogens

Diseases in plants generally occur due to their interacting activity among three major components of disease-causing triangle which includes the host, environment and pathogen. When included the Biocontrol agents in this triangle, it results in the inhibition of the disease. The mechanism of action is important to understand by which bio control agents work to inhibit the disease in plants. The most common mechanisms of interspecies antagonism comprises direct and indirect antagonism (Junaid, *et al.,* 2013; Pal and Gardener 2006; Parveen, *et al.,* 2016), which leads to bio-control of phytopathogen.

The action is generally classified into 2 major categories which includes;

1. Direct Antagonism
2. Indirect Antagonism

Direct Antagonism

Direct antagonism involves the direct interaction among the BCA and pathogen. This mechanism includes Antibiosis, Enzyme production, siderophore and Parasitism.

Table 14.1: Classification of Biocontrol Agents on the Basis of their Action Mechanisms (Adapted from Nega *et al.*, 2014; Agriculture and Healthcare 2014 and modified accordingly)

Type	*Mechanism*	*Examples*
Direct antagonism	Enzymes	Proteases, Chitinases, Glucanases
	Antibiotics	Cyclic lipopeptides 2,4-diacetylphloroglucinol, Phenazines
	Siderophore	Catecholates (*i.e.* enterobactin), carboxylates (*i.e.* rhizobactin) hydroxamates (*i.e.* ferrioxamine
	Parasitism	Lytic or nonlytic mycoviruses, *Lysobacter enzymogenes*, *Trichoderma virens*, *Ampelomyces quisqualis, Pasteuria penetrans*
Indirect antagonism	Induction of Systemic Resistance	Fungal cell wall contact,Pathogen associated detection, molecular patterns, Phyto-hormone-mediated induction
	Competition	Exudate or leachate consumption
	Plant Growth Promotion and Nutrition	*Erwinia, Pseudomonas, Azospirillum*

Antibiosis

Antibiosis process is mediated by the production of antibiotics which has lower molecular weight and inhibits the pathogen growth. Antibiotics are the metabolites produced by microbe for example Endophytes is microbial agent which uses antibiosis mechanism and have which are isolated from plants have capability to produce antimicrobial compounds such as VOC's, polyketides, steroids, phenols, flavonoids, terpenoids are specific metabolites which suppress the growth of aggressive pathogens (Fravel 1988; Howell and Gliocladium 1998; Raaijmakers, *et al.*, 2002).

Majorly 3 types of antibiotics are produced by biocontrol agents that are; water soluble, non-polar volatile, polar/non-volatile among which volatile biocontrol agents are most effective as they have power to act upon sites other than the site of their production (Khasa 2017). Researchers have shown that the genetic stability of microbes have been altered by carbon sources and numerous trace metals like zinc (Zn), which makes them capable for the production of antibiotics (Rokem, *et al.*, 2007).

Trichoderma and *Gliocladioum* are majorly seen in production of antibiotics that exhibits biocontrol activity. For example, Gliotoxin produce by *G. virens* is considered as key factor in its biocontrol activity against *Phythium ultimum* and *R. solani*. There is large number of bacterial BCA which produces antibiotics and other secondary metabolites which controls the different fungal pathogens. The well-known example is the phenazine-carboxylic acid production by *Pseudomonas fluorescens* strain which is involved in biocontrol activity against take-all of wheat (Adetunji and Varma 2020; Rajput, *et al.*, 2019).

One of the best examples of antibiosis is bio-control of crown gall disease caused by *Agrobacterium tumefaciens* using agrocin 84 which is produced by *Agrobacterium radiobacter*. There are certain strains of *Pseudomonas putida* WCS358r, which was genetically engineered in order to produce 2,4-diacetylphloroglucinol (2,4-DAPG) and phenazine, which causes inhibition of the plant pathogens. *Pseudomonas fluorescens* strain 2-79 controls all the diseases of wheat caused by *Gaeumannomyces graminis* var. Also, P. florescent CHA0 can produce numerous antibiotic products including pyoluteorin and 2, 4-diacetylphloroglucinol helps in suppression of fungal pathogen which causes root diseases (McManus, *et al.*, 2002; Raaijmakers, *et al.*, 2002).

Table 14.2: Mechanisms of Specific Biocontrol Agents for Controlling Plant Disease (Adapted from Nega *et al.*, 2014, Agriculture and Healthcare 2014 and modified accordingly).

Sl.No.	*Antibiotics*	*Source*	*Target pathogen*	*Disease*
1	Bacillomycin D	*Bacillius subtilis* AU195	*Aspergillus flavus*	Aflotoxin contamination
2	Bacillomycin, fengycin	*Bacillus amylolique-faciens* FZB42	*Fusarium oxysporium*	Wilt
3	Xanthobaccin A	*Lysobacter* sp. Strain SB-K88	*Aphanomyces cochlioides*	Damping off
4	Gliotoxin	*Trichoderma virens*	*Rhizoctonia solani*	Root rots
5	Mycosubtilin	*Bacillus Subtilis* BBG 100	*Pythium aphanidermatum*	Damping off

Enzymes

Enzymes not only work as a biocatalyst in the essential metabolic reactions but also in the survival of microbes in the specific niche. Some microbes produce enzymes which can hydrolyse cellulose, chitin, hemicellulose, proteins and they also suppresses the plant pathogens (Poshina *et al.*, 2018; Veliz, *et al.*, 2017).

Production of lytic enzymes is the other mechanism of action through many microbial antagonists. The efficiency of these enzymes is enough to suppress the activity of pathogen. For example, β-1,3-glucanase dependent *Lysobacter enzymogenes* strain C3, *Serratia marcescens* also acts against causal agent of southern blight *Sclerotium rolfsii* in bean crop, by producing chitinase enzyme. These enzymes basically obtain their nutrition by degrading complex cell wall of pathogen and hence, inhibiting the pathogenic activity on the crop.

There are 2 major organisms which produces plethora of enzymes for lysis of pathogenic cell wall *i.e., Myxobacteria* and *Lysobacteri.* Also, *Rhizosphere* microbes is the best example which includes the ability to promote growth of plant as well as control the pathogens, this also enhances the plant tolerance against pathogens through several means. Among, all known ways of biocontrol secretion of enzymes are considered as the most efficient way to deter the growth of plant pathogen that are residing in the close vicinity of the rhizospheres.

Experimentally, the role of enzyme in disease inhibition was checked using the mutant which does not contain the gene which codes for the specific enzyme, say for example in the *Serratia marcescens* the gene containg the codes for ChiA production was struck off and it results in the lower efficiency for controlling *Fusarium* wilt under controlled conditions in greenhouses and with introduction of the gene in other microbe, resulted in an increased biocontrol activity. When *Escherichia coli* strain was engineered to insert ChiA gene it turned up with amazing results in reduction of southern blight of bean. Recently it was studied that the product of lytic enzymes shows indirect effect on plant pathogen. It is considered as best choice for biocontrol agents however, it has certain constrains such as stability, quality, shelf life, production cost and formulation design (Haran, *et al.,* 1996; Viterbo, *et al.,* 2002).

Siderophore

Siderophore are low-molecular weight chelating agents having strong and specific affinity towards Fe (III). Microorganisms as a BCAs having the ability of siderophore production inhibits the, metabolic activity, growth and development of plant pathogens by iron chelation (Ahmed and Holmström 2014; Chincholkar, *et al.,* 2005). Growth and development of other fungal plant pathogens is prevented by different species of *Trichoderma* as biocontrol antagonist that releases more effective siderophores with chelate iron (Naher, *et al.,* 2014). *Pseudomonas fluorescens* having siderophore production ability play a vital role in the prevention of plant pathogen. Postharvest pathogens like *B. cinerea* and *P. expansum* are inhibited by *Rahnella aquatilis* with siderophore production. *Metschnikowia pulcherrima* and *Monilinia fructicola* yeasts produce siderophore *pulcherrimin* that can be applied for bio- control of postharvest pathogens *A.alternate, B. cinerea,* and *P. expansum* in apple fruits (Calvo, *et al.,* 2007; Costa and Loper 1994; Saravanakumar, *et al.,* 2008).

Parasitism

Mycoparasitism is the capability of fungi to parasitize other fungal strains for utilizing them as food and also acts as antagonistic agents. Hyperparasitism of a plant phytopathogen by an obligate parasite is direct type of antagonism. In this case, the specific biocontrol agent directly attacks the pathogen. Hyperparasitism involves tropic growth of biocontrol agent towards the organism in the target causing them to coil, attack them and end up in dissolution of the target pathogen's cell wall or membrane through the activity of enzymes. This is important mechanism involved in *Trichoderma.* Also, *Trichoderma harzianum* exhibits amazing activity of mycoparasitic against *Rhizoctonia solsni* hyphae (Shalini and Kotasthane 2007). Pathogens that have biotrophic nature (depends on other organism for nutrition) are attacked by fungal antagonists through antibiosis and mycoparasitism only. Mycoparasitism uses the mechanism of hyperparasitism in which the host organism is fungi (Viterbo, *et al.,* 2010).

There are 4 major steps involved in mycoparasitism (Steyaert, *et al.,* 2003):

1. Chemotropism: It is the process through which biocontrol agents are attracted by chemical stimulus released from pathogen for example, pathogen releases volatile or water-soluble substances which work as the chemoattractant for their parasite.
2. Recognition: It is the process of recognising the specific pathogenic fungi and is attacked by specific antagonistic microbe. It involves the interaction of carbohydrate-surface receptors on the biocontrol agents and the lectins which are produced by pathogens.
3. Attachment and degradation: The next stem involves two processes one is the attachment to the cell wall using the receptor ligand interaction and in turn degrading it. For example, fungal hyphae can either coil around it and produce enzymes for degrading the cell-wall which may include chitinases and β-1, 3-glucanase or grow alongside the host hyphae.
4. Penetration: This includes penetration of special structures such as appressoria into the fungal cell wall.

It has been observed that single fungal pathogen is attacked by plethora of hypo parasites such as *Acrodontium crateriforme, Acremonium altenatum, Gliocladium virens* and *Ampelomyces quisqualis* are few fungal strains that are having the capacity to parasitize powdery mildew pathogens. The 4 major groups of hyperparasites are obligate bacterial pathogens, parasites, predators and hypo-parasites. Hypovirus can infect the fungi *Cryphonectria parasitica* which causes blight in chestnut. The Hypovirus causes hypo virulence (decrease of pathogenicity of the phytopathogen). This in turn controls the chestnut blight at many places. Some more examples of mycoparasitism by fungi include the parasitism of rust causing fungi that is *Uromyces* by *Spharellopsis filum* and *Puccinia* (Sri, 2019).

Indirect Antagonism

An indirect antagonist is a substance that enhances action or release of an endogenous neurotransmitter but does not have any receptor. This method as the name suggest, off course doesn't involve through direct interaction among BCA and pathogen (Pal and Gardener, 2006). This process includes systemic resistance induction in plant, plant growth promotion, the competition for space, and nutrient.

Induced Systemic Resistance

The responses in plants are developed towards the stimuli such as abiotic stress *i.e.* iron deficiency and biotic stress which are provoked by biological agents for example pathogens or insects. Not all responses are systemically spread but some are limited towards the damaged organ. The former responses include the Induced Systemic Resistance (ISR) and Systemic Acquired Resistance (SAR) (Shoresh, *et al.*, 2010). ISR is induced by beneficial microbes which are residing in the Rhizosphere, like fungi and bacteria. These root- related mutualistic relationship impacts plant growth and nutrition and also can boost defence system in plants causing whole to develop the resistance against pests and pathogens. SAR is induced by the local infections such as insects and pathogen attack on some organs and remains systemic in the body and hence provides the long-term resistance to pathogen (Khan *et al.*, 2012).

In recent times, the abundance of compounds that have been discovered that develops the resistance in plants *e.g.* proteins that have enzymatic activity such as cellulases, endochitinase and xylanases. Another, protein group that work to induce resistant are the Avirulence genes' products results in the activation of hypersensitive response. Rhizobacteria also, triggers the salicylic acid- dependent SAR pathway and these pathways are studied in *Arabidopsis thaliana* which in turn is dependent on jasmonic acid (Bakker, *et al.*, 2003; Montejano-Ramírez, *et al.*, 2020). Some *Pseudomonas* strains are also known for inducing systemic resistance. Combination of both SAR and ISR has the capability to increase the protection against pathogens (Choudhary, *et al.*, 2007). In defence response to SAR usually thickening of cell walls occurs through lignifications, deposition of callose, and synthesis of lytic enzymes such as glucanases or chitinases and synthesis of light weight antimicrobial compounds for example phytoalexins this in turned employed by variety of *Rhizobacteria* which are involved in plant growth-promoting *Rhizobacteria* (Dukare *et al.*,2020; Wani, *et al.*, 2018).

Competition

The growth of plant and plant pathogen both depends upon the nutrient uptake. The pathogens which take the nutrition exclusively from the infected host cell are known as obligate biotrophic pathogens and they are independent on any exogenous nutrient from the environment.

Another class of plant pathogen known as Necrotrophic plant pathogenic fungi, bacteria or oomycetes obtain their nutrient by killing and invading the tissue of host plant and taking up plant nutrient by colonizing itself with in the tissue. Saprophytic non-pathogenic microorganisms also start residing and colonizing itself on the necrotic tissue that has been induced by the pathogenic organisms which induces the competition between these two colonies in order to get enough space and nutrients for their healthy survival.

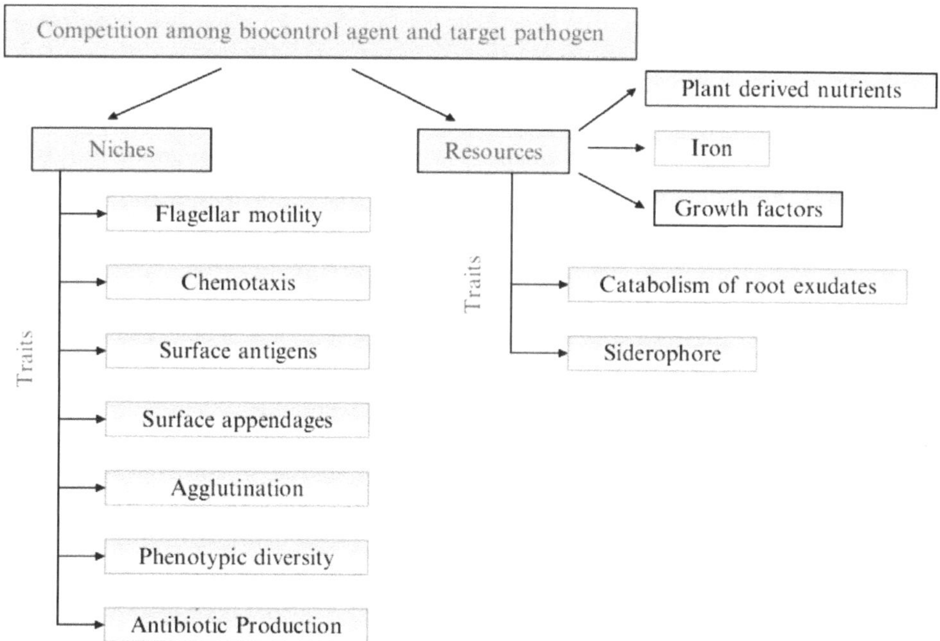

Figure 14.1: Showing Competition among Biological Agent and Target Pathogen (Adapted from Prashar, *et al.*, 2013).

The competition among them is advantageous as the Necrotrophic pathogens first colonize directly after they incited death. Also, similar advantage is taken up by the Non-pathogenic endophytes that are latently present in the affected tissue and this may play an role in competitive substrate colonization.

The biocontrol agents and pathogens usually compete for the survival conditions such as space, nutrient and essential micronutrients such as iron, manganese and specific growth substances which stimulants germination. The examples of competition can include the usage of high-affinity iron chelators say siderophores producing BCA results in the starvation of pathogenic microbes that need iron as the major constituent for survival. The importance of siderophore was first described by (Kloepper *et al.*, 1980) as mechanism of suppression of Erwinia carotovora by *P. flurescenes* strains A1, BK1 and TL3B1.

Table 14.3: Shows the Bioassays Used to Select the Antagonist Nutrient Competition

Sl.No.	Mode of Action	Methods for Screening	Pathogen Specificity
1	Competition	Simplified bioassay	Broad
2	Induced resistance	Complex bioassay on plant	Specific to broad
3	Antimicrobial metabolite produced *in situ*	Simplified bioassay	Specific to broad
4	Hyperparasitism	Simplified bioassay	Pathogen specific interaction
5	Antimicrobial metabolites in product	*In vitro* assay	Broad
6	Helper strains	Complex bioassay	Depends on MBCA
7	Assembled consortia combined with different modes of action	*In silico* design followed by complex boassay	Broad
8	Modulation of indigenous microbiota	Complex site specific bioassay	Broad

Plant Growth Promotion and Biofortification

The increase in solubilisation and uptake of variety of nutrients has been shown by BCA. Thus, inhibition of disease-causing phyto pathogens by BCA can also enhances the plant growth and acid produced (2-carboxy-methyl-3-hexyl-maleic anhydride and2-methylene-3-hexyl-butanedioic acids) by *Aspergillus niger,* helps to increase the shoot and root length in the crop. It is reported that treatment of seed T. *viride* which in turn results in the increment of dry and fresh weight of shoot, root and nodules of broad beans. Some, bacterial species can affect the hormones associated with plant for example many bacterias' has the capability to produce ACC (1-aminocyclopropane-1-carboxylate) deaminase which in turn decreases the level of ethylene by degrading the ACC which is known as the precursor of ethylene synthesis. A number of PGPR is known such *Erwinia, Pseudomonas, Azospirillum, Bacillus, Rhizobium* and *Serratia* have the capability to solubilise phosphate. They work to convert non-soluble to soluble phosphate (Hamed *et al.,* 2011; Jha *et al.,* 2015; Kumar *et al.,* 2014; Suleria *et al.,* 2015).

The rhizosphere is generally expressed as the volume of soil under the physical, chemical and biological influence of plant roots and is the area of microbial activity and diversity. This activity of microbe is utmost important for plant nutrition and health as it contributes towards the uptake of nutrients through the plant roots and offers resistance against the microbes that causes diseases in the plants. The ability to improve the nutritional status of plants and production of phytohormones are the derivatives of the direct effect of PGPR on plant's growth (Bhattacharyya *et al.,* 2012; Borriss *et al.,* 2019; Axel *et al.,* 2012; Datta *et al.,* 2015).

Various Categories of Biocontrol Agents

Not only insects such as ladybird and dragonflies are employed to be wiped out the plant pathogens such as aphids and mosquitos respectively but also some microbes are involved in the process and act as biocontrol agents, these may include bacteria, viruses, fungi and even protozoans. They work as BCA in 3 basic ways that are either by causing disease in the pests or compete with them for basic survival needs or do kill them as discussed in earlier part of the chapter (Bejarano and Puopolo 2020).

In this section, we would be discussing the mechanism and examples of individual microbe as BCA.

Bacteria as Biocontrol Agent

Certain strains of *Pseudomonas* spp., *Enterobacter cloacae* and *Bacillus subtilis* have been extensively studied agents of bacterial family for bio-controlling plants against pathogens and its research have been extended towards vegetable crops also. The damping-off diseases have shown the significant decrease which are being caused by *Sclerotium rolfsii, Rhizoctonia solani Kuhn and Pythium spp.* The bacterias were employed either as drench to the growth medium or seed treatment. Recent studies done on bacterial agents such as *Pseudomonas aureofaciens, P. Fluorescens, Serratia marcescens, P. putida, E. Cloacae and B. Subtilis* shows reduction in diseases and in some cases plant growth complimenting it and yield mature plants. The positive reports were seen in vegetable crops such as celery, cucumber, onion and tomato were grown under the action of bacterial agents in the field or the glasshouse conditions (Erdogan and Benlioglu 2010).

Application of P. Fluorescens has level up not only in the cucumber production but also has caused the fruit growth and has given the increase in yield of around 18 per cent when compared to plants which were infected by *P. Aphanidermatum* and were treated with agents.

There is enormous number of bacterial endophytes which has the ability to work as biocontrol agents. Endophytes are nothing but organism that is associated with plant tissues and infection or symptoms are seen. The endophytic species influence the physiology of the host plants to the greater extend. Endophytes have the capability to be used as BCA because of their special capability to carry nutrient assimilation and production of many secondary metabolites such as antibiotics. Bacterial endophytes work as the competition to the plant pathogens by having the capability to colonise themselves in the same niche as that of pathogens which turn causes vascular wilt diseases, and works as biopesticides against wilt diseases which are observed to occur due to *Ralstonia, Erwinia, Clavibacterb* and *Pseudomonas.* Vascular wilts have the capapbility to affect annual crops and woody perennials which end up causing huge food losses. These wilts do nothing just invades the xylem vascular tissue thus creating problem in transport of water and minerals. The most commonly employed endophytes that have the ability to inhibit the adverse effect of phyto pathogens are the *Enterobacter, Bacillus, Agrobacterium* and *Pseudomonas*. They use multiple ways to inhibit the action of pathogens such as using antibiotics, growth promotion, inducing host resistance, competition, signal interference and parasitism (Ardakani *et al.*, 2010; cottyn *et al.*, 2009).

Sl.No.	***Stages of Fungal Pathogen Development Affected***	***Biocontrol Agent***
1	Survival of propagules	*Coniothyrium minitans, Gliocladium virens, Sporidesmium sclerotivorum, Talaromyces flavus, Trichoderma* spp.
2	Germination and infection	*Coniothyrium minitans, Gliocladium virens, Pseudomonas* spp., *Sporidesmium sclerotivorum, Trichoderma* spp., *Talaromyces flavus*
3	Tissue colonization	*Epicoccum purpurascens, Rhizoctonia* sp., *Trichoderma harzianum*
4	Growth and sporulation	*Ampelomyces quisqualis, Tilletiopsis* spp., *Sporothrix flocculosa*

Effects of Biological Control Agents on the Development of Fungal Plant Pathogen

Four different bacterial strains of *Pseudomonas i.e.* FP25, FP30, FP23 and FP35 isolated from the roots of various cotton plants and their effects were studied on the phyto pathogen verticillium causing wilt. It was elucidated that by treating cotton seeds with strains of *Pseudomonas* caused significant reduction in disease severity and also improves growth of the cotton plant. They also show positive results on the bacterial wilt diseases including *Bacillus amyloliquefaciens* Bg-C31 and *Streptomyces virginiae* which controls the disease caused by *Ralstonia solanacearum* on tomato and capsicum plant irrespectively (Dandurishvili *et al.*, 2011; Eljounaidi *et al.*, 2016).

It was also, shown that *Pseudomonas fluorescens* PICF7 has the capability to suppress the growth of Verticillium wilt on olive trees. It takes entry in the plant through the root hair and delays the colonization of plant pathogens giving competition for their niche. This has been environmentally observed in the greenhouse condition through treatment of olive trees with the *Pseudomonas fluorescens* PICF7 and was observed the significant reduction in the severity of disease by 96 per cent. The bacteria work through the process of inducing of systemic resistance in roots and in other distant tissues. Also, the efficiency of *Paenibacillus* K165 strain against the soil borne phyto pathogen *V. Dahlia* was observed in reducing the severity or chances of disease in eggplant and potato.

Most bacterial strains are being used as biopesticides which belongs to *Bacillus, Pseudomonas* and *Agrobacterium* genera. Enormous *Bacillus* strains have the great potential as biofertilizers due to their special ability of colonizing rhizosphere which induces resistance in plants and compete for nutrition with plant pathogens and also produce antibiotics, endospores which promote growth of plants. Not only this, *Bacillus* strains have the capability to produce the variety of antimicrobial compounds such as non-ribosomal synthesized lipopeptides, ribosomal synthesized bacteriocins and polyketides. *Bacillus* strains are also being used to control the crown gall diseases which are caused by *Agrobacterium tumefaciens* strains.

The potential of lactic acid bacteria as BCA and variety of mechanisms employed by *Bacillus* are discussed in following sections;

a) Use of lactic acid bacteria as biocontrol agent to control the disease caused by *Phytophthora infestans i.e.*, late blight of potato: Active metabolites produced by lactic acid bacteria such as 3-hydroxy fatty acids, hydrogen peroxide, cylic dipeptides, carbon dioxide, organic acids and proteinaceous compounds are produced by lactic acid bacteria that have the antifungal and antibacterial properties. LAB are also tested for their potential against *P. Infestans* and variety of other fungus which are pathogenic to plant say *Sclerotium, Rhizopus, Microsporum, Sclerotium, Penicillium, Endomyces* due to their capability to produce enormous antifungal compounds. There are 2 commercially available LAB-based products namely EM5 and AgroMos™. However, they are still not successfully able to control the late blight. EM5 was used as BCA for the *P. Infestans* and shows up 30 per cent of pathogen inhibition (Axel *et al.*, 2012).

b) Molecular mechanisms employed by *Bacillus*: In order, to enhance the ability of the bacteria to work more efficiently as BCA one should be aware of the detailed mechanism involved at the genetic level. *B. amyloliquefaciens* FZB42 strain is generally used as a model to find the genetic feature linked with their biocontrol activity. The presence of 10 giant gene clusters was seen in the computational analysis of genome of this model organism which are related with the production of several compounds that are responsible for the biocontrol activity exhibited by them. The genes which are responsible for the synthesis of nonribosomal and ribosomal associated secondary metabolites were identified having antibacterial properties. Also, many genes whose products inhibit the action of pathogen and prevent plant from the damage were also recognised by the chemical mass spectroscopy and were rechecked using the creation of knockouts. Ten clusters of genes were found using this approach which is in turn involved in Sfp (4'-phosphopantetheine transferase)-dependent nonribosomal synthesis of polyketides and cyclic lipopeptides (cLPs), Sfp-independent nonribosomal synthesis of bacilysin are highly modified. The growth of *Erwinia amylovora*, was inhibited by Sfp-dependent nonribosomal cyclic lipopeptides and bacilysin was demonstrated by the construction of a double mutant, which does not have the capability to synthesise both bacilysin and nonribosomal peptides and it was concluded that the double mutant was incapable of inhibiting the growth of *E. Amylovora*. Genes involved in aromatic acid and bacilysin synthesis are involved in the inhibition of *Microcystis aeruginosa* the causative agent of algal bloom. The over expression of genes which are involved in the secondary metabolite production say degU enhances the biocontrol activity significantly. For enhancing the production of the surfactin, biosurfactant and inducible promoter P_{spac} is generally used in the *B. subtilis*. In the similar way genes were engineered to enhance the production of 2, 3-butanediol dehydrogenase. Genetic engineering using Cre/Lox site combination with PCR is also used (Borriss 2015; Erdogan and Benlioglu 2010; Singh *et al.*, 1999; Wu *et al.*, 2015).

Fungi as Biocontrol Agent

The fungi as bio control agents almost constitute the 10 per cent of the total BCA. The fungal Biocontrol agents were evaluated for the ability to demolish the diseases that occur in vegetable crops and was concluded that *Trichoderma spp.* and *Gliocladium virens* were major contributors. The list also contains the species such as *Chaetomium globosum, Sporidesmium sclerotivorum Coniothyrium minitans, Talaromyces flavus* and some sterile fungi. They have been evaluated through number of stages against the soil borne diseases and resulted in reducing the diseases with significant amount which were caused by pathogens such as *Rhizoctonia solani, Pythium spp., Sclerotinia sclerotivorum* and *rolfsii*. Certain positive results were also seen in the epiphytic fungi, such as *Cladosporium cladosporoides, Alternaria alternate* and *Epicoccum purpurascenes are* successful in reducing the infections caused by *S. sclerotiorum. Glomus versiforme* is the effective BCA against harmful wilt causing plant pathogen namely Ralstonia solanacearum. It initiates the response of ISR *i.e.*, induced systemic resistance in the pathogen attacked plants. *Pythium oligannacearum* is also found to be successful in treating bacterial wilt. The leachate of Shiitake mycelia contains certain antibiotic ingredients which show the ability to inhibit the plant pathogen namely R. *solanacearum.* Some fungi have the capability to the demolish variety of pesticides, say for example *Trichoderma harzianum* can degrade the organochlorine pesticides, mostly endosulfan and aromatic hydrocarbon which includes anthracene (Azcón-Aguilar *et al.*, 2002).

The fungal biocontrol agents reduce the diseases in the plants by producing the toxins and lytic enzymes which results in the pathogen growth or the its survival, occupation of infection sites on the host plant, direct parasitism of pathogen and induction of host resistance mechanisms. The secretion of toxins such as Gliotoxin, enzymes chitinases, glucanases and proteases by *Gliocladium and Trichoderma* species and other antibiotics by *G. virens* play a major role in reducing inoculum potential of fungal plant pathogens. The protection against root diseases of vegetable crops *e.g.* cucumber, bean and tomato is achieved by weakly virulent strains of pathogenic species say *Pythium oligandrum and F. oxysporum,* and non-pathogenic species say binucleate *Rhizoctonia and Pythium nunn*(Banerjee and Mandal2020).

Trichoderma is the most capable fungal antagonist is, extensively used for the suppression of diseases in vegetable crops. It uses its exclusive ability to hyperparasitize other fungal strains, with the production of the antibiosis and cell wall lytic enzymes. Fungal agents also produce a variety of secondary metabolites and bring about resistance in the plants. This is the most valuable feature for the plant protection (Butt *et al.,* 2001).

Trichoderma stimulates a variety of defence mechanisms in plants such as ISR *i.e.* induced systemic resistance, SAR *i.e.* Systemic acquired resistance and hypersensitive response. The evidence, involved is that expression of phenylalanine and hydroperoxidase lyase is induced by *T. asperellum* which causes the accumation of phytoalexinin cucumber plants against *Pseudomonas syringae pv. lachrymans.* Trichoderma invasion is reacted by plants through rapid invasion of the ion fluxes, oxidative burst, synthesis of polyphenols and deposition of callose followed by activation of ethylene and salicylate signalling which in turn results in the inducing of ISR in the plants. Various initiators of the plant immune system are known to be produced by *Trichoderma* such as xylanases, trichovirin and alamethicin.

The enzymes and secondary metabolites tend to play the major and important role in the activity of the *Trichoderma* against plant pathogen. Trichoderma tend to produce enormous amount of antibiotics which posses the antibacterial and antifungal properties as well. The pyrone- like antibiotic which is produced by *T. harzianum* suppresses the growth of *Gaeumannomyces graminis.* Secondary metabolites, that are produced by *Trichoderma spp.* Assemble into 3 different types; (1) volatile compounds like 6-pentyl-alpha-pyrone (2) water-soluble compounds like heptelidic acid (3) peptaibol compounds.

Some fungal biocontrol agents, such as *G. virens and T. Harzianum* have been extensively researched and lead its pathway towards commercial utility. Though, many other organisms yet to come in the market have to go additional research.

The fungal Biocontrol agent serves many advantages such as they are easy to grow and many formulations have been studied to enhance their efficacy. The most widespread utilization of fungal agents is mainly against the soil borne plant pathogens, granular formulations can be directly applied to the soil. The major problem with fungal bio control agent used in the foliar diseases is that it reduces their survival in the phyllosphere (Melo and Faull 2000; Urquhart *et al.,* 1994; De Silva *et al.,* 2019; Goudjal *et al.,* 2014).

Yeast as Biocontrol Agent

In order, to protect the plant majorly vegetable crops from postharvest diseases the usage of yeast organisms as the biological control has been harvested intensively. Most of the research on yeast has been carried to control the vegetable disease which is generally emphasized on the powdery mildew diseases on the cucurbits however, a significant reduction has been observed in tomato diseases caused by *Botrytis cinerea.* The most used yeast organisms all around the world are

Ampelomyces quisqualis Ces, tilletiopsis spp. and *Sporothrix* spp. were extensively evaluated. There should be optimum condition for the yeast to work efficiently say it should have the high relative humidity (70 per cent) and the inclusion of protectant formulation is necessary to enhance their survival rate to control disease at low humidity. This includes the use of vegetable oils, liquid paraffin, lecithin or mineral oil in the process of soil suppression.

The mechanism by which yeast tend to work against pathogens include mycoparasitism of pathogen structure by *A. quisaqualis,*which produce lytic enzymes say β-1,3-glucanases, by *Tilletiopsis* sp. and secrete antibiotics such as heptadecenoic and methyl-heptadecenoic acids by *Sporothrix.* Also, in addition in some cases competition for nutrients were also seen which contributes in the reduction of infection by foliar pathogens.Some, other mechanisms which have the potential to protect against diseases were seen but they were not studied in detail which includes the stimulation of host defensive mechanism stimulated by yeast which is somehow similar to those described above in the fungi section.

The potential of the yeast used commercially in vegetable crops is effective, as the organism can be easily grown and formulated to enhance its survival in the environment were tested. The most appropriate utility of yeast is to control powdery mildew diseases and other foliar fungal plant pathogens. Also, may studies have successfully completed when the yeast was combined or altered with the fungicides, the level of disease-control was achieved far more when the yeast was used alone. This observation in turn suggested that the use of bio-control agents could be a successful strategy if they are used along with some chemical additives (Chaube *et al.,* 2004; Errakhi *et al.,* 2007; Jiskani *et al.,* 2007).

Interactions Contributing towards Biological Control

It is important to understand the interactions contributing to their biocontrol activity in order to suppress the activity of pathogens. These crucial interactions may include commensalism, competition, predation, parasitism and mutualism. These interactions between pathogen and bio control agents are complex in nature and it is known that plant, biocontrol agentand environmental factors play an important role in control of plant disease.Microbes and plants had observed to have the different forms of mutualistic interactions during co-evolution. Mutualism in normal language can be defined as the association between two or more species through which both or all the species are being benefitted. The classical example of the mutualism in plants is the association of roots of leguminous plants and *Rhizobium,* where nutrients are providedby the host plant to the bacteria and in return bacteria gives the NH_3 to the plants for amino acidand protein synthesis. Most microbes are considered as the facultative mutualists, that are been used as BCA. Facultative mutualism is the special type of mutualism in which the species interacting derive the benefit from each other but they are not fully dependent on each other for their survival. Occasionally, mutualism involves long-term biochemical and physical interactions between the partners that are been interacted for example interaction between the plants and mycorrhizal fungi. The fungi here prevent the roots infection by initiating the host defences and it also provides tolerance towards the plants in the plants. The colonization of the mycorrhizal fungi is used to reduce the damage caused by *Pseudomonas syringae*on the tomato plant.

Another, most common type of interaction is the commensalism,a symbiotic relationship between organisms where one organism is benefitted and the other is unaffected. Majority of the microbes associated with the plants are considered as commensals but their presence is the challenge to phytopathogens. The competition for various factors such as space, nutrients and growth stimulators between the BCA's and pathogens resulted in suppression of plant disease.

Another, interaction is the parasitism *i.e.* interaction between two organisms where the smaller one is benefitted and the host or larger organism is harmed. Somewhere the hyperparasitism are also, in this is the type of interaction; one organism is parasitizedby the other, particularly the pathogen, and further results in the biocontrol for example, parasitism of pathogen *Rhizoctonia solani*. Here, in this type of interaction, the parasite which is the antagonistic microbe would harm the host *i.e.*, the plant pathogen thus acting as BCA.

In contrast, the last type of interaction is predation which refers to killing of one organism by the other in order to take nutrition from killed one, such as the pathogen biomass for substance is consumed by the fungus-feeding nematode (Carmen and Roberto 2011; Chailleux *et al.*, 2014; Choudhary and Johri 2009; Hofstetter *et al.*, 2006; Kiss 2003; Massart and Jijakli 2007).

Engineered Bacteriophages as Biocontrol Agent

Bacteriophages are studied to be the potent biocontrol agents but somehow their potential is completely realized owing to number of ethical issues of natural pathogen. With, the significant increase in the technology there has been inclination towards the usage of self-engineered natural phages to overcome the limitations associated with natural microorganisms. Engineered phages are being used as targeted detection and removal of pathogens, *in situ* microbiome editing, gene delivery and programmable control of phage-bacterial interactions.

Traditional way of controlling pathogens has relied upon the broad-spectrum approaches like pasteurization or antibiotics that may differ in effectiveness can negatively affect food quality and can impact even natural microflora of food. Bacteriophages are nothing but the viruses that have capability to kill bacteria and can be promising alternative for biocontrol of bacterial diseases and are cheaper in production. Phages are ever-present natural predators of bacteria which have the capability to precisely target and kill bacterial pathogens without affecting the food quality.

Although, phages based agents have been commercializing since decade, but the adaptation through industry is quite at lower side and their game changing ability is not fulfilled. Synthetic biology offers new techniques to build engineered phages by various recombination and in vitro genome assemblies. Phages without any natural limitation could lead to better opportunity towards controlling diseases.A major obstacle in the usage of natural phages as biocontrol agents is their poor efficacy. However, initial application of natural phage has reduced the target bacteria significantly in the log manner but the residual bacterial load remains the same. However, this limited efficacy is achieved with low bacteria to phage ratio which is not possible to use outside the labs. It is seen that bacterias tend to grow enormously and quickly recover after the phage application which indicated the swift emergence of bacterial resistance and/or low phage susceptibility(Frampton *et al.*, 2012; Gerbore *et al.*, 2014; Harper 2021; Huss and Raman 2020).

Application Technologies

Biocontrol agents can be applied to the target plants through seed inoculation, soil or foliar application depends on its mode of action, type of formulation and stage of plant growth. Therefore, formulations for the control of soil-borne and above-ground phytopathogens require a different application approach. Similarly, the method of application of liquid formulations, wettable powders, and water-dispersible granules differs from dry powder and granule applications (Raguchander *et al.*, 2003).

Seed Application

Microorganisms when used as biological control agents for seed treatments protect seeds and seedlings from a variety of plant pathogens. Seed treatment is a common and convenient application method. Slurry method is generally used for seed coating with granules and powders. In this method solid seed treatment agents are mixed by means of a strapping adhesive solution, to form slurry, which is then applied to seeds. Liquid inoculants are applied by spraying on seeds. Vegetable oils, wheat flour, sucrose, starch, arabic gum, hydroxypropyl methylcellulose, CMC and non-toxic preparations are commonly used adhesives for seed coating. Seed coating is done either by mechanical tumbling machines, rotating drums, cement mixers, hand or automated seed coaters and subsequently drying is done by seed drier using forced air. Coating and drying can also be done using fluidized beds and seed coaters, where the seeds are dispersed on a cushion of pressurized air.

A major constraint in seed treatment technique is that the seeds are coated with a limited amount of BCA needed for successful bio control. Sometimes the inoculants are not properly attached to the seed and lost during sowing. There are a number of factors which influence microbial survival on seed such as toxins released by the seed coat, incompatibility between the bioinoculant and seed-applied chemicals (Kumari *et al.*, 2018; Tripathi *et al.*, 2020).

Soil Application

Soil application is direct application method of BCAs in soil against soil-borne plant pathogens, generally applied in the seeding furrows. In this method

BCAs are applied in the form of liquid inoculants, wettable powders, granular inoculants or water-dispersible granules. In soil they are applied by using granular applicators, under, above or alongside the seed. BCAs in powder form are also used for soil application; they are dustier than granular form and are less user-friendly. Liquid form of BCAs *i.e.* water-dispersible granules, wettable powders and liquid formulations can be applied in the soil by hand or by using mechanical spraying equipments, with even distribution of the formulation on the cropped area. These can be delivered directly to the root zone of individual plants by drenching furrows, hydroponic systems or drip irrigation. Soil application is generally practised where large populations of BCAs are needed to be applied into the soil. This avoids damage to small, fragile seeds and overcomes the adverse effects of chemicalsapplied on the seed(Mercado-Blanco *et al.*, 2004; Monda 2004; Musheer *et al.*, 2020; Noumavo *et al.*, 2016).

Aerial Application

Aerial application of BCAs is especially convenient to treat above-ground plant pathogens, mainly applied on the leaves. This technique is used for multiple application of BCA formulation during crop growth and also controls the location and application rate. The dose and frequency of application can be standardized based on each pathogens life cycle. They can be effectively applied by aerial application using spray equipment such as an aerosol can, hand or mechanical equipment and aircraft. In warmer regions aerial application of BCA is done in early morning or in late evening, when the leaves are turgid,high relative humidity and low temperature. The major challenge in foliar application is to maintain the viability and threshold inoculums potential for efficient control of plant pathogens (Punja 1997; Punja and Yip 2003).

Commercial Products

The bio control of phyto pathogens is relatively a new approach as compared to bio control of insect/pests. *Agrobacterium radiobacter* strain K 84 was the first bacterium registered with the United States Environmental Protection agency (EPA) for the control of crown gall in ear 1979. Ten years later *Trichoderma harzianumi* ATCC 20476 the first fungus was registered with the EPA for the control of plant diseases. Till date a total of 14 bacteria and 12 fungal strains have been registered as bio control agents with the EPA for the control of plant diseases (Fravel, 1988). Most of these are sold commercially as one or other product. The technology for commercialization of these products as bio control agents is still in its initial phase and 65 per cent of the organisms registered with EPA were registered in the past 10 years while the remaining 36 per cent were registered in the past 5 years (Ramanujam *et al.*, 2010; Rotem *et al.*, 1999; Sharma *et al.*, 2019).

REFERENCES

Abdel-Kader, M.M., El-Mougy., Nehal, S., Aly, M.D.E. and Lashin, S.M. (2012). Long activity of stored formulated bio-agents against some soil-borne plant pathogenic fungi causing root rot of some vegetables. J. Appl. Sci. Res. 8(4), 1882-1892.

Adetunji, C.O. and Varma, A. (2020). Biotechnological Application of Trichoderma: A Powerful Fungal Isolate with Diverse Potentials for the Attainment of Food Safety, Management of Pest and Diseases, Healthy Planet, and Sustainable Agriculture. In Trichoderma: Agricultural Applications and Beyond (pp. 257-285): Springer. Ahmed, E. and Holmström, S.J.M. (2014). Siderophores in environmental research: roles and applications. Microbial biotechnology. 7(3), 196-208.

Ardakani, S.S., Heydari, A., Khorasani, N. and Arjmandi, R. (2010). Development of new bioformulations of Pseudomonas fluorescens and evaluation of these products against damping-off of cotton seedlings. Journal of Plant Pathology. 83-88.

Axel, C., Zannini, E., Coffey, A., Guo, J., Waters, D.M., Arendt, E.K. (2012). Eco friendly control of potato late blight causative agent and the potential role of lactic acid bacteria: a review. Applied microbiology and biotechnology. 96(1), 37-48.

Azcón-Aguilar, C., Jaizme-Vega, M.C. and Calvet, C. (2002). The contribution of arbuscular mycorrhizal fungi to the control of soil-borne plant pathogens. In Mycorrhizal technology in agriculture (pp. 187-197): Springer.

Bakker, P., Ran, L.X., Pieterse, C.M. and Van Loon, L.C. (2003). Understanding the involvement of rhizobacteria-mediated induction of systemic resistance in biocontrol of plant diseases. Canadian Journal of Plant Pathology. 25(1), 5-9.

Banerjee, S. and Mandal, N.C. (2020). Fungal Bioagents in the Remediation of Degraded Soils. In Microbial Services in Restoration Ecology (pp. 191-205): Elsevier.

Bejarano, A. and Puopolo, G. (2020). Bioformulation of Microbial Biocontrol Agents for a Sustainable Agriculture. In How Research Can Stimulate the Development of Commercial Biological Control Against Plant Diseases (pp. 275-293): Springer.

Bhattacharyya, P.N. and Jha, D.K. (2012). Plant growth-promoting rhizobacteria (PGPR): emergence in agriculture. World Journal of Microbiology and Biotechnology. 28(4), 1327-1350.

Borriss, R. (2015). Bacillus, a plant-beneficial bacterium. In Principles of plant-microbe interactions (pp. 379-391): Springer.

Borriss, R., Wu, H., and Gao, X. (2019). Secondary metabolites of the plant growth promoting model rhizobacterium Bacillus velezensis FZB42 are involved in direct suppression of plant pathogens and in stimulation of plant-induced systemic resistance. In Secondary Metabolites of Plant Growth Promoting Rhizo microorganisms (pp. 147-168): Springer.

Butt, T.M., Jackson, C. and Magan, N. (2001). Fungi as biocontrol agents: progress problems and potential: CABI.

Calvo, J., Calvente, V., de Orellano, M.E., Benuzzi, D. and de Tosetti, M.I.S. (2007). Biological control of postharvest spoilage caused by Penicillium expansum and Botrytis cinerea in apple by using the bacterium Rahnella aquatilis. International Journal of Food Microbiology. 113(3), 251-257.

Carmen, B. and Roberto, D. (2011). Soil bacteria support and protect plants against abiotic stresses. Abiotic stress in plants-Mechanisms, and daptations, Italy. 143-170.

Chailleux, A.M., Emily, K., Teixeira, A., Mickaël, M., Gerben, J. and Desneux, N. (2014). Natural enemy mediated indirect interactions among prey species: potential for enhancing biocontrol services in agroecosystems. Pest management science. 70(12), 1769-1779.

Chaube, H.S., Mishra, D.S., Varshney, S. and Singh, U.S. (2004). Biocontrol of plant pathogens by fungal antagonists: historical background, present status and future prospects. Annual Review of Plant Pathology. 2, 1-42.

Chincholkar, S.B., Rane, M.R. and Chaudhari, B.L. (2005). Siderophores: their biotechnological applications. Biotechnological Application of Microbes. 177-198.

Choudhary, D.K. and Johri, B.N. (2009). Interactions of Bacillus spp. and plants–with special reference to induced systemic resistance (ISR). Microbiological research. 164(5), 493-513.

Choudhary, D.K., Prakash, A. and Johri, B.N. (2007). Induced systemic resistance (ISR) in plants: mechanism of action. Indian Journal of Microbiology. 47(4), 289-297.

Cook, R.J. (1993). Making greater use of introduced microorganisms for biological control of plant pathogens. Annual review of phytopathology. 31(1), 53-80.

Costa, J.M. and Loper, J.E. (1994). Characterization of siderophore production by the biological control agent Enterobacter cloacae. MPMI-Molecular Plant Microbe Interactions. 7(4), 440-448.

Cottyn, B., Debode, J., Regalado, E., Mew, T.W. and Swings, J. (2009). Phenotypic and genetic diversity of rice seed associated bacteria and their role in pathogenicity and biological control. Journal of Applied Microbiology. 107(3), 885-897.

Damalas, C.A., Eleftherohorinos. and Ilias, G. (2011). Pesticide exposure, safety issues, and risk assessment indicators. International journal of environmental research, and health, public. 8(5), 1402-1419.

Dandurishvili, N., Toklikishvili, N., Ovadis, M., Eliashvili, P., Giorgobiani, N., Keshelava, R., and Szegedi, E. (2011). Broad range antagonistic rhizobacteria Pseudomonas fluorescens and Serratia plymuthica suppress Agrobacterium crown gall tumours on tomato plants. Journal of applied microbiology. 110(1), 341-352.

Datta, A., Singh, R.K., Kumar, S. and Kumar, S. (2015). An effective and beneficial plant growth promoting soil bacterium "Rhizobium": a review. Ann Plant Sci. 4(1), 933-942.

De Silva., Nimali I., Brooks, S., Lumyong, S, and Hyde, K.D. (2019). Use of endophytes as biocontrol agents. Fungal Biology Reviews. 33(2), 133-148.

Dhall, RK. (2016). Application of edible coatings on fruits and vegetables. Biobased Environ. Benign Coatings. 87-119.

Dordas, C. (2008). Role of nutrients in controlling plant diseases in sustainable agriculture. A review. Agronomy for sustainable development. 28(1), 33-46.

Dukare, A., Kamble, U. and Koulagi, R. (2020). Biological Management of Agricultural Crops Diseases Using Microbial Biocontrol Agents. Biotechnology in plant protection and crop improvement. Pg 57-67.

Eljounaidi, K., Lee, S.K. and Bae, H. (2016). Bacterial endophytes as potential biocontrol agents of vascular wilt diseases–review and future prospects. Biological Control. 103, 62-68.

Erdogan, O. and Benlioglu, K. (2010). Biological control of Verticillium wilt on cotton by the use of fluorescent Pseudomonas spp. under field conditions. Biological Control. 53(1), 39-45.

Errakhi, R., Bouteau, F., Lebrihi, A., Barakate, M. (2007). Evidences of biological control capacities of Streptomyces spp. against Sclerotium rolfsii responsible for damping-off disease in sugar beet (Beta vulgaris L.). World Journal of Microbiology, and Biotechnology. 23(11), 1503-1509.

Frampton, R.A., Pitman, A.R. and Fineran, P.C. (2012). Advances in bacteriophage-mediated control of plant pathogens. International journal of microbiology. 2012.

Fravel, D.R. (1988). Role of antibiosis in the biocontrol of plant diseases. Annual review of phytopathology. 26(1), 75-91.

Gentz, M.C, Murdoch, G. and King, G.F. (2010). Tandem use of selective insecticides and natural enemies for effective, reduced-risk pest management. Biological Control. 52(3), 208-215.

Gerbore, J., Benhamou, N., Vallance, J., Le, F.G., Grizard, D. and Regnault-Roger. (2014). Biological control of plant pathogens: advantages and limitations seen through the case study of Pythium oligandrum. Catherine, Research, Pollution. 21(7), 4847-4860.

Gnanamanickam., Samuel, S.V., Preeti, R., Munagala, S., Defago, G. and Kloepper, J.W. (2002). Principles of biological control. Biological control of crop diseases. 1-9.

Goudjal, Y., Toumatia, O., Yekkour, A., Sabaou, N., Mathieu, F. and Zitouni, A. (2014). Biocontrol of Rhizoctonia solani damping-off and promotion of tomato plant growth by endophytic actinomycetes isolated from native plants of Algerian Sahara. Microbiological Research. 169(1), 59-65.

Hajek, A.E. and Eilenberg, J. (2018). Natural enemies: an introduction to biological control: Cambridge University Press.

Hamed, H.A., Moustafa, Y.A. and Abdel-Aziz, S.M. (2011). In vivo efficacy of lactic acid bacteria in biological control against Fusarium oxysporum for protection of tomato plant. Life Science Journal. 8(4), 462-468.

Haran, S., Schickler, H. and Chet, I. (1996). Molecular mechanisms of lytic enzymes involved in the biocontrol activity of Trichoderma harzianum. Microbiology. 142(9), 2321-2331.

Harper, D.R. (2021). Bacteriophage as Biocontrol Agents. Bacteriophages: Biology, Technology, Therapy. 751-766.

Hofstetter, R.W., Cronin, J.T., Klepzig, K.D., Moser, J.C. and Ayres, M.P. (2006). Antagonisms, mutualisms and commensalisms affect outbreak dynamics of the southern pine beetle. Oecologia. 147(4), 679-691.

Howell, C.R. (1998). The role of antibiosis in biocontrol. Trichoderma, and Gliocladium. 2, 173-184.

Huss, P. and Raman, S. (2020). Engineered bacteriophages as programmable biocontrol agents. Current opinion in biotechnology. 61, 116-121.

Jensen, D.F., Karlsson, M., Sarrocco, S. and Vannacci, G. (2016). Biological control using microorganisms as an alternative to disease resistance. Plant pathogen resistance biotechnology. 341-363.

Jha, C.K. and Saraf, M. (2015). Plant growth promoting rhizobacteria (PGPR): a review. Journal of Agricultural Research, and Development. 5(2), 108-119.

Jiskani, M.M., Pathan, M.A., Wagan, K.H., Imran, M. and Abro, H. (2007). Studies on the control of tomato damping-off disease caused by Rhizoctonia solani Kuhn. Pak. J. Bot. 39(7), 2749-2754.

Junaid, J.M., Dar, N., Ahmad, B., Tariq, A., Bhat, A.H. and Bhat, M.A. (2013). Commercial biocontrol agents and their mechanism of action in the management of plant pathogens. International Journal of Modern Plant, and Sciences, Animal. 1(2), 39-57.

Khan, N., Mishra, A. and Nautiyal, C.S. (2012). Paenibacillus lentimorbus B-30488r controls early blight disease in tomato by inducing host resistance associated gene expression and inhibiting Alternaria solani. Biological Control. 62(2), 65-74.

Khasa, Y.P. (2017). Microbes as biocontrol agents. In Probiotics and Plant Health (pp. 507-552): Springer.

Khattak, K.F. and Rahman, T.U. (2017). Analysis of vegetable's peels as a natural source of vitamins and minerals. International Food Research Journal.24(1), 292.

Kiss, L. (2003). A review of fungal antagonists of powdery mildews and their potential as biocontrol agents. Pest Management Science: formerly Pesticide Science.59(4), 475-483.

Kloepper, J.W., Leong, J., Teintze, M. and Schroth, M.N. (1980). Enhanced plant growth by siderophores produced by plant growth-promoting rhizobacteria. Nature. 286(5776), 885-886.

Köhl, J., Kolnaar, R. and Ravensberg, W.J. (2019). Mode of action of microbial biological control agents against plant diseases: relevance beyond efficacy. Frontiers in Plant Science. 10, 845.

Koike, S.T., Gladders, P. and Paulus, A.O. (2007). Vegetable diseases: a color handbook: Gulf Professional Publishing.

Kumar, S., Thakur, M. and Rani, A. (2014). Trichoderma: Mass production, formulation, quality control, delivery and its scope in commercialization in India for the management of plant diseases. African Journal of Agricultural Research. 9(53), 3838-3852.

Kumari, R., Ashraf, S., Bagri, G.K., Khatik, S.K., Bagri, D.K and Bagdi, D.L. (2018). Impact of seed treatment from bio-agents and fungicides on growth, biomass and yield of lentil (Lens culinaris Medik). Journal of Pharmacognosy, and Phytochemistry. 7(3), 251-253.

Lemaire, A. and Limbourg, S. (2019). How can food loss and waste management achieve sustainable development goals?, Journal of cleaner production. 234, 1221-1234.

Mahmood, I.I., Sameen, R.S., Kanwal, G.A. and Hakeem, K.R. (2016). Effects of pesticides on environment. In Plant, soil and microbes (pp. 253-269): Springer.

Massart, S. and Jijakli, H.M. (2007). Use of molecular techniques to elucidate the mechanisms of action of fungal biocontrol agents: a review. Journal of Microbiological Methods. 69(2), 229-241.

McGrath, M.T. (2001). Fungicide resistance in cucurbit powdery mildew: experiences and challenges. Plant disease. 85(3), 236-245.

McManus, P.S., Stockwell, V.O., Sundin, G.W. and Jones, A.L. (2002). Antibiotic use in plant agriculture. Annual review of phytopathology. 40(1), 443-465.

Melo, Itamar., Soares, d. and Faull, J.L. (2000). Parasitism of Rhizoctonia solani by strains of Trichoderma spp. Scientia Agricola. 57(1), 55-59.

Mercado-Blanco, J., Rodrýguez-Jurado, D., Hervás, A. and Jiménez-Dýaz, R.M. (2004). Suppression of Verticillium wilt in olive planting stocks by root-associated fluorescent Pseudomonas spp. Biological control. 30(2), 474-486.

Monda, E. (2002). Biological control of Fusarium wilt of tomato–a review. Journal of Tropical Microbiology, and Biotechnology. 1(1), 74-78.

Montejano-Ramírez, V., García-Pineda, E. and Valencia-Cantero, E. (2020). Bacterial compound N, N-dimethylhexadecylamine modulates expression of iron deficiency and defense response genes in Medicago truncatula independently of the jasmonic acid pathway. Plants. 9(5), 624.

Musheer, N., Ashraf, S., Choudhary, A., Kumar, M. and Saeed, S. (2020). Role of Microbiotic Factors Against the Soil-Borne Phytopathogens. In Phytobiomes: Current Insights and Future Vistas (pp. 251-280): Springer.

Naher, L., Yusuf, U.K., Ismail, A. and Hossain, K. (2014). Trichoderma spp.: a biocontrol agent for sustainable management of plant diseases. Pak. J. Bot. 46(4), 1489-1493.

Nega, A. (2014). Review on concepts in biological control of plant pathogens. Journal of Biology, Agriculture, and Healthcare. 4(27), 33-54.

Newell, D.G., Koopmans, M., Verhoef, L., Duizer, E., Aidara-Kane, A., Sprong, H. and Scheutz, F. (2010). Food-borne diseases—the challenges of 20 years ago still persist while new ones continue to emerge. International journal of food microbiology. 139, S3-S15.

Noumavo, P., Agossou, A., Nadège A., Baba-Moussa, F., Adjanohoun, A. and Baba-Moussa, Lamine. (2016). Plant growth promoting rhizobacteria: Beneficial effects for healthy and sustainable agriculture. African Journal of Biotechnology. 15(27), 1452-1463.

Oerke, E-C. and Dehne, H-W. (2004). Safeguarding production—losses in major crops and the role of crop protection. Crop protection. 23(4), 275-285.

Pal, K.K. and Gardener, B.M. (2006). Biological control of plant pathogens. The plant health instructor. 2, 1117-1142.

Parveen., Shazia, W., Abdul, H., Bhat, M.Y. and Koka, J.A. (2016). Biological control of postharvest fungal rots of rosaceous fruits using microbial antagonists and plant extracts—a review. Czech Mycology. 68(1).

Porat, R., Lichter, A., Terry, L.A., Harker, R. and Buzby, J. (2018). Postharvest losses of fruit and vegetables during retail and in consumers' homes: Quantifications, causes, and means of prevention. Postharvest Biology and Technology. 139, 135-149.

Poshina, D.N., Raik, S.V., Poshin, A.N and Skorik, Y.A. (2018). Accessibility of chitin and chitosan in enzymatic hydrolysis: A review. Polymer Degradation, and Stability. 156, 269-278.

Powell, K.A and Jutsum, A.R. (1993). Technical and commercial aspects of biocontrol products. Pesticide Science. 37(4), 315-321.

Prashar, P., Kapoor, N. and Sachdeva, S. (2013). Biocontrol of plant pathogens using plant growth promoting bacteria. In Sustainable Agriculture Reviews (pp. 319-360): Springer.

Punja, Z.K. (1997). Comparative efficacy of bacteria, fungi, and yeasts as biological control agents for diseases of vegetable crops. Canadian Journal of Plant Pathology. 19(3), 315-323.

Punja, Z.K. and Yip, R. (2003). Biological control of damping-off and root rot caused by Pythium aphanidermatum on greenhouse cucumbers. Canadian Journal of Plant Pathology. 25(4), 411-417.

Raaijmakers, J.M., Vlami, M. and De Souza, J.T. (2002). Antibiotic production by bacterial biocontrol agents. Antonie van leeuwenhoek. 81(1), 537-547.

Raguchander, T., Rajappan, K. and Samiappan, R. (1997). Evaluating methods of application of biocontrol agent in the control of mungbean root rot. Indian Phytopathology. 50(2), 229-234.

Rajput, R.S., Ram, R.M., Vaishnav, A. and Singh, H.B. (2019). Microbe-based novel biostimulants for sustainable crop production. In Microbial Diversity in Ecosystem Sustainability and Biotechnological Applications (pp. 109-144): Springer.

Ramanujam, B., Prasad, R.D., Sriram, S. and Rangeswaran, R. (2010). Mass production, formulation, quality control and delivery of Trichoderma for plant disease management. The Journal of Plant Protection Sciences. 2(2), 1-8.

Rokem, J.S., Lantz, A.E. and Nielsen, J. (2007). Systems biology of antibiotic production by microorganisms. Natural product reports. 24(6), 1262-1287.

Rotem, Y., Yarden, O. and Sztejnberg, A. (1999). The mycoparasite Ampelomyces quisqualis expresses exgA encoding an exo-β-1, 3-glucanase in culture and during mycoparasitism. Phytopathology. 89(8), 631-638.

Saravana, K., Duraisamy, C., Annalisa, S., Davide, G., Angelo, G. and Lodovica, M. (2008). Metschnikowia pulcherrima strain MACH1 outcompetes Botrytis cinerea, Alternaria alternata and Penicillium expansum in apples through iron depletion. Postharvest Biology, and Technology. 49(1), 121-128.

Schmitt, A. and Seddon, B. (2005). Biocontrol of plant pathogens with microbial BCAs and plant extracts-advantages and disadvantages of single and combined use. Paper presented at the Modern fungicides and antifungal compounds IV: 14th International Reinhardsbrunn Symposium, Friedrichroda, Thuringia, Germany, April 25-29, 2004.

Shalini, S. and Kotasthane, A.S. (2007). Parasitism of Rhizoctonia solani by strains of Trichoderma spp. EJEAF Chemistry. 6, 2272-2281.

Sharma, S., Kour, D., Rana, K.L., Dhiman, A., Thakur, S., Thakur, P., Yadav, N. (2019). Trichoderma: biodiversity, ecological significances, and industrial applications.

In Recent advancement in white biotechnology through fungi (pp. 85-120): Springer.

Shoresh, M., Harman, G.E. and Mastouri, F. (2010). Induced systemic resistance and plant responses to fungal biocontrol agents. Annual review of phytopathology. 48, 21-43.

Singh, P.P., Shin, Y.C., Park, C.S. and Chung, Y.R. (1999). Biological control of Fusarium wilt of cucumber by chitinolytic bacteria. Phytopathology. 89(1), 92-99.

Singh, S., Kumar, V., Dhanjal, D.S. and Singh, J. (2020). Biological control agents: diversity, ecological significances, and biotechnological applications. In Natural Bioactive Products in Sustainable Agriculture (pp. 31-44): Springer.

Smith, H.S. (1939). Insect populations in relation to biological control. Ecological Monographs. 9(3), 311-320.

Sri, A. (2019). Occurrence of mycoparasitic fungi Sphaerellopsis paraphysata on pearl millet rust pathogen, Puccinia substriata in India. Madras Agricultural Journal. 106(march (1-3)), 1.

Steyaert, J.M., Ridgway, H.J., Elad, Y. and Stewart, A. (2003). Genetic basis of mycoparasitism: a mechanism of biological control by species of Trichoderma. New Zealand Journal of Crop, and Science, Horticultural. 31(4), 281-291.

Suleria, H.A.R., Butt, M.S., Anjum, F.M., Saeed, F. and Khalid, N. (2015). Onion: Nature protection against physiological threats. Critical reviews in food science, and nutrition. 55(1), 50-66.

Suman., Yadav, A., Nath, A. and Verma, P. (2016). Endophytic microbes in crops: diversity and beneficial impact for sustainable agriculture. In Microbial inoculants in sustainable agricultural productivity (pp. 117-143): Springer.

Sundh, I. and Goettel, M.S. (2013). Regulating biocontrol agents: a historical perspective and a critical examination comparing microbial and macrobial agents. BioControl. 58(5), 575-593.

Tripathi, A.N., Meena, B.R., Pandey, K.K. and Singh, J. (2020). Microbial bioagents in agriculture: current status and prospects. In New Frontiers in Stress Management for Durable Agriculture (pp. 331-368): Springer.

Urquhart, E.J., Menzies, J.G. and Punja, Z.K. (1994). Growth and biological control activity of Tilletiopsis species against powdery mildew (Sphaerotheca fuliginea) on greenhouse cucumber. Phytopathology. 84(4), 341-351.

van Lenteren, J.C., Bolckmans, K., Köhl, J., Ravensberg, W.J. and Urbaneja, A. (2018). Biological control using invertebrates and microorganisms: plenty of new opportunities. BioControl. 63(1), 39-59.

Veliz, E.A., Martínez-Hidalgo, P. and Hirsch, A.M. (2017). Chitinase-producing bacteria and their role in biocontrol. AIMS microbiology. 3(3), 689.

Vey, A., Hoagland, R.E., Butt, T.M. (2001). Toxic metabolites of fungal biocontrol agents. Fungi as biocontrol agents: progress, problems, and potential. 1, 311-346.

Viterbo, A., and Horwitz, B.A. (2010). Mycoparasitism. Cellular, and fungi, molecular biology of filamentous. 676-693.

Viterbo, A., Ramot, O., Chernin, L. and Chet, I. (2002). Significance of lytic enzymes from Trichoderma spp. in the biocontrol of fungal plant pathogens. Antonie Van Leeuwenhoek. 81(1), 549-556.

Wani, M.Y., Mehraj, S., Rather, R.A., Rani, S., Hajam, O.A., Ganie, N.A. and Kamili, A.S. (2018). Systemic acquired resistance (SAR): A novel strategy for plant protection with reference to mulberry. Int. J. Chem. Stud. 2, 1184-1192.

Wilson, M. and Backman, P.A. (1999). Biological control of plant pathogens. Handbook of pest management. 309-335.

Wu, L., Wu, H., Qiao, J., Gao, X., and Borriss, R. (2015). Novel routes for improving biocontrol activity of Bacillus based bioinoculants. Frontiers in Microbiology. 6, 1395.

Chapter 15

Probiotics in Human Health

Sakshi Khurana[1], Sangeeta C. Sindhu[2]*, Ajeet Kumar[3], Anil Sindhu[4] and Anju Kumari[5]

[1]Department of Biotechnology, DCRUST, Murthal, Haryana
[2]Department of Foods and Nutrition, CCSHAU Hisar, Haryana
[3]Department of Animal Biotechnology, NDRI, Karnal, Haryana
[4]Department of Molecular Biology, Biotechnology and Bioinformatics,
[5]Department Centre for Food Science and Technology,
CCSHAU, Hisar, Haryana
*e-mail: chahalsangeeta@hau.ac.in

ABSTRACT

Last few decades have seen a lot of progress in the field of probiotics due to rising awareness of the dietary supplements and also for the curative healthcare concerns. According to the FAO/WHO, Probiotics are "live microorganisms that provide a health benefit on the host when supplied in sufficient numbers." Certain minimal criteria have been set for probiotic strains; such as - well characterized, safe for their use, support the generally accepted scientific standards by at least one human clinical trial and active in massive quantities of the product throughout the shelf life. There are different type of bacteria, yeast and other microorganisms that are heavily involved in the development of the probiotic related products. But most commonly used microbes are the lactic acid bacteria (LAB). Lactobacillus and Bifidobacterium genus are among the LAB, that help in the fermentation of food and provide health benefits. Probiotics strain can act through competitive exclusion, production of inhibitory substances, adhesion of cells, immune system modulation or by other means, gives potential to be used as therapeutic purpose. Now, beside the several dairy products, probiotics are also used in the non-dairy products and plant-based products, and regarded to be an effective natural way alternative to different pharmaceutical treatments. Probiotics improve digestion while also providing various health advantages such as improved immunity, intestinal health, weight control, obesity prevention, and many more.

Keywords: *Probiotics, Selection criteria, Lactic acid bacteria, Health benefits, Mechanisms of Action, Probiotics in Covid-19*

Defining Probiotics

Over the last decade, with increasing scientific evidence; focus has increased on use of food and dietary supplements as a means of sustaining one's health. The concept of probiotics is not new to mankind. Its history dates back to era when humans started yeast fermentation of beverages. Yoghurt is perhaps the most common form of probiotic fermented food used in almost every household.

Elie Metchnikovis regarded as the father of the idea that probiotics are beneficial to human health (Ozen and Dinleyici, 2015). In 1908, Metchnikov published a book 'The prolongation of life; optimistic studies.' He suggested that "phagocytosis in intestinal flora produces toxic substances, leading to the ageing process, and if these proteolytic microbes could be prevented, the result would be longer life expectancy." In the same year, he was awarded a Nobel Prize for his work on phagocytosis. The term 'Probiotic' was first coined by Lilley and Stillwell (1965) and has been redefined by several authors (Table 15.1).

Table 15.1: Tracing History of Probiotic Definitions

Year	*Author*	*Definition*
1965	Lilly and Stillwell	"Substances secreted by one microorganism which stimulated the growth of another". This defined probiotics as exact opposite of 'antibiotic'(Lilly and Stillwell, 1965)
1971	Sperti	"Tissue extracts which stimulated microbial growth"(Mihich, 1972)
1974	Parker	"Organisms and substances which contribute to intestinal microbial balance"(PARKER and RB, 1974)
1989	Fuller	"A live microbial feed supplement which beneficially affects the host animal by improving its intestinal microbial balance"
2001	FAO/WHO	"Live microorganisms which when administered in adequate amounts confer a health benefit on the host"(FAO and WHO, 2002)
2014	Hill *et al.*	"Live microorganisms that, when administered in adequateamounts, confer a health benefit on the host"(Hill *et al.*, 2014)

Selection of Probiotics

Probiotics include a wide range of microorganisms, mainly bacteria, yeasts and fungi. Going by the definition, a large number of microbial species have been identified with probiotic properties. Various criteria have been postulated for selecting suitable probiotics for intended use (Binda *et al.,* 2020; Koirala and Anal, 2021; Santacroce *et al.,* 2019). Common selection criteria (Shewale *et al.,* 2014) for use as 'probiotics' include the following: ability to exert a beneficial effect on the host; withstand incorporation into a foodstuff at high cell counts, and remain viable throughout the shelf life of product, withstand transit through a GI tract; adhere to intestinal epithelium cell lining and colonize the lumen of the tract; produce antimicrobial substances towards the pathogens; technologically suitable for industrial process.

The various genera of bacteria most commonly used in probiotic formulations are *Lactobacillus, Bifidobacterium, Enterococcus* and *Streptococcus* as well as some fungal strains belonging to the genus *Saccharomyces* such as the *Saccharomyces boulardii.* The most widely applied microorganisms that are utilised as probiotics in food and feed are *Lactobacillus* and *Bifidobacterium* species. Many commonly exploited probiotic strains have a generally recognized as safe (GRAS) status in the USA or belong to species with qualified presumption of safety (QPS) status with the European Food Safety Authority (EFSA). However, novel probiotic species with no history of use need to be evaluated for safety. Developments in affordable complete genome sequencing and powerful cultivation methods have allowed isolation and characterisation of a new range of microorganisms from human microbiomes with potential health benefits(Cunningham *et al.,* 2021). The probiotic potential is known to be directly connected to particular strains, not to the genus or species of a microorganism (Hill *et al.,* 2014). Also, data is important about the minimum effective values. It is generally accepted that probiotic products should have a minimum concentration of 10^6 CFU per milliliter or gram and that a total of some 10^8 to 10^9 probiotic microbial amount must be consumed each day for the probiotic benefits to be transferred to the consumer (Kechagia *et al.,* 2013).

Table 15.2: Commonly Used Probiotic Microorganisms*

Lactobacillus Species	*Bifidobacterium Species*	*Others*
Lactobacillus acidophilus	*Bifidobacterium adolescentis*	*Bacillus cereus*
Lactobacillus brevis	*Bifidobacterium breve*	*Enterococcus faecalis*[a]
Lactobacillus amyloyorus	*Bifidobacteriumanimalis*	*Clostridium botyricum*
Lactobacillus casei	*Bifidobacterium bifidum*	*Enterococcus faecalis*[a]
Lactobacillus rhamnosus	*Bifidobacteriuminfantis*	*Escherichia coli*
Lactobacillus crispatus	*Bifidobacterium lactis*	*Lactobacillus lactis* subsp. *Cremoriss*
Lactobacillus delbrueckii subsp. *Bulgaricus*	*Bifidobacterium longum*	*Lactobacillus lactis* subsp. *Lactis*

Lactobacillus Species	*Bifidobacterium Species*	*Others*
Lactobacillus fermentum		*Leuconostocmes enteroides* subsp. *Dextranicum*
Lactobacillus gasseri		*Propionibacterium freudenreichii*[a]
Lactobacillus helyeticus		*Pediococcus acidilactici*
Lactobacillus johnsonii		*Saccharomyces boulardii*
Lactobacillus lactis		*Streptococcus salivarius* subsp. *Thermophilus*
Lactobacillus paracasei		*Sporolacto bacillusinulinus*[a]
Lactobacillus plantarum		
Lactobacillus reuteri		
Lactobacillus salivarius		
Lactobacillus gallinarum[a]		

a: Mostly used on animals.

*Adapted from (Prado *et al.*, 2008) and (Leroy *et al.*, 2008).

For use of identified probiotics in foods as well as dietary supplements, certain requirements apply (Binda *et al.*, 2020). The probiotic strains must be (i) adequately characterized; (ii) safe for the intended use; (iii) supported by at least one positive human clinical trial conducted according to generally accepted scientific standards; and (iv) alive in sufficient numbers in the product at an efficacious dose throughout the shelf life.

Over the years certain selection criteria have been established for selection of identified probiotics for therapeutic applications (Table 15.3).

Table 15.3: Suggested Traits in Probiotics for Specific Applications

Therapeutic Use	*Suggested Traits*	*Reference*
Atopic disease	☆ Ability to modulate *in vitro* production of cytokines by peripheral blood mononuclear cells (PBMCs)	(Niers *et al.*, 2005)
Hypocholes-terolemia	☆ Tolerance to Simulated Gastric Juice ☆ Bile Resistance ☆ Assimilation of Cholesterol ☆ Bile salt deconjugation ☆ Cholesterol co-precipitation ☆ Cholesterol adhesion to probiotic cell wall ☆ Miceller sequestration of cholesterol	(Bhat and Bajaj, 2019)(Romero-Luna *et al.*, 2021)

Therapeutic Use	Suggested Traits	Reference
Anticarcinogenic ability	✰ Production of compounds with anticarcinogenic activity ✰ The binding and degradation of potential mutagens ✰ The reduction of the activity of enzymes involved in carcinogen formation ✰ Ability to reduce genotoxic, immunesuppressive and nephrotoxic mycotoxins ✰ Inhibition of tumor cell proliferation ✰ Induction of the apoptosis in cancer cells.	(de Melo Pereira *et al.*, 2018; Kumar *et al.*, 2013; Tjalsma *et al.*, 2012)
Anti-depression and anti-anxiety	✰ Use of rat models for measuring behaviors in the forced swim test ✰ Evaluation of immune (*e.g.*, plasma cytokines measurement) and neuroendocrine and central monoaminergic activities (analysis of monoamine, such as serotonin, noradrenaline and dopamine, in the brain, as well as serotonin-precursor, tryptophan in the plasma)	(Dinan and Cryan, 2013; Pirbaglou *et al.*, 2016; R.A. and J.A., 2015)
Anti-obesity and anti-diabetic activities	✰ Abilities to breakdown lipopolysaccharides ✰ Ability to modulate the production of cytokines by peripheral blood mononuclear cells ✰ Inhibit mast cell activation ✰ Bring desirable change in biochemical parameters such as glucose, insulin, triglycerides, C-peptide, total cholesterol, HDL cholesterol, LDL cholesterol, alanine, lowering of Homeostatic Model Assessment for Insulin Resistance (HOMA-IR) values during clinical trials	(Alokail *et al.*, 2013; Hu *et al.*, 2017; Niers *et al.*, 2005)

Mechanism of Action of Probiotics

The intestinal microbial has a vital role in maintaining immune homeostasis. Probiotic bacteria have been demonstrated to exhibit a beneficial effect in treating gastrointestinal diseases in a number of controlled clinical trials. This serves as the proof of principle that probiotic bacteria have a potential to be used as a therapeutic strategy in order to overcome gastrointestinal diseases(Ng *et al.*, 2009). However, different probiotic strains exhibit a variety of mechanisms by which they exert their beneficial effects and are largely influenced by dosage, route as well as frequency employed (Sherman *et al.*, 2009). Different mechanisms by which probiotics exert their beneficial effects include competitive exclusion, production of inhibitory substances, production of mucus in intestine, cell adhesion, modulation of immune system, production of short chain fatty acids (SCFA) and growth stimulation of indigenous lactic acid bacteria. Various proposed mechanisms have been discussed below and gist is shown in Figure 15.1.

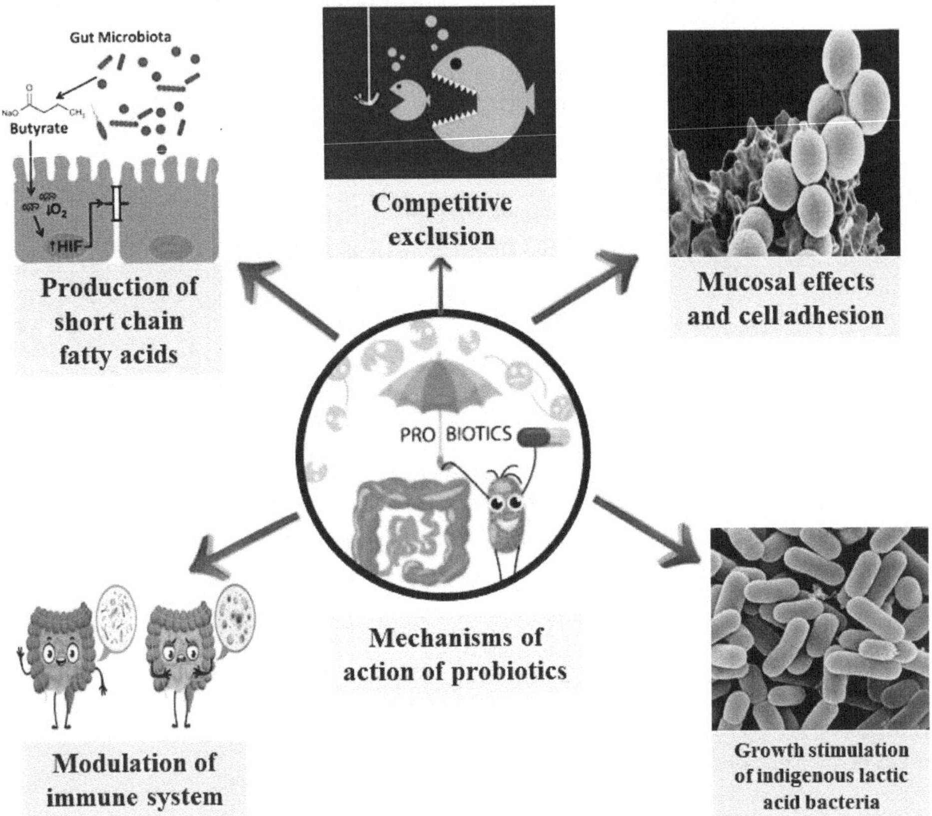

Figure 15.1: Representing Mechanisms of Action of Probiotics.

Competitive Exclusion and Production of Inhibitory Substances

In order to affect host, microbial strain must be attached to receptor sites in the intestinal tract. Competitive exclusion is a condition in which one bacterial species competes for receptor sites in intestinal tract more dominantly than other species (Bermudez-Brito *et al.,* 2012). Mechanism underlying and specific pathway of this phenomenon is still largely unknown. Competition for nutrients, resultant decrease in luminal pH and production of inhibitory substances such as bacteriocins and related compounds are few of the proposed mechanisms (Collado *et al.,* 2010). Majority of the studies are conducted on *Salmonella typhi* and *E. coli* as these two are considered as major microbial strains associated with contaminated foods (Bermudez-Brito *et al.,* 2012). A number of metabolic products of probiotics such as organic acids, hydrogen peroxide, bacteriocin, amines, *etc.* have been reported to combine with multiple targets in metabolic pathways that regulate cell proliferation, inflammation, apoptosis and differentiation. Proliferation of selected pathogens is prevented by bacteriocins which is well known antimicrobial peptides produced by some *lactobacilli* and *bifidobacteria*.

Mucosal Effect and Cell Adhesion

Any microbial strain to be termed as probiotic must meet some specific prerequisites. Cell adhesion to the intestinal mucosa for colonization and further interaction with host is one such criteria. Microbial strain such as *Lactobacillus plantarum* possess ability to elevate the production and secretion of mucin (MUC2 and MUC3) from human intestinal epithelial cells (David R. Mack *et al.,* 1999). The enhanced mucus layer on lining of gastrointestinal tract act as an antimicrobial shield thus prevents binding of pathogenic microorganisms such *as E. coli* (Mack *et al.,* 2003). Mucosal surface also helps in clearance of pathogens from gastrointestinal tract. Mucin producing cells also secretes antimicrobial peptides in response to harmful stimuli. Above mentioned phenomenon act synergistically and reduce the binding of pathogenic microorganisms in the epithelial cell surface (Clyne *et al.,* 2004).

Modulation of Immune System

A number of microbial strains have been known to produce molecules that possess immunomodulatory and anti-inflammatory functions that are capable of stimulating immune cells. Immunomodulatory effects of probiotic are a result of interaction of probiotic bacteria with monocytes/macrophages and lymphocytes(D'Amelio and Sassi, 2018). Regulation of host immune system is one of the mechanisms of probiotic. Thus, immune system is divided into two broad classes a) adaptive system and b) innate system. In adaptive immune system, response is characterized by B and T lymphocytes which shows affinity to specific antigens. While innate system responds to common structures known as pathogen-associated molecular patterns (PAMPs), shared by majority of pathogens. The primary response to pathogen is achieved by pattern recognition receptors (PRRs) which binds to PAMPs. PRRs comprises of transmembrane proteins known as toll like receptors (TLRs) that are expressed on various immune and non-immune cells such as epithelial cells, endothelial cells, fibroblast cells, B-cells, natural killer cells, macrophages and dendritic cells. Furthermore, PRRs also possess nucleotide binding oligomers, lectins and adhesive molecules. Through above mentioned mechanisms probiotic helps in preservation of intestinal homeostasis by modulation immune response (Gómez-Llorente *et al.,* 2010).

Production of Short Chain Fatty Acids (SCFA)

Organic acids, short chain fatty acids in particular are produced in small amounts (millimolar) in gastrointestinal tract and higher amount is also produced when anaerobic microorganisms are predominantly present (Markowiak-Kopeæ and Œli¿ewska, 2020). The most common short chain fatty acids include acetic acid, propionic acid and butyric acid in (3:1:1 molar ratio) and represents about 90-95 per cent of total SCFAs in human colon while formic acid is present in small proportion. SCFA plays a significant role in maintaining intestinal and immune homeostasis. Thus, presence of these acids is essential for health and well-being of the host. SCFA helps in inhibiting prospective contraction in human colon that derive the digesta downward. This inhibitory effect is one of the proposed mechanisms as intestinal bacteria do not retain their population against fast digesta movement (Ohashi and Ushida, 2009).

Growth Stimulation of Indigenous Lactic Acid Bacteria

Most of the researchers throughout the world have demonstrated the positive effect of probiotics solely through administrated probiotics strains. However, few of the studies also suggests that administration of various *Lactobacilli* of different origin such as *Lactobacillus Casei Shirota* (LcS), *L. delbruekii subsp. bulgaricus* or *L. plantarum* helps in increasing *Lactobacilli* in pigs. In case of human, fermented milk containing *Bifidobacterium* have been reported to increase number of *Lactobacillus* in faecal matter. These results represent potential of probiotics to stimulate growth of indigenous *Lactobacillus* or *Bifidobacterium*.

Regulations and Guidelines for Probiotics

Regulatory restrictions vary considerably depending on whether a probiotic is used as a medication or a food supplement. If a probiotic is to be used as a medication, it must go through the same FDA approval process as any other novel therapeutic substance. Before an experimental or biological product may be delivered to people, an Investigational New Drug application must be submitted and approved by the Food and Drug Administration. Before a probiotic medication can be marketed, it must be proved to be safe and effective for its intended application(Lahtinen *et al.*, 2011). Probiotic products are sold as foods or dietary supplements in the United States to a usually healthy population. Premarket proof of safety and effectiveness, as well as FDA clearance, are not necessary for dietary supplements; only premarket notice is required. Manufacturers of dietary supplements are allowed to make structure/function or health claims with regard to assertions on essential micronutrients under the legislation. Around the world, efforts are being undertaken to develop relevant probiotic product standards or guidelines (Table 15.2). Guidelines developed by the Joint Food and Agriculture Organization of the United Nations/World Health Organization Expert Consultation on Evaluation of Health and Nutritional Properties of Probiotics could be used as global standards for evaluating probiotics in food, allowing health claims to be substantiated.

The strain must be identified using accepted scientific procedures, named using valid current nomenclature, and given a strain designation that can be retrieved. Methods will differ based on the probiotic species and are likely to alter as technology advances. It should also be put in a worldwide cultural collection. Furthermore, the strain must have shown safety for its intended use as well as a documented health benefit based on at least one research that satisfies widely recognized scientific criteria or as per the recommendations and regulations of local/national authorities, if applicable. To provide the stated (and evidence-based) health benefit, sufficient quantities of the probiotic strain(s) must be present in the finished product throughout its shelf life. Products should be made in accordance with the applicable good manufacturing practices to ensure safety, purity, and stability(Jackson *et al.*, 2019) and should be labelled in a way that conveys to the end-user important information about the product's composition (particular strains, level of live probiotic given at end of shelf life, and health benefits assertions where authorized). Following these guidelines will ensure that there are no goods on the market that utilize the term "probiotic" incorrectly.

Table 15.2: Various Organizations Working in Order to Establish Probiotics Standards in Commercially Goods.

Organizations	*Region Affected*	*Action taken*
Joint committee organization of Food Agriculture Organization (FAO)/World Health Organization (WHO)	Throughout the world	Prescribed guidelines for assessment of probiotics in foods
International Dairy Federation (IDF)	Throughout the world	Started developing criteria for determining particular functional and safety qualities stated in the FAO recommendations for probiotics in food assessment.
European Food and Feed Culture Association	European countries	Suggested regulations for the use of probiotics in food.
Codex Standard for Fermented Milks (Codex Stan 243-2003)	Throughout the world	Apart from composition consideration, standards also specify counts of microorganisms and their additional labelled in fermented milk products such as kefir, acidophilus milk, kumys and yoghurt.
National Yogurt Association	United State America	Grant of petition under suitable criteria by the FDA which would ultimately result in alteration of the standard of identity of yoghurt, particularly the requirement of minimum levels of live cultures that should be present in yoghurt.
International Scientific Association for Probiotics and Prebiotics	Throughout the world	Board of Directors as well as Industry Advisory Committee in order to take into consider method of validation and establishment of laboratory sites to assess microbiological aspects of foods containing probiotic.

Source: Food Science and Technology Bulletin: Fu - R. G Prof. Gibson - Google Books, n.d.; Sanders and Heimbach, 2003.

Applications of Probiotics in Human Health Diseases

At present, functional foods and beverages have become common in people's diets, and they have been shown to provide health advantages in addition to the well-known nutritional impacts. Consumption of functional foods and drinks is currently one of the most popular eating habits, worldwide. Probiotics improve digestion while also providing various health advantages such as improved immunity, intestinal health, weight control, obesity prevention, and many more.

Probiotics in Dental Caries

Tooth decay is caused by the deterioration of the tooth enamel, often known as **dental caries** or **cavities.** Dental caries is a bacterial-caused multifactorial illness marked by acid demineralization of the tooth enamel (Dallal *et al.,* 2021). Dental caries continues to affect both children and adults around the world, putting a substantial financial burden on them (Angarita-Díaz *et al.,* 2020). As there are extensive studies going on the beneficial effects of probiotics, there is ample evidence showing the significant role of probiotics in oral care. Probiotics latch to dental surfaces, causing a modification in the mouth's environment. The inflammatory response is reduced, and antimicrobial compounds are produced in response to oral infections (X. Cheng *et al.,* 2021). Probiotics can be found in toothpastes, mouthwashes, powders, tablets, lozenges, and chewing gums(Ramanujam *et al.,* 2019).Thermophili 110 is produced by *S. thermophilus* B59671. Thermophilin 110, it is Bacteriocin that stops the oral pathogen *Streptococcus* mutants from growing and was manifest to prevent biofilm development by *S. mutants* UA159 (Renye and Steinberg, 2021). Sandoval and co-workers (2021) reported that frequent consumption of *Lactobacillus rhamnosus* SP1-supplemented milk to preschool children with a high caries risk reduced the incidence of caries. Further, according to many clinical trials, daily consumption of probiotic-rich yoghurt, cheese or milk resulted in a decrease in the number of pathogenic *Streptococci* in the saliva and a reduction in dental plaque (Ahola *et al.,* 2002; Çaglar *et al.,* 2007; Nikawa *et al.,* 2004; Petti *et al.,* 2008). *Streptococcus thermophilus* and *Lactobacillus lactis* subsp. *lactis* were the only bacterial strains with the ability to integrate into a biofilm present on a hydroxyapatite surface and interfere with the formation of cariogenic species, according to Comelli and colleagues ("*The Role of Prebiotics and Probiotics in Human Health,*" 2013). Another study by Siyuan and his team conducted a systematic review and meta-analysis and it was found that *Bifidobacterium* isn't really a suitable probiotic choice for caries prevention. According to the report, there was no statistically significant reduction in *Streptococcus Mutans*or *Lactobacilli* counts in saliva in the presence of *Bifidobacterium* (Zhao *et al.,* 2021).

Effects of Probiotics on Urinary Tract Infection (UTI)

UTI is a frequent bacterial infection that can cause abscess formation, bacteraemia, fistula, pyelonephritis, stricture and kidney kaliuretic are linked to a high rate of morbidity and mortality, and they have a negative impact on patients' quality of life. Comparing to men, women suffer more from UTI especially during pregnancy and in peri- and postmenopausal women (Gupta and Trautner, 2015). UTI is usually associated with vaginal infections and is caused by microorganisms that originate in the digestive tract (Czajkowski *et al.*, 2021). The more complicated one happens when there is an irregular urinary system, which increases infection susceptibility (Fazly Bazzaz *et al.*, 2021). Various databases were used to find research that reported on natural solutions such as vitamins, herbal medicines, microelement, carbohydrates, and probiotics that were not time limited (Khameneh *et al.*, 2019; Yarnell, 2002; Akour *et al.*, 2021).

With the discovery of a manifold, symbiotic microorganism in the healthy urinogenital tract, the entrenched diagnostic and treatment for UTIs has indeed been called into issue. Therefore, the study of the gut microbiota's importance in the creation of bacteriuria and urinary tract infections (UTI) is an important area of research (Meštroviæ *et al.*, 2021). *Lactobacilli* have been shown to have clinical usefulness as a non-chemotherapeutic strategy of reinstating a normal urogenital microbiota, and that probiotic bacteria can most effectively normalise vaginal microbiota, heal existing infections, and prevent UTI recurrence (Somayeh Ziyadi *et al.*, 2016). Herbal medications can be used to prevent infection as soon as they appear, as well as for short-term prophylaxis (Table 15.3) (Devi *et al.*, 2021). Probiotics offer a lot of promise in mitigating antibiotic overuse and the spread of antibiotic-resistant bacteria. For instance, Diebold and co-workers (2021) concluded that treatment with lactic acid gel may improve uropathogen resistance, reducing the requirement for antibiotics to treat recurrent urinary tract infections.

Table 15.3: Probiotic Microorganisms Used as Herbal medicines in the Treatment of UTI

Herbal Medicine	*Functioning Constituent*	*Mechanism of Action*	*References*
Berberis aquifolium	Berberine	Bacterial adherence to the bladder epithelium is inhibited.	J Barker *et al.*, 2006
Armoracia rusticana	Isothiocyanates	Preventing the invasion of human cells	S Manuguerra *et al.*, 2020
Agathosma betulina	Phenolic compounds	Anti-adhesive, antiseptic, and diuretic characteristics	B Banerjee *et al.*, 2021
Vaccinium macrocarpon	Proanthocyanidins	Bacterial adhesion to the bladder walls is reduced.	ES Andreani *et al.*, 2021
Juniperus communis	Terpenoids	Antimicrobial and diuretic properties	KA El-Tarabily *et al.*, 2021

Recently the gut microbiota–UTI axis was described by Magruder and co-workers (2019). The scientists found that an increase in *E. coli* in the gut was linked to the development of *E. coli* bacteriuria and *E. coli*-induced UTI in the future.

Moreover, the *E. coli* strains in the gut were more similar to the *E. coli* strain in the urine of the same patient, indicating that the gut microbiota is a source of urinary tract colonisation and UTI (Magruder *et al.,* 2019). Furthermore, study by Magruder and co-workers (2020) observed that in kidney transplant patients, large relative abundances of two bacterial taxa—*Faecali bacterium* and *Romboutsia*—have been associated with a lower incidence of *Enterobacteriaceae* bacteriuria and UTI.

Probiotic in Allergic Diseases

Allergic diseases (AD) are a complex category of diseases characterised by an abnormal hypersensitivity response to antigen stimulation, which is an inflammatory disease induced by IgE-mediated reaction (Boonpiyathad *et al.,* 2020; Bousquet *et al.,* 2008). According to the World Health Organization (WHO), AD affects 30–40 per cent of the global population and has a substantial influence on general health, medical care, and also affect the economy of nations (Biagioni *et al.,* 2020; Feng *et al.,* 2021). Till now there is no effective cure for AD, patients can only count on drugs to relieve AD (Tuzimski and Petruczynik, 2021).

In recent years, probiotics have gained global attention, as a preventive and therapeutic treatment for food allergies.When consumed in sufficient amounts, probiotics can strengthen immunity, modulate the gut microbes, alleviate irritable colon syndrome, and lower blood sugar levels (García-Ruiz *et al.,* 2014; Weichselbaum, 2009). Probiotics beneficial impacts are derived from their cell metabolism and cell components, such as lipoteichoic acid, genetic material, and peptidoglycan, according to studies(Lei *et al.,* 2021; Lopez santamarina *et al.,* 2021; Patil *et al.,* 2021). LAB has the ability to downregulate histamine and the cytokines TNF- and IL-4, which are the major components associated with allergy and inflammatory disorders (J. H. Cheng *et al.,* 2016; Lei *et al.,* 2021). Anania and co-workers (2021) studied Probiotic Mixture Containing *Bifidobacterium animalis* Subsp. *Lactis* BB12 and *Enterococcus faecium* L3 statistically decreased signs and symptoms as well as requirement for drug use in children with Allergic rhinitis

Moreover, Co-relation between allergic asthma and microbiome has been explored (Zhang *et al.,* 2021). In the observations, the continuous administration of mixed probiotics reduced costimulatory molecules on the surface of dendritic cells (DCs) and increased mucosal gut-primed Tregs triggered by mesenteric CD103+DCs, decreasing OVA-induced allergic inflammation. Also, the study found that the stains with *Lactobacillus* genus are dominant in showing this protective effect. Additionally, Probiotic supplementation throughout pregnancy and after birth lowered the risk of Allergic disease in newborns and children. The findings showed that initiating probiotic treatment during pregnancy and continuing for the first six months of a child's life may be beneficial in preventing Allergic disease (Li *et al.,* 2019).

Probiotics in Cardiovascular Disease

Cardiovascular diseases (CVDs) are the leading cause of mortality worldwide, killing an estimated 17.9 million people each year, according to the World Health Organization (WHO) (Nongrum *et al.,* 2021). Coronary artery disease, stroke, venous thrombosis, hypertensive heart disease, arrhythmia, cardiomyopathy, and thromboembolic disease are some of the most well-known cardiovascular ailments. There are various factors that have been linked to a higher risk of CVD such as (Olas, 2020; Oniszczuk *et al.,* 2021; Suganya *et al.,* 2021). However, emerging data suggests that the intestinal microbiota and metabolites play a key role in the course of many disorders associated with CVD. There is substantial changes in the bacterial makeup between patients with chronic heart failure and the control group. During their research, in patients with chronic heart failure, there was a drop in the level of *Faecalibacterium prausnitzii* and an increase in the level of *Ruminococcus gnavus* (Cui *et al.,* 2018). According to another research by Kummen and his team, Patients with heart failure exhibited greater amounts of Prevotella, Hungatella, and Succinclasticum and relatively low levels of the Lachnospiraceae family (*Faecalibacterium* and *Bifidobacterium*) than control group. Moreover pathogens including *Candida, Yersinia, Shigella,* and *Campylobacter* were also shown to be more prevalent in the faeces of CVD patients (Pasini *et al.,* 2016). Through a number of mechanisms, the gut microbiota interacts with distant organs, including the heart. Trimethylamine (TMA)/trimethylamine N-oxide (TMAO), short-chain fatty acids, bile acids, lipopolysaccharides (LPS), and peptidoglycans are among them(Gerdes *et al.,* 2020). Many studies have shown that short-chain fatty acids (SCFA) play a function in cardiovascular disease (CVD). SCFAs are generated primarily by *Bifidobacterium, Provotella, Clostridium, Bacteroides,* and *Ruminococcus,* and are involved in sympathetic system activation and lymphocyte influx to the gut epithelial maintenance. SCFAs can also regulate renin secretion and blood pressure by activating G-protein coupled receptors (Gpr41, Gpr43, Gpr109a, and Olfr78) (Hosseinkhani *et al.,* 2021; Marras *et al.,* 2021). Additionally, study by Marques and co-workers (2017) reported that in hypertensive mice, a high-fiber diet and acetate supplementation were found to change the gut microbiota composition, resulting in the prevention of hypertension and heart failure. Hypercholesterolemia is becoming a more common cause of CVD. The role of probiotics in reducing cholesterol levels has been studied by various researchers (J. Jiang, Wu, *et al.,* 2020; Jung *et al.,* 2015; J. Jiang, Feng, *et al.,* 2020). According to the clinical trial by Asemi and his team, supplementing 70 pregnant women in the third trimester of pregnancy with 200 g/day of a probiotic yoghurt containing *Streptococcus thermophiles, L. bulgaricus, L. acidophilus* LA-5, and *B. animalis* BB12 for nine weeks resulted in significant reductions in total cholesterol, low-density lipoprotein cholesterol, and high-density lipoprotein cholesterol, as well as serum triglyceride concentrations (Asemi *et al.,* 2012).

Probiotics against COVID-19 Pandemic

Virus development and emergence have grown dramatically in the previous two decades as a result of their fast mutation. Viruses arise as a result of a variety of circumstances, including an increase in the number of immunocompromised individuals, climatic change, a lack of antiviral medicines, greater geographical movement of people and commodities, and virus genetic mutation (Zolnikova *et al.,* 2018). Both developing and developed nations have experienced a major cause of mortality and disability due to respiratory infections. Furthermore, they have a significant economic and societal effect because of the high prevalence of hospitalisation, high medical expenditures, and lost productivity due to time away from work or school.

SARS-CoV-2 (Severe Acute Respiratory Syndrome Coronavirus 2) is a novel coronavirus that causes coronavirus disease (COVID-19) in humans, which is related to a member of the zoonotic beta-coronavirus family. It was initially detected in Wuhan, China in December 2019(Rodriguez-Morales *et al.,* 2020; Xie and Chen, 2020). SARS-CoV-2 is a single-stranded positive sense RNA virus with an enclosed genome (Zhu *et al.,* 2020). Coronaviruses get their name from their crown-like forms and lengthy surface spikes. COVID-19 was thought to have been first transmitted by bats and subsequently transferred to people via wild animals; nevertheless, the virus spread later through human-to-human transmission(Zu *et al.,* 2020). Humans, as well as other vertebrate reservoirs such as camels, bats, masked palm civets, mice, dogs, and cats, host coronaviruses(F. Jiang *et al.,* 2020; Lu *et al.,* 2020).

Probiotics have been shown to have a direct or indirect beneficial effect on ACE enzymes(Robles-Vera *et al.,* 2017). Probiotics create bioactive peptides during food fermentation that can block the active sites of ACE enzymes, inhibiting them(Ayyash *et al.,* 2020). Furthermore, ACE inhibitors were found in the detritus of the dead probiotic cells (Miremadi *et al.,* 2014). These findings imply that probiotics might be used to inhibit the ACE receptor, which is used by SARS-CoV-2 to infect GI cells. Fernández-Fernández (2020) presented the idea of utilising medicines to inhibit ACE receptors as a therapy for COVID-19. According to Imai *et al.* (2005) utilising an ACE blocker to treat respiratory distress syndrome has a favourable effect.

By increasing probiotic development and survival, probiotics may have an outstanding potential impact against COVID-19. Furthermore, through inhibiting the ACE enzymes, probiotics may have a direct influence on GI symptoms induced by COVID-19. Yeh *et al.* (2018) conducted a comprehensive analysis of 12 research on the effects of prebiotic and probiotic supplements on influenza illness and concluded that supplementing with probiotics and prebiotics after an influenza vaccine might increase hemagglutination inhibition antibody titters.

SARS-CoV-2 is a novel virus for which there are presently no cures or less efficacy for vaccinations. There has been no research on the use of prebiotics and probiotics to treat or prevent COVID-19, although using probiotics in the therapeutic treatment or prevention of COVID-19 might be a suitable strategy(Infusino *et al.,* 2020). Intestinal microbial dysbiosis was seen in certain COVID-19 patients, defined by low levels of various probiotic species such as Bifidobacterium and Lactobacillus. This might be a sign of a weakened immune system; thus, it's been advised that these individuals need dietary assistance as well as prebiotic or probiotic supplements to re-establish a healthy gut flora balance to decrease the risk of infections (Xu *et al.,* 2020). Meanwhile, a patient's food pattern is a critical determinant in GI microbiota diversity, structure, and function. As a result, a well-balanced diet rich in probiotics and immunity-boosting micronutrients including polyphenols, vitamins A, C, and D, and minerals (especially selenium and zinc) may reduce the incidence of COVID-19 infection (Gasmi *et al.,* 2020). Probiotic-rich foods, such as fermented foods, have a high potential for preventing COVID-19. Consumption of fermented milk containing probiotic microorganisms has been shown to lower the incidence of upper respiratory tract infections in healthy babies, children, adults, and the elderly in earlier research (Shida *et al.,* 2017; Taipale *et al.,* 2011).

Probiotics might help reduce COVID-19 by maintaining the human GI or lung microbiota because dysbiosis has a key role in people's vulnerability to infectious illnesses. To investigate the potential preventative and therapeutic benefits of probiotics against SARS-CoV-2 infection, *in vitro* and clinical research are needed.

Economical Prospects

"The global probiotics market size was USD $48.88 billion in 2019 and is projected to reach USD $94.48 billion by 2027, exhibiting a CAGR of 7.9 per cent during the forecast period." The market is expected to be driven by increased consumer awareness of preventive healthcare measures, as well as rising demand for natural and safe health-beneficial goods. The growing popularity of probiotic strains as a natural, cost-effective, and safe alternative to pharmaceutical treatments, as well as the expanding data on the health advantages associated with their intake, are expected to favourably affect market growth. Consumers are becoming more health-conscious, and the increasing trend of veganism is tempting them to a plant-based diet, which is projected to drive demand for plant-based products. For examples, Chobani, an American food company, announced the introduction of a new probiotic drink in August 2020. Lemon Ginger, Pineapple, Turmeric, Peach Mint, and Cherry Hibiscus Tea are the four flavours available in the new product. This report was published by "Fortune Business Insight" titled as "Market research report-2020".

With an eye on personalised nutrition and precision medicine, the probiotics industry is all set to grow with development of new interventions such as prebiotics, synbiotics as well as postbiotics.

REFERENCES

Ahola, A. J., Yli-Knuuttila, H., Suomalainen, T., Poussa, T., Ahlstrom, A., Meurman, J. H., and Korpela, R. (2002). Short-term consumption of probiotic-containing cheese and its effect on dental caries risk factors. Archives of Oral Biology. https: //doi.org/10.1016/S0003-9969(02)00112-7

Akour, A., Abuloha, S., Mulakhudair, A. R., Kasabri, V., and Al-Tammemi, A. B. (2021). Complementary and alternative medicine for urinary tract illnesses: A cross-sectional survey in Jordan. Complementary Therapies in Clinical Practice. https: //doi.org/10.1016/j.ctcp.2021.101321

Alokail, M. S., Sabico, S., Al-Saleh, Y., Al-Daghri, N. M., Alkharfy, K. M., Vanhoutte, P. M., and McTernan, P. G. (2013). Effects of probiotics in patients with diabetes mellitus type 2: Study protocol for a randomized, double-blind, placebo-controlled trial. Trials. https: //doi.org/10.1186/1745-6215-14-195

Anania, C., Di Marino, V. P., Olivero, F., De Canditiis, D., Brindisi, G., Iannilli, F., De Castro, G., Zicari, A. M., and Duse, M. (2021). Treatment with a probiotic mixture containing bifidobacterium animalis subsp. Lactis bb12 and enterococcus faecium l3 for the prevention of allergic rhinitis symptoms in children: A randomized controlled trial. Nutrients. https: //doi.org/10.3390/nu13041315

Angarita-Díaz, M. P., Forero-Escobar, D., Cerón-Bastidas, X. A., Cisneros-Hidalgo, C. A., Dávila-Narvaez, F., Bedoya-Correa, C. M., Freitas, S. C., Cabrera-Arango, C. L., and Melo-Colina, R. (2020). Effects of a functional food supplemented with probiotics on biological factors related to dental caries in children: a pilot study. European Archives of Paediatric Dentistry. https: //doi.org/10.1007/s40368-019-00468-y

Asemi, Z., Samimi, M., Tabasi, Z., Talebian, P., Azarbad, Z., Hydarzadeh, Z., and Esmaillzadeh, A. (2012). Effect of daily consumption of probiotic yoghurt on lipid profiles in pregnant women: A randomized controlled clinical trial. Journal of Maternal-Fetal and Neonatal Medicine. https: //doi.org/10.3109/14767058.2011.640372

Ayyash, M., Olaimat, A., Al-Nabulsi, A., and Liu, S. Q. (2020). Bioactive properties of novel probiotic lactococcus lactis fermented camel sausages: Cytotoxicity, angiotensin converting enzyme inhibition, antioxidant capacity, and antidiabetic activity. Food Science of Animal Resources. https: //doi.org/10.5851/kosfa.2020.e1

Bermudez-Brito, M., Plaza-Díaz, J., Muñoz-Quezada, S., Gómez-Llorente, C., and Gil, A. (2012). Probiotic mechanisms of action. In Annals of Nutrition and Metabolism. https: //doi.org/10.1159/000342079

Bhat, B., and Bajaj, B. K. (2019). Hypocholesterolemic potential and bioactivity spectrum of an exopolysaccharide from a probiotic isolate Lactobacillus paracasei M7. Bioactive Carbohydrates and Dietary Fibre. https: //doi.org/10.1016/j.bcdf.2019.100191

Biagioni, B., Annesi-Maesano, I., D'Amato, G., and Cecchi, L. (2020). The rising of allergic respiratory diseases in a changing world: from climate change to migration. In Expert Review of Respiratory Medicine. https: //doi.org/10.1080/17476348.2020.1794829

Binda, S., Hill, C., Johansen, E., Obis, D., Pot, B., Sanders, M. E., Tremblay, A., and Ouwehand, A. C. (2020). Criteria to Qualify Microorganisms as "Probiotic" in Foods and Dietary Supplements. In Frontiers in Microbiology. https: //doi.org/10.3389/fmicb.2020.01662

Boonpiyathad, T., Sözener, Z. C., Akdis, M., and Akdis, C. A. (2020). The role of treg cell subsets in allergic disease. In Asian Pacific Journal of Allergy and Immunology. https: //doi.org/10.12932/AP-030220-0754

Bousquet, J., Khaltaev, N., Cruz, A. A., Denburg, J., Fokkens, W. J., Togias, A., Zuberbier, T., Baena-Cagnani, C. E., Canonica, G. W., Van Weel, C., Agache, I., Aït-Khaled, N., Bachert, C., Blaiss, M. S., Bonini, S., Boulet, L. P., Bousquet, P. J., Camargos, P., Carlsen, K. H., ... Williams, D. (2008). Allergic Rhinitis and its Impact on Asthma (ARIA) 2008 update (in collaboration with the World Health Organization, GA2LEN and AllerGen). In Allergy: European Journal of Allergy and Clinical Immunology. https: //doi.org/10.1111/j.1398-9995.2007.01620.x

Çaglar, E., Kavaloglu, S. C., Kuscu, O. O., Sandalli, N., Holgerson, P. L., and Twetman, S. (2007). Effect of chewing gums containing xylitol or probiotic bacteria on salivary mutans streptococci and lactobacilli. Clinical Oral Investigations. https: //doi.org/10.1007/s00784-007-0129-9

Cheng, J. H., Tsai, C. L., Lien, Y. Y., Lee, M. S., and Sheu, S. C. (2016). High molecular weight of polysaccharides from Hericium erinaceus against amyloid beta-induced neurotoxicity. BMC Complementary and Alternative Medicine. https: //doi.org/10.1186/s12906-016-1154-5

Cheng, X., He, F., Sun, P., and Chen, Q. (2021). Identification of unknown acid-resistant genes of oral microbiotas in patients with dental caries using metagenomics analysis. AMB Express. https: //doi.org/10.1186/s13568-021-01199-4

Clyne, M., Dillon, P., Daly, S., O'Kennedy, R., May, F. E. B., Westley, B. R., and Drumm, B. (2004). Helicobacter pylori interacts with the human single-domain trefoil protein TFF1. Proceedings of the National Academy of Sciences of the United States of America. https: //doi.org/10.1073/pnas.0308489101

Collado, M. C., Gueimonde, M., and Salminen, S. (2010). Probiotics in Adhesion of Pathogens: Mechanisms of Action. In Bioactive Foods in Promoting Health. https: //doi.org/10.1016/B978-0-12-374938-3.00023-2

Cui, X., Ye, L., Li, J., Jin, L., Wang, W., Li, S., Bao, M., Wu, S., Li, L., Geng, B., Zhou, X., Zhang, J., and Cai, J. (2018). Metagenomic and metabolomic analyses unveil dysbiosis of gut microbiota in chronic heart failure patients. Scientific Reports. https: //doi.org/10.1038/s41598-017-18756-2

Cunningham, M., Azcarate-Peril, M. A., Barnard, A., Benoit, V., Grimaldi, R., Guyonnet, D., Holscher, H. D., Hunter, K., Manurung, S., Obis, D., Petrova, M. I., Steinert, R. E., Swanson, K. S., van Sinderen, D., Vulevic, J., and Gibson, G. R. (2021). Shaping the Future of Probiotics and Prebiotics. In Trends in Microbiology. https: //doi.org/10.1016/j.tim.2021.01.003

Czajkowski, K., Broœ-Konopielko, M., and Teliga-Czajkowska, J. (2021). Urinary tract infection in women. In Przeglad Menopauzalny. https: //doi.org/10.5114/ pm.2021.105382

D'Amelio, P., and Sassi, F. (2018). Gut Microbiota, Immune System, and Bone. In Calcified Tissue International. https: //doi.org/10.1007/s00223-017-0331-y

Dallal, M. M. S., Yazdi, M. K. S., Afradi, M. R., Rajabi, Z., Samimirad, K., and Erfani, Y. (2021). Effect of bacteriocin produced by lactobacillus casei with probiotic potential isolated from milk on the ability of biofilm production by streptococcus salivarius obtained from dental plaque. Journal of Mazandaran University of Medical Sciences.

de Melo Pereira, G. V., de Oliveira Coelho, B., Magalhães Júnior, A. I., Thomaz-Soccol, V., and Soccol, C. R. (2018). How to select a probiotic? A review and update of methods and criteria. In Biotechnology Advances. https: //doi.org/10.1016/j. biotechadv.2018.09.003

Dinan, T. G., and Cryan, J. F. (2013). Melancholic microbes: A link between gut microbiota and depression? In Neurogastroenterology and Motility. https: // doi.org/10.1111/nmo.12198

FAO and WHO. (2002). WHO working group guidelines for the evaluation of probiotics in food. Guidelines for the Evaluation of Probiotics in Food.

Fazly Bazzaz, B. S., Darvishi Fork, S., Ahmadi, R., and Khameneh, B. (2021). Deep insights into urinary tract infections and effective natural remedies. In African Journal of Urology. https: //doi.org/10.1186/s12301-020-00111-z

Feng, Z., Yi, X., and Hajavi, J. (2021). New and old adjuvants in allergen-specific immunotherapy: With a focus on nanoparticles. In Journal of Cellular Physiology. https: //doi.org/10.1002/jcp.29941

Fernández-Fernández, F. J. (2020). COVID-19, hypertension and angiotensin receptor-blocking drugs. In Journal of Hypertension. https: //doi.org/10.1097/ HJH.0000000000002468

Food Science and Technology Bulletin: Fu - R. G Prof. Gibson - Google Books. (n.d.). Retrieved August 17, 2021, from https: //books.google.co.in/books?hl=en and lr= and id=2lNMZrslFK8C and oi=fnd and pg=SA8-PA1 and dq=+probiotics+ per cent 2B+Sanders+and+Heimbach and ots=aQzM6oYqhw and sig

=qkqpbXdsAIasWfmHTF6xiSUQd38 and redir_esc=y#v=onepage and q=probiotics per cent 2B Sanders and Heimbach and f=false

García-Ruiz, A., González de Llano, D., Esteban-Fernández, A., Requena, T., Bartolomé, B., and Moreno-Arribas, M. V. (2014). Assessment of probiotic properties in lactic acid bacteria isolated from wine. Food Microbiology. https: //doi.org/10.1016/j.fm.2014.06.015

Gasmi, A., Noor, S., Tippairote, T., Dadar, M., Menzel, A., and Bjørklund, G. (2020). Individual risk management strategy and potential therapeutic options for the COVID-19 pandemic. In Clinical Immunology. https: //doi.org/10.1016/j.clim.2020.108409

Gerdes, V., Gueimonde, M., Pajunen, L., Nieuwdorp, M., and Laitinen, K. (2020). How strong is the evidence that gut microbiota composition can be influenced by lifestyle interventions in a cardio-protective way? In Atherosclerosis. https: //doi.org/10.1016/j.atherosclerosis.2020.08.028

Gómez-Llorente, C., Muñoz, S., and Gil, A. (2010). Role of Toll-like receptors in the development of immunotolerance mediated by probiotics. Proceedings of the Nutrition Society. https: //doi.org/10.1017/S0029665110001527

Gupta, K., and Trautner, B. W. (2015). Urinary Tract Infections, Pyelonephritis, and Prostatitis. Harrison's Principles of Internal Medicine.

Hill, C., Guarner, F., Reid, G., Gibson, G. R., Merenstein, D. J., Pot, B., Morelli, L., Canani, R. B., Flint, H. J., Salminen, S., Calder, P. C., and Sanders, M. E. (2014). Expert consensus document: The international scientific association for probiotics and prebiotics consensus statement on the scope and appropriate use of the term probiotic. Nature Reviews Gastroenterology and Hepatology. https: //doi.org/10.1038/nrgastro.2014.66

Hosseinkhani, F., Heinken, A., Thiele, I., Lindenburg, P. W., Harms, A. C., and Hankemeier, T. (2021). The contribution of gut bacterial metabolites in the human immune signaling pathway of non-communicable diseases. In Gut Microbes. https: //doi.org/10.1080/19490976.2021.1882927

Hu, Y. meng, Zhou, F., Yuan, Y., and Xu, Y. cheng. (2017). Efectos del suplemento de probióticos en pacientes con diabetes mellitus tipo 2: metaanálisis de ensayos aleatorizados. In Medicina Clinica. https: //doi.org/10.1016/j.medcli.2016.11.036

Imai, Y., Kuba, K., Rao, S., Huan, Y., Guo, F., Guan, B., Yang, P., Sarao, R., Wada, T., Leong-Poi, H., Crackower, M. A., Fukamizu, A., Hui, C. C., Hein, L., Uhlig, S., Slutsky, A. S., Jiang, C., and Penninger, J. M. (2005). Angiotensin-converting enzyme 2 protects from severe acute lung failure. Nature. https: //doi.org/10.1038/nature03712

Infusino, F., Marazzato, M., Mancone, M., Fedele, F., Mastroianni, C. M., Severino, P., Ceccarelli, G., Santinelli, L., Cavarretta, E., Marullo, A. G. M., Miraldi, F., Carnevale, R., Nocella, C., Biondi-Zoccai, G., Pagnini, C., Schiavon, S., Pugliese, F., Frati, G., and D'Ettorre, G. (2020). Diet supplementation, probiotics, and nutraceuticals in SARS-CoV-2 infection: A scoping review. In Nutrients. https: //doi.org/10.3390/nu12061718

Jackson, S. A., Schoeni, J. L., Vegge, C., Pane, M., Stahl, B., Bradley, M., Goldman, V. S., Burguière, P., Atwater, J. B., and Sanders, M. E. (2019). Improving end-user trust in the quality of commercial probiotic products. Frontiers in Microbiology. https: //doi.org/10.3389/fmicb.2019.00739

Jiang, F., Deng, L., Zhang, L., Cai, Y., Cheung, C. W., and Xia, Z. (2020). Review of the Clinical Characteristics of Coronavirus Disease 2019 (COVID-19). In Journal of General Internal Medicine. https: //doi.org/10.1007/s11606-020-05762-w

Jiang, J., Feng, N., Zhang, C., Liu, F., Zhao, J., Zhang, H., Zhai, Q., and Chen, W. (2020). Lactobacillus reuteri A9 and Lactobacillus mucosae A13 isolated from Chinese superlongevity people modulate lipid metabolism in a hypercholesterolemia rat model. FEMS Microbiology Letters. https: //doi.org/10.1093/femsle/fnz254

Jiang, J., Wu, C., Zhang, C., Zhao, J., Yu, L., Zhang, H., Narbad, A., Chen, W., and Zhai, Q. (2020). Effects of probiotic supplementation on cardiovascular risk factors in hypercholesterolemia: A systematic review and meta-analysis of randomized clinical trial. In Journal of Functional Foods. https: //doi.org/10.1016/j.jff.2020.104177

Jung, S., Lee, Y. J., Kim, M., Kim, M., Kwak, J. H., Lee, J. W., Ahn, Y. T., Sim, J. H., and Lee, J. H. (2015). Supplementation with two probiotic strains, Lactobacillus curvatus HY7601 and Lactobacillus plantarum KY1032, reduced body adiposity and Lp-PLA2 activity in overweight subjects. Journal of Functional Foods. https: //doi.org/10.1016/j.jff.2015.10.006

Kechagia, M., Basoulis, D., Konstantopoulou, S., Dimitriadi, D., Gyftopoulou, K., Skarmoutsou, N., and Fakiri, E. M. (2013). Health Benefits of Probiotics: A Review. ISRN Nutrition. https: //doi.org/10.5402/2013/481651

Khameneh, B., Iranshahy, M., Soheili, V., and Fazly Bazzaz, B. S. (2019). Review on plant antimicrobials: A mechanistic viewpoint. In Antimicrobial Resistance and Infection Control. https: //doi.org/10.1186/s13756-019-0559-6

Koirala, S., and Anal, A. K. (2021). Probiotics-based foods and beverages as future foods and their overall safety and regulatory claims. Future Foods. https: //doi.org/10.1016/j.fufo.2021.100013

Kumar, M., Nagpal, R., Verma, V., Kumar, A., Kaur, N., Hemalatha, R., Gautam, S. K., and Singh, B. (2013). Probiotic metabolites as epigenetic targets in the prevention of colon cancer. Nutrition Reviews. https: //doi.org/10.1111/j.1753-4887.2012.00542.x

Lahtinen, S., Ouwehand, A. C., Salminen, S., and von Wright, A. (2011). Lactic acid bacteria: Microbiological and functional aspects, fourth edition. In Lactic Acid Bacteria: Microbiological and Functional Aspects, Fourth Edition. https: //doi.org/10.3920/bm2012.x003

Lei, W., Liu, C., Pan, L., Peng, C., Wang, J., and Zhou, H. (2021). Screening of probiotic Lactobacilli with potential anti-allergic activity based on hyaluronidase inhibition and degranulation of RBL-2H3 cells in vitro. LWT. https: //doi.org/10.1016/j.lwt.2020.110707

Leroy, F., Falony, G., and De Vuyst, L. (2008). Latest developments in probiotics. In Meat Biotechnology. https: //doi.org/10.1007/978-0-387-79382-5_10

Li, L., Han, Z., Niu, X., Zhang, G., Jia, Y., Zhang, S., and He, C. (2019). Probiotic Supplementation for Prevention of Atopic Dermatitis in Infants and Children: A Systematic Review and Meta-analysis. In American Journal of Clinical Dermatology. https: //doi.org/10.1007/s40257-018-0404-3

Lilly, D. M., and Stillwell, R. H. (1965). Probiotics: Growth-promoting factors produced by microorganisms. Science. https: //doi.org/10.1126/science.147.3659.747

Lopez santamarina, A., Gonzalez, E.G., Lamas, A., Mondragon, A. del C., Regal, P., and Miranda, J. M. (2021). Probiotics as a possible strategy for the prevention and treatment of allergies: A narrative review. In Foods. https: //doi.org/10.3390/foods10040701

Lu, C. wei, Liu, X. fen, and Jia, Z. fang. (2020). 2019-nCoV transmission through the ocular surface must not be ignored. In The Lancet. https: //doi.org/10.1016/S0140-6736(20)30313-5

Mack, D. R., Ahrne, S., Hyde, L., Wei, S., and Hollingsworth, M. A. (2003). Extracellular MUC3 mucin secretion follows adherence of Lactobacillus strains to intestinal epithelial cells in vitro. Gut. https: //doi.org/10.1136/gut.52.6.827

Mack, David R., Michail, S., Wei, S., McDougall, L., and Hollingsworth, M. A. (1999). Probiotics inhibit enteropathogenic E. coli adherence in vitro by inducing intestinal mucin gene expression. American Journal of Physiology - Gastrointestinal and Liver Physiology. https: //doi.org/10.1152/ajpgi.1999.276.4.g941

Markowiak-Kopeæ, P., and Œli¿ewska, K. (2020). The effect of probiotics on the production of short-chain fatty acids by human intestinal microbiome. In Nutrients. https: //doi.org/10.3390/nu12041107

Marques, F. Z., Nelson, E., Chu, P. Y., Horlock, D., Fiedler, A., Ziemann, M., Tan, J. K., Kuruppu, S., Rajapakse, N. W., El-Osta, A., Mackay, C. R., and Kaye, D. M. (2017). High-fiber diet and acetate supplementation change the gut microbiota and prevent the development of hypertension and heart failure in hypertensive mice. Circulation. https: //doi.org/10.1161/CIRCULATIONAHA.116.024545

Marras, L., Caputo, M., Bisicchia, S., Soato, M., Bertolino, G., Vaccaro, S., and Inturri, R. (2021). The role of bifidobacteria in predictive and preventive medicine: A focus on eczema and hypercholesterolemia. In Microorganisms. https: //doi.org/10.3390/microorganisms9040836

Mihich, E. (1972). Probiotics, G. S. Sperti (editor), 120 pages, 1971, avi publishing company, Westport, Connecticut. Journal of Surgical Oncology. https: //doi.org/10.1002/jso.2930040211

Miremadi, F., Ayyash, M., Sherkat, F., and Stojanovska, L. (2014). Cholesterol reduction mechanisms and fatty acid composition of cellular membranes of probiotic Lactobacilli and Bifidobacteria. Journal of Functional Foods. https: //doi.org/10.1016/j.jff.2014.05.002

Ng, S. C., Hart, A. L., Kamm, M. A., Stagg, A. J., and Knight, S. C. (2009). Mechanisms of action of probiotics: Recent advances. In Inflammatory Bowel Diseases. https: //doi.org/10.1002/ibd.20602

Niers, L. E. M., Timmerman, H. M., Rijkers, G. T., Van Bleek, G. M., Van Uden, N. O. P., Knol, E. F., Kapsenberg, M. L., Kimpen, J. L. L., and Hoekstra, M. O. (2005). Identification of strong interleukin-10 inducing lactic acid bacteria which down-regulate T helper type 2 cytokines. Clinical and Experimental Allergy. https: //doi.org/10.1111/j.1365-2222.2005.02375.x

Nikawa, H., Makihira, S., Fukushima, H., Nishimura, H., Ozaki, Y., Ishida, K., Darmawan, S., Hamada, T., Hara, K., Matsumoto, A., Takemoto, T., and Aimi, R. (2004). Lactobacillus reuteri in bovine milk fermented decreases the oral carriage of mutans streptococci. International Journal of Food Microbiology. https: //doi.org/10.1016/j.ijfoodmicro.2004.03.006

Ohashi, Y., and Ushida, K. (2009). Health-beneficial effects of probiotics: Its mode of action. In Animal Science Journal. https: //doi.org/10.1111/j.1740-0929.2009.00645.x

Olas, B. (2020). Probiotics, prebiotics and synbiotics-a promising strategy in prevention and treatment of cardiovascular diseases? International Journal of Molecular Sciences. https: //doi.org/10.3390/ijms21249737

Oniszczuk, A., Oniszczuk, T., Gancarz, M., and Szymañska, J. (2021). Role of gut microbiota, probiotics and prebiotics in the cardiovascular diseases. In Molecules. https: //doi.org/10.3390/molecules26041172

Ozen, M., and Dinleyici, E. C. (2015). The history of probiotics: The untold story. In Beneficial Microbes. https: //doi.org/10.3920/BM2014.0103

PARKER, and RB. (1974). Probiotics, the other half of the antibiotic story. Anim Nutr Health, 29, 4–8. https: //ci.nii.ac.jp/naid/10014898669

Pasini, E., Aquilani, R., Testa, C., Baiardi, P., Angioletti, S., Boschi, F., Verri, M., and Dioguardi, F. (2016). Pathogenic Gut Flora in Patients With Chronic Heart Failure. JACC: Heart Failure. https: //doi.org/10.1016/j.jchf.2015.10.009

Patil, S. V., Mohite, B. V., and Patil, V. S. (2021). Probiotics for Allergic Airway Infection and Inflammations. In Probiotic Research in Therapeutics. https: // doi.org/10.1007/978-981-33-6236-9_13

Petti, S., Tarsitani, G., and Simonetti D'Arca, A. (2008). Antibacterial activity of yoghurt against viridans streptococci in vitro. Archives of Oral Biology. https: //doi.org/10.1016/j.archoralbio.2008.04.009

Pirbaglou, M., Katz, J., de Souza, R. J., Stearns, J. C., Motamed, M., and Ritvo, P. (2016). Probiotic supplementation can positively affect anxiety and depressive symptoms: a systematic review of randomized controlled trials. In Nutrition Research. https: //doi.org/10.1016/j.nutres.2016.06.009

Prado, F. C., Parada, J. L., Pandey, A., and Soccol, C. R. (2008). Trends in non-dairy probiotic beverages. In Food Research International. https: //doi.org/10.1016/j.foodres.2007.10.010

R.A., L., and J.A., F. (2015). Gut brain axis: Diet microbiota interactions and implications for modulation of anxiety and depression. Current Opinion in Biotechnology.

Ramanujam, P., Poorni, S., Srinivasan, M. R., and Sureshbabu, N. M. (2019). Probiotics in Dental Caries Prevention. The Indian Journal of Nutrition and Dietetics. https: //doi.org/10.21048/ijnd.2019.56.1.22609

Renye, J. A., and Steinberg, D. H. (2021). Thermophilin 110 inhibits growth and biofilm formation of Streptococcus mutans. Biotechnology Reports. https: //doi.org/10.1016/j.btre.2021.e00647

Robles-Vera, I., Toral, M., Romero, M., Jiménez, R., Sánchez, M., Pérez-*Viz.*caíno, F., and Duarte, J. (2017). Antihypertensive Effects of Probiotics. In Current Hypertension Reports. https: //doi.org/10.1007/s11906-017-0723-4

Rodriguez-Morales, A. J., Cardona-Ospina, J. A., Gutiérrez-Ocampo, E., Villamizar-Peña, R., Holguin-Rivera, Y., Escalera-Antezana, J. P., Alvarado-Arnez, L. E., Bonilla-Aldana, D. K., Franco-Paredes, C., Henao-Martinez, A. F., Paniz-Mondolfi, A., Lagos-Grisales, G. J., Ramírez-Vallejo, E., Suárez, J. A., Zambrano, L. I., Villamil-Gómez, W. E., Balbin-Ramon, G. J., Rabaan, A. A., Harapan, H., ... Sah, R. (2020). Clinical, laboratory and imaging features of COVID-19: A systematic review and meta-analysis. In Travel Medicine and Infectious Disease. https: //doi.org/10.1016/j.tmaid.2020.101623

Romero-Luna, H. E., Peredo-Lovillo, A. G., and Jiménez-Fernández, M. (2021). Probiotic and Potentially Probiotic Bacteria with Hypocholesterolemic Properties. Food Reviews International. https: //doi.org/10.1080/87559129.2021.1926481

Sanders, M. E., and Heimbach, J. T. (2003). Functional foods in the USA: emphasis on probiotic foods. Food Science <html_ent Glyph="@amp; " Ascii=" and amp; "/> Technology Bulletin: Functional Foods. https: //doi.org/10.1616/1476-2137.13616

Sandoval, F., Faleiros, S., Cabello, R., Díaz-Dosque, M., Rodríguez, G., and Escobar, A. (2021). The consumption of milk supplemented with probiotics decreases the occurrence of caries and the salivary concentration of hβD-3 in children. Clinical Oral Investigations. https: //doi.org/10.1007/s00784-020-03712-8

Santacroce, L., Charitos, I. A., and Bottalico, L. (2019). A successful history: probiotics and their potential as antimicrobials. In Expert Review of Anti-Infective Therapy. https: //doi.org/10.1080/14787210.2019.1645597

Sherman, P. M., Ossa, J. C., and Johnson-Henry, K. (2009). Invited review: Unraveling mechanisms of action of probiotics. In Nutrition in Clinical Practice. https: //doi.org/10.1177/0884533608329231

Shewale, R. N., Sawale, P. D., Khedkar, C. D., and Singh, A. (2014). Selection criteria for probiotics: A review. In International Journal of Probiotics and Prebiotics.

Shida, K., Sato, T., Iizuka, R., Hoshi, R., Watanabe, O., Igarashi, T., Miyazaki, K., Nanno, M., and Ishikawa, F. (2017). Daily intake of fermented milk with Lactobacillus casei strain Shirota reduces the incidence and duration of upper respiratory tract infections in healthy middle-aged office workers. European Journal of Nutrition. https: //doi.org/10.1007/s00394-015-1056-1

Suganya, K., Son, T., Kim, K. W., and Koo, B. S. (2021). Impact of gut microbiota: How it could play roles beyond the digestive system on development of cardiovascular and renal diseases. In Microbial Pathogenesis. https: //doi.org/10.1016/j.micpath.2020.104583

Taipale, T., Pienihkkinen, K., Isolauri, E., Larsen, C., Brockmann, E., Alanen, P., Jokela, J., and Söderling, E. (2011). Bifidobacterium animalis subsp. lactis BB-12 in reducing the risk of infections in infancy. British Journal of Nutrition. https: //doi.org/10.1017/S0007114510003685

The Role of Prebiotics and Probiotics in Human Health. (2013). Journal of Probiotics and Health. https: //doi.org/10.4172/2329-8901.1000108

Tjalsma, H., Boleij, A., Marchesi, J. R., and Dutilh, B. E. (2012). A bacterial driver-passenger model for colorectal cancer: Beyond the usual suspects. In Nature Reviews Microbiology. https: //doi.org/10.1038/nrmicro2819

Tuzimski, T., and Petruczynik, A. (2021). Review of new trends in the analysis of allergenic residues in foods and cosmetic products. Journal of AOAC International. https: //doi.org/10.1093/JAOACINT/QSAA015

Weichselbaum, E. (2009). Probiotics and health: A review of the evidence. In Nutrition Bulletin. https: //doi.org/10.1111/j.1467-3010.2009.01782.x

Xie, M., and Chen, Q. (2020). Insight into 2019 novel coronavirus — An updated interim review and lessons from SARS-CoV and MERS-CoV. In International Journal of Infectious Diseases. https: //doi.org/10.1016/j.ijid.2020.03.071

Xu, K., Cai, H., Shen, Y., Ni, Q., Chen, Y., Hu, S., Li, J., Wang, H., Yu, L., Huang, H., Qiu, Y., Wei, G., Fang, Q., Zhou, J., Sheng, J., Liang, T., and Li, L. (2020). [Management of corona virus disease-19 (COVID-19): the Zhejiang experience]. Zhejiang Da Xue Xue Bao. Yi Xue Ban = Journal of Zhejiang University. Medical Sciences.

Yarnell, E. (2002). Botanical medicines for the urinary tract. In World journal of urology. https: //doi.org/10.1007/s00345-002-0293-0

Yeh, T. L., Shih, P. C., Liu, S. J., Lin, C. H., Liu, J. M., Lei, W. Te, and Lin, C. Y. (2018). The influence of prebiotic or probiotic supplementation on antibody titers after influenza vaccination: A systematic review and meta-analysis of randomized controlled trials. Drug Design, Development and Therapy. https: //doi.org/10.2147/DDDT.S155110

Zhang, J., Ma, J., Li, Q., Su, H., and Sun, X. (2021). Exploration of the effect of mixed probiotics on microbiota of allergic asthma mice. Cellular Immunology. https: //doi.org/10.1016/j.cellimm.2021.104399

Zhao, S., Zhou, Q. yu, Huang, Y. yan, Nan, S. gang, and Liu, D. mei. (2021). Comparative analysis of physico-chemical, rheological, sensory and flavour properties of yoghurts using a new probiotic Bacillus coagulans 13002 with traditional yoghurt starter. International Journal of Food Science and Technology. https: //doi.org/10.1111/ijfs.14795

Zhu, N., Zhang, D., Wang, W., Li, X., Yang, B., Song, J., Zhao, X., Huang, B., Shi, W., Lu, R., Niu, P., Zhan, F., Ma, X., Wang, D., Xu, W., Wu, G., Gao, G. F., and Tan, W. (2020). A Novel Coronavirus from Patients with Pneumonia in China, 2019. New England Journal of Medicine. https: //doi.org/10.1056/nejmoa2001017

Zolnikova, O., Komkova, I., Potskherashvili, N., Trukhmanov, A., and Ivashkin, V. (2018). Application of probiotics for acute respiratory tract infections. In Italian Journal of Medicine. https: //doi.org/10.4081/itjm.2018.931

Zu, Z. Y., Di Jiang, M., Xu, P. P., Chen, W., Ni, Q. Q., Lu, G. M., and Zhang, L. J. (2020). Coronavirus Disease 2019 (COVID-19): A Perspective from China. In Radiology. https: //doi.org/10.1148/radiol.2020200490

Chapter 16

Noble Role of Phytochemicals in Nanoparticle Synthesis and as Proven Antimicrobial Agents: A Review

Manjit Kaur[1], Sunita Khatak[2], Harinder Singh[2]*

[1]CSIR-CBRI, Roorkee – 247 667, Uttarakhand
[2]Department of Biotechnology, University Institute of Engineering and Technology, Kurukshetra University, Kurukshetra – 136 119, Haryana
**e-mail: sunitakhatak2019@gmail.com*

ABSTRACT

In medicinal plants, Phytochemicals exhibit a vital role in nanoparticles (NPs) synthesis. Nanoparticles are essential for growth and development of plant as well as for increasing its photosynthesis capabilities. Phytochemicals act as reducing agent and stabilizer during nanoparticle synthesis. Antimicrobial potency of phytochemicals has been reported in varied plant extracts. Antimicrobial medicines in combination with phytochemicals reviewed to be more effective than antimicrobial drugs alone. This study signifies the importance of medicinal phytochemicals through review of its important characteristics.

Keywords: *Phytochemical, Nanoparticles, Plant growth, Photosynthesis, Antimicrobial, Synergetic interactions.*

Introduction

Bioactive phytochemicals or bio nutrients are plentiful in medicinal plants. These secondary metabolites evolved as a necessity of plant for its interaction with its environment (Michael,2002) and are indispensable part of phyto-immune process benefitting both to the plant as well as to its consumer (Ku *et al.,* 2020). Chemical structures, botanical sources, biosynthetic routes, and biological characteristics may all be used to classify phytochemicals into several categories. Chemical structures such as polyacetylenes, limonoids, phenolics, alkaloids, saponins, terpenoids among others are used to classify phytochemicals. The therapeutic effects of medicinal plants have been shown to be related to these phytochemicals which are synthesised and accumulated with marked physiological activities (Doughariet, 2012). Phyto-chemicals synthesize metallic nano-particles. This green synthesis is carried out through the ability of phytochemicals to reduce metal ions. In result, there stabilized formulation produces nanoparticles (NPs) (Marslin *et al.,* 2018). In addition to it, Phytochemicals are also found to exhibit antimicrobial potency. Flavonoids, alkaloids, steroids, tannins and saponins are phytochemical substances that are physiologically active and hence responsible for antimicrobial effect, which is why they are used in traditional medicine (Nethathe, 2011).

Table 16.1: Explain Main Classes of Phytochemicals

Sl.No.	*Phyto Content*	*Group Responsible*	*Disease/Role*
1.	Tannins	Phenolic groups	Diarrheal
2.	Terpenes	Lactones, Sesquiterpenes	Antibacterial, antiprotozoal
3.	Saponins	Sapogenin (hydrolysis), saponins (Poisonous)	Irritation to mucosal membrane, haemolysis of blood.
4.	Phenols/Polyphenols	Natural colour pigments	Plant defence against pathogens, herbivore/ predators. Cause cattle poisoning
5.	Flavonoids (Flavans derived)	Quercetin, Kaempferol, quercitrin+ dihydro flavones, flavones, anthocyanidins, calchones, catechin, leucoanthocyanidins, pro anthocyanidins	Antioxidants and free radical scavengers, as pigments in higher plants.
6.	Glycosides (Bitter taste)	Trophanthidin, digitoxin, barbaloin, salicin, cantharidin,pruwasin	Promote-petite, aid digestion, increased flow of saliva and gastric juices, astringent, antiprotozoal.
7.	Alkaloids (Largest group of secondary plant metabolites	Nitrogen bases	Act in plant defence, widely used as pharmaceuticals, pscycho stimulants, Narcotics, poisons.

Role of Nanoparticles in Plants

Nanoparticles are small particles with a dimension of 1–100 nm. Compared to the volumetric shape, NPs have higher biological activities, higher catalytic reactivity, non-linear optical overall performance and chemical stability due to their huge surface area to volume ratio. NPs are utilized in the cosmetic industries, feed, food, chemical, space, health due to their qualities (Agarwal *et al.*, 2017; Prasad *et al.*, 2017; Dimkpa *et al.*, 2017). Depending on the qualities of nanoparticles; they interact with plants and give rise to the diversity in genotype and phenotype of plant. The chemical composition, size, surface coverage, reactivity, and most crucially amount at which NPs are efficient(Khodakovskaya *et al.*, 2012).Nanoparticles covers persuasive role in plant growth and photosynthesis (Siddiqui *et al.*, 2015).

Nanoparticles Role on Plant Growth

The initial indicators of plant development and boom are observed via seed germination, way of shoot emergence and root elongation. As a result, it's critical to comprehend plant boom and development in connection to NPs. Gold Nanoparticles (AuNPs) improves seed germination (Gopinath *et al.*, 2014), leaf area, number of leaf, chlorophyll concentration, plant height, and content of sugar are all factors which influence crop production (Arora *et al.*, 2012; Kumar *et al.*, 2013). Found that AuNPs affects the antioxidant system in *Arabidopsis thaliana*, seed germination and, as well as changed levels of microRNA expression, which controls metabolic, morphological, physiological processes in plants (Kumar *et al.*, 2013).Silver Nanoparticles (AgNPs) exhibit to decrease phenol content and increases peroxides activities in plant (Krishnaraj *et al.*, 2012). *Brassica juncea*, common bean, and maize plants had their growth profiles (root and shoot length, leaf area) and characteristics associated with biochemical (antioxidant enzymes, carbohydrate, protein and chlorophyll contents) improved by AgNPs (Salama, 2012). Titanium Dioxide Nanoparticles (TiO_2NPs) regulates the activity of enzymes involved in metabolism of nitrogen, like glutamate dehydrogenase, nitrate reductase, glutamic-pyruvic transaminase, and glutamine synthase, which aids in the nitrate absorption and favours the conversion of inorganic to organic nitrogen aschlorophyll and protein, potentially increasing the dry and fresh weight of plants (Yang *et al.*, 2006; Mishra *et al.*, 2014). Carbon Nanotubes (CNTs) possess the capacity to penetrate in the cell walls or membranes, as well as offer a chemical delivery mechanism to cells. SWCNTs are carbons with single wall nanotubes that function as nano transporters for Deoxyribonucleic acid and dye molecules inside the plant cells (Srinivasan and Saraswathi, 2010). Zinc Oxide Nanoparticles (ZnONPs) supplementation with MS medium improved stress association with (living organism) biotic tolerance by enhancing plantlet regeneration, shooting, somatic embryogenesis as well as inducing peroxidase, proline synthesis, superoxide dismutase activity (Helaly *et al.*, 2014). Silicon Dioxide Nanoparticles (nano-Sio_2) improved germination of seed by increasing the accessibility of nutrients to maize seeds and increasing the pH and permeability of the growth medium (Suriyaprabha *et al.*, 2012).

Nanoparticles Role in Photosynthesis

Only 2–4 per cent of the energy available in radiation as a result of radiation is converted into plant growth by plants (Kirschbaum, 2011). *Rubisco* is an enzyme that catalyses the incorporation of carbon dioxide into biological molecules during the photosynthesis process. This replaced with two genes of cyanobacterium *Synechococcus elongates* in tobacco plant. The photosynthetic efficiency of these novel designed plants is higher than that of native plants (Lin *et al.*, 2014). *Nano-anatase* TiO_2 has a photocatalytic property that enhances absorption of light and energy converted from light to chemical and electrical energy, as well as causing carbon dioxide assimilation. For long-term illumination, TiO_2NPs protect chloroplasts against ageing (Hong *et al.*, 2005; Yang *et al.*, 2006). In photosynthetic systems, nanoparticles of metals can increase the effectiveness of chemical energy generation. Because of plasmon resonance, the chlorophyll inside the photosynthetic reaction centre bonds to the Ag nanocrystals and AuNPs, producing a new hybrid system which could create electrons that are ten times more excited (Govorov *et al.*, 2007).

Nanoparticles Synthesis using Phytochemicals

Phytochemicals may act both as stabilizing and reducing agents in the process of synthesis of nanoparticles (Kumar *et al.* 2010). Nanoparticles may be created in a variety of ways, either a "top down" or a "bottom up" strategy (Sepeur, 2008). Top-down synthesis produces NPs by reducing the size of appropriate initiating material (Meyers *et al.*, 2006). In bottom-up synthesis, NPs are assembled by compact components and smaller particles (Mukherjee *et al.*, 2001). Several plants are utilized in the production of nanoparticles (Li *et al.* 2007).

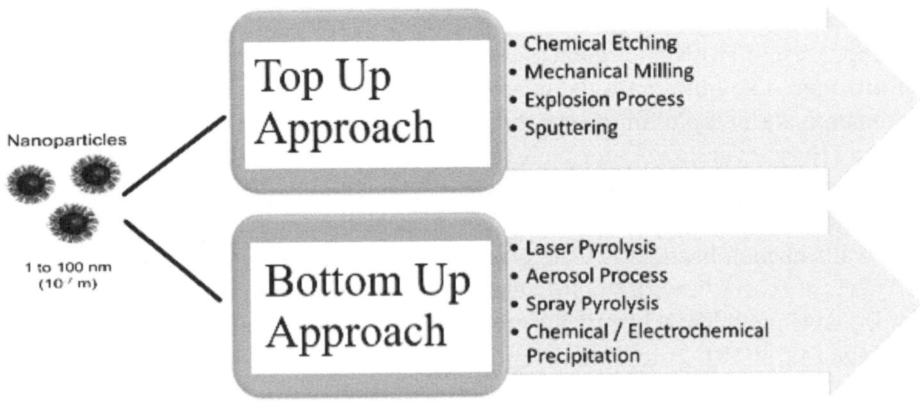

Figure 16.1: Nanoparticle Synthesis Methods and Cofactor-Dependent Bio-reduction.

Mechanism of Nanoparticles Synthesis using Phytochemicals

When a metallic salt breaks down into anion and cation, the cations become saturated and form hydroxyl complexes. Crystallite formation of metal with oxygen species begins shortly after the supersaturation of hydroxyl complexes. This causes the development of crystalline planes with various energy levels. Heat is an important source of energy for the reaction system. This process takes place till the phytochemical activate the capping agent, which stops the formation of high-energy atomic growth planes. Specific type NPs are formed as a result of this (Figure 16.2) (Markov *et al.*, 2014).

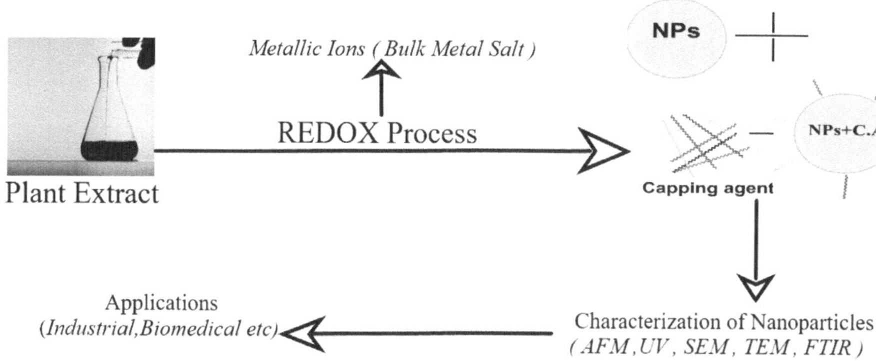

Figure 16.2: Figure Illustrating the Nanoparticle Synthesis through Phytochemicals Activation.

Phytochemical as Bioreductant and Stabilizer while Formulating Nanoparticle

Several phytochemicals have been shown to stimulate the production of metallic nanoparticles from ionic compounds (Dubey *et al.*, 2010). Numerous functional groups may be found in phytochemicals including O=H (phenolic and alcohol), CH and COO- (carboxylic group), N-H (amine), C=N (amide) and C=C (Alkenyl). It's mostly represented by phytochemicals, although it might also include micro and macro macromolecules (Jha *et al.*, 2009). These chemical compounds are totally involved in the creation of nanoparticles. Likewise, *R. hymenosepalus* phytochemical extracted from medicinal plant *Rumex,* known to exhibit antioxidant property (Vanderjagt *et al.*, 2002) enables the production of nanoparticles at room temperature with rapid reaction kinetics. As a result, the solvent extract of *R. hymenosepalus* is excessive in polyphenols such as stilbenes and catechins, which function as stabilising agents and reducing agents in the synthesis of silver NPs (Awwad *et al.*, 2013).

Additionally to the bio-reduction, phytochemicals provide stability to nanoparticle synthesises acting as a stabilizer or capping agent through the implication of alkaloids, terpenoids, polyphenols while reduction of metal ions into NPs (Makarov *et al.*, 2014). The phytochemical content of Ag NPs made with 3 distinct natural polyphenols, resveratrol (RSV), epigallocatechin-3-gallate (EGCG), and fisting are substantially associated with their scavenging capabilities (Sanna *et al.*, 2014). In streptozotocin (STZ)-induced diabetic mice, ZnO nanoparticles that had been stabilised by plant metabolites, varies in their anti-diabetic potency depending on their size (Bala *et al.*, 2015). The antioxidant activity of Au NPs made from Azolla microphyll methanolic extract found to be significant (Kunjiappan *et al.*, 2014).

Table 16.2: Phytochemical Tested in available Plants in Vicinity of Kurukshetra University

Plant (Part)	*Phytochemicals Present*	*Phytochemicals Absent*	*Extracted with Solvent*
Achyranthes asper (Leaves)	Tannins, Coumarins, Terpenoids	Quinones, Phlobatinins, Aminoacids, Leucoanthocyanin	Methanolic
Eriobotrya japonica (Leaves)	Couarins, Terpenoids	Tannins,Quinones, Anthocyanins	Methanolic
Terminalia chebula (Leaves)	Coumarin, Terpenoids, Anthocyanin	Tannins, Quinones, Leucoanthocyanins	Methanolic
Trigonella corniculata	Tannins, Coumarins and anthocyanins	Quinones, Phlobatinins, Terpenoids, Leucoanthocyanins	Methanolic

Phytochemicals Exhibit Antimicrobial Potency

Flavonoids, saponins, alkaloids, steroids and others are phytochemical substances that are physiologically active and hence largely responsible for plants' antimicrobial activities, which is why they are used in traditional medicine (Nethathe *et al.*, 2011). With the emergence of multidrug resistance (MDR) importance of using phytochemicals as a natural alternate has emerged.

Individual Antimicrobial Compounds Derived from Phytochemicals

Phenolic Compound

A single substituted phenolic ring makes up the most basic bioactive phytochemicals with highest oxidation state known to have antibacterial properties (Brantner *et al.*, 1996). Both the hydroxylated phenols pyrogallol and catechol have been proven to be harmful for microorganisms. Enhanced microbial toxicity is directly proportional to the enhanced hydroxylation of the phenol group (Geissman *et al.*, 1963).

Alkaloids

They are nitrogen molecules that are heterocyclic. Morphine, first medicinally useful isolation was done from opium poppy. Diterpenoid alkaloids are often obtained from the *Ranunculaceae* family of plants (Atta-ur-Rahman *et al.*, 1995) are usually found to possess antibacterial characteristics (Omulokoli *et al.*, 1997). From the berries of *Solanum khasianum*, Solamargine, a glycoalkaloid an alkaloid, is derived which might be beneficial in the fight against HIV (McMahon *et al.*, 1995).

Flavonoids

Flavonoids are naturally occurring chemicals, but plants also produce them by microbial infection (Dixon,1986). Flavonoids have been discovered to exhibit antibacterial action *in vitro* against a variety of pathogens. Their capacity to bind with soluble and extracellular proteins, as well as cell walls of bacteria, has been ascribed to their action (Cowan, 1999).

Quinones

In *Bacillus subtilis,* anthraquinones, the biggest category of quinones (Harborne *et al.*, 1999), have been reported to exert antibacterial effects *via* blocking nucleic acid production (Levin *et al.*, 1988).

Coumarins

Coumarin are phenolic compounds consist of a-pyrone rings which are bonded together and benzene (Kennedy *et al.*, 1997).Gram-positive bacteria appear to be inhibited by hydroxycinnamic acids, which are linked to coumarins (Fernandez *et al.*, 1996). Carrots also generate hydroxylated derivatives of coumarins, such as phytoalexins, in reaction to fungal infection and are thought to have antifungal action (Hoult *et al.*, 1996).

Essential Oils and Terpenoids

Terpenes are a broad collection of chemicals that give plants their scent and make up the so-called essential oil fraction. Terpenes and terpenoids have been discovered to have a variety of properties antibacterial properties (Eloff *et al.*, 2006; Amaral *et al.*, 1998; Barre *et al.*, 1997; Himejima *et al.*, 1992) antifungal properties (Ayafor *et al.*, 1994; Kubo *et al.*, 1993) and antiviral properties (Pengsuparp *et al.*, 1994; Sun *et al.*, 1996).

Tannins

Tannins are a class of polyphenolic chemicals that are found in variety of plants. The capacity of tannins inactivate microbial adhesins, enzymes, and cell envelope transport proteins is connected to their antimicrobial mechanism of action (Cowan,1999).Tannins also hinder the development of bacteria by depriving them of substrate and metal ions due to their capacity to bind to proteins and metals (Scalbert *et al.*, 1991).The antifungal (filamentous fungus) and antibacterial efficacy of hydrolysable and condensed tannins found to be equal, while the hydrolysable tannins shown to be more effective against yeasts (Cowan, 1999).

Other Phytochemicals

Antimicrobial activities have been discovered in a variety of phytochemicals not mentioned above. This review paper has sought to focus on reports of compounds that have been confirmed to be active in numerous cases. Bibenzyls and stilbenes are Stilbenoids made up of two benzene rings connected by an ethane or ethene bridge. The antifungal and antibacterial effects of several Stilbenoids are well recognised (Eloff *et al.,* 2006). Also, antimicrobial potency of polyamines (Flayeh *et al.,* 1987), isothiocyanates (Iwu *et al.,* 1991) and glycosides (Rucker *et al.,* 1992) has been identified as potent alternative antimicrobial agents.

Synergistic Interactions of Antimicrobial Drugs and Phytochemicals

Plants have variety of phytochemicals effective in antimicrobial propensity. But phytochemicals exhibit weaker antimicrobial activity than common-chemically synthesized compounds (Chung *et al.,* 2011). Thus, synergetic interactions of phytochemicals with known compounds are required (Sanhueza *et al.,* 2017).

Antibacterial Drugs and Phytochemical Interaction

Synergistic combinations of antibacterial and natural compounds provide some beneficial pharmacological effects (Hooton *et al.,* 1984). 52 -methoxyhydnocarpin and berberine, two phytochemicals often found in plants as barberry, exhibit energy by accumulating inside the bacteria and inhibiting the pump called MDR (Stermitz *et al.,* 2000). This interaction inhibits active site modifications, efflux pump overexpression and neutralizing resistant mechanism (Muhammad *et al.,* 2019).

Antiviral Drugs and Phytochemical Interaction

In conjunction with interferon, glycyrrhizin, a sweet-tasting component of licorice (*Glycyrrhiza glabra*), suppresses the hepatitis C virus (Ashfaq *et al.,* 2011). The use of a sub-optimal accumulation of oseltamivir in combination with *Hedera helix* extract or its component hedrasaponin F offers improved defence against influenza infection, whereas oseltamivir alone was not enough to eradicate the illness. The synergy between the plant and its active component accounts for oseltamivir's strong antiviral efficacy when used in combination (Hong *et al.,* 2015).

Antifungal Drugs and Phytochemical Interaction

Combining natural products as phytomedicines with antifungal drugs can minimize toxicity (Ngo *et al.,* 2016) and overcome fungal resistance (Movahed *et al.,* 2016). On the other hand, extract of *Rosae Chinensis* is behind for the conversion of lanosterol to eburicol. Both the medicines in proper proportion resulted in a decrease in ergosterol levels, which resulted in cell death. Fresh garlic extracts have been shown to have synergistic effect against *Candida albicans* when used in combination with fluconazole in other studies (Salih *et al.*, 2016).

Limiting Factor

The natural active ingredients isolated from living organisms and plants are limiting just because the metabolite composition in these natural resources changes according to environmental factors and stress conditions prevalent in India.

In spite of all these limitations scientist is synthesizing artificial or synthetic molecules which are less popular in market and people again have inclination towards natural products. The most of the natural products are already in market having originated from higher plants.

Table 16.3: Explain Drugs in Plant Extracts and their Respective Treatments

Sl.No.	*Drug in Use*	*Plant*	*Treating*
1.	Artemotil	*Artemisia annua* L.	Malaria
2.	Reminyl	*Galanthus woronowii* L.	Alzheimers disease
3.	Paclitaxel R	*Taxus brevifolia*	Cancer treatment
4.	Vinblastin and Vincristine	*Catharanthus roseus*	Cancer treatment
5.	Camptothecin	*Camptotheca acuminate* Decne	Cancer treatment
6.	Silymarin	*Silybum marianum*	Liver disorders

Conclusion

Most of the green route synthesized nanoparticles are capped with secondary metabolites which further enhances bioactivities of these nanoparticles. This conjugation of nanoparticles with secondary metabolites offers great opportunity in different fields of industrial biotech. Nanoparticles mediated delivery of medicinally important flavonoids and other biomolecules will increase their therapeutic efficacy. Hesperitin conjugated gold nanoparticles enhanced the treatment of hepatocellular carcinoma by minimizing side effects and reduced the dose of chemotherapy drugs. Titanium dioxide nanoparticles have been synthesized using a flavonoid assisted method having applicability in food samples. Although nanoparticles synthesized using green route has widespread occurrence but medical and industrial application is still restricted. The nanoparticles owing to non-homogeneous and differential size and morphology along with factors affecting like temperature, pH, capping agents, concentrations need to be standardized for obtaining spherical, mono-dispersed and stable nanoparticles. The phytochemicals or secondary metabolites have immense potential to generate active ingredients which are novel and have almost null side effects. Indians' flora and fauna is a huge resource being 17th mega biodiversity centre and ancient knowledge needs to be documented for future discovery of novel drugs as medicinal healthcare. Although secondary metabolites have main factor for redox reaction in green route synthesis and inhibiting bacteria (as antimicrobial agents). In spite of different cell wall structure, alkaloids, flavonoids, terpenes, phenols, terpenoids, and saponins are major bioactive constituents in natural plant products. They have been still restricted and least concerned by pharmaceuticals industry because of reduced bioavailability and efficacy. Moreover, their size and high molecular structure is major hurdle in their absorption capacity *via* lipid bilayer. Moreover, the number of doses and high systemic clearance further add to their usage. The present review focussed on collaborative effects of both phytochemicals in concern with nanoparticle synthesis and antimicrobial activity.

REFERENCES

Agarwal H., Venkat K., S. Rajesh K.S.(2017) A review on green synthesis of zinc oxide-nanoparticles—An eco-friendly approach. *Resour. Effic. Technol.*, 3, 406–413.

Amaral J.A., Ekins A., Richards S.R., and Knowles R. (1998). Effect of selected monoterpenes on methane oxidation, denitrification, and aerobic metabolism by bacteria in pure culture. *Applied and Environmental Microbiology*, 64: 520-525.

Arora S., Sharma P., Kumar S., Nayan R., Khanna P.K., Zaidi MGH (2012) Gold-nanoparticle induced enhancement in growth and seed yield of Brassica juncea. *Plant Growth Regul*,66: 303–310

AshfaqU.A., MasoudM.S., NawazZ., RiazuddinS.(2011) Glycyrrhizin as antiviral agent against hepatitis C virus, *J. Transl. Med.* 9: 112.

Atta-ur-Rahman and Choudhary M.I. (1995). Diterpenoid and steroidal alkaloids. *Natural Products Rep.*, 12: 361-379.

Awwad A.M., Salem N.M., Abdeen A.O.(2013). Green synthesis of silver nanoparticles using carob leaf extract and its antibacterial activity. *Int. J. Ind. Chem.* 4, 29-35.

Ayafor, J.F., Tchuendem, M.H.K. and Nyasse, B. (1994). Novel bioactiveditерpenoids from Aframomumaulacocarpos. *Journal of Natural Products*, 57: 917-923.

Bala N., Saha S., Chakraborty M., Maiti M., Das S., Basu R., Nandy P. (2015) Green synthesis of zinc oxide nanoparticles using Hibiscus subdariffa leaf extract: Effect of temperature on synthesis, anti-bacterial activity and anti-diabetic activity. *RSC Adv.*, 5, 4993–5003.

Barre J.T., Bowden B.F., Coll J.C., Jesus J., Fuente V.E., Janairo G.C., and Ragasa C.Y. (1997). A bioactive triterpene from Lantana camara. *Phytochemistry*, 45: 321- 324.

Brantner A.Z., Males S., Pepeljnak S., and Antolic A. (1996). Antimicrobial activity of Paliurus spina-christi Mill. *Journal of Ethnopharmacology*, 52: 119-122.

Chung P.Y., NavaratnamP., ChungL.Y.(2011) Synergistic antimicrobial activity between pentacyclic triterpenoids and antibiotics against *Staphylococcus aureus* strains, *Ann.Clin. Microbiol. Antimicrob.* 10: 25

Cowan M.M. (1999). Plant products as antimicrobial agents. *Clinical Microbiology Reviews,* 12: 564-582.

Cowan M.M. (1999). Plant products as antimicrobial agents. *Clinical Microbiology Reviews,* 12: 564-582

Dimkpa C.O., BindrabanP.S.(2018)Nanofertilizers: New Products for the Industry? *J. Agric. Food Chem.*66, (26): 6462–6473

Dixon R.A. (1986). The phytoalexin response: elicitation, signaling and control of host gene expression. *Biology Reviews*, 61: 239-291.

Doughari J.H. (2012) Phytochemicals: extraction methods, basic structures and mode of action as potential chemotherapeutic agents. chemical.report Available from: http: //cdn.intechopen. com/pdfs-wm/32936.pdf

Dubey S.P., Lahtinen M., Särkkä H., Sillanpää M. (2010). Bioprospective of Sorbus aucuparia leaf extract in development of silver and gold nanocolloids. *Colloid Surface B.* 80, 26–33.

Eloff J.N., and McGaw L.J. (2006). Plant Extracts Used to Manage Bacterial, Fungal and Parasitic Infections in Southern Africa, In*: Modern Phytomedicine. Turning Medicinal Plantsinto Drugs*, 97-121

Eloff J.N., and McGaw L.J. (2006). Plant Extracts Used to Manage Bacterial, Fungal and Parasitic Infections in Southern Africa. *Modern Phytomedicine. Turning Medicinal Plantsinto Drugs*. pp. 97-121

Fernandez M. A., Garcia M.D., and Saenz M.T. (1996). Antibacterial activity of the phenolic acids fraction of Scrophularia frutescens and Scrophulariasambucifolia. *Journal of Ethno-pharmacology,* 53: 11-14.

Flayeh K.A., and Sulayman K.D. (1987). Antimicrobial activity of the amine fraction of cucumber (Cucumis sativus) extract. *Journal Applied Microbiology*, 3: 275-279.

Geissman T.A. (1963). Flavonoid compounds, tannins, lignins and related compounds, In: Pyrrole pigments, isoprenoid compounds and phenolic plant constituents, Ed. By Florkin, M. and Stotz E.H., *Elsevier, New York, N.Y.*,9: 265.

Gopinath K., Gowri S., Karthika V., Arumugam A. (2014) Green synthesis of gold nanoparticles from fruit extract of Terminalia arjuna, for the enhanced seed germination activity of Gloriosa superba. *J of Nanostruct Chem* 4: 1–11

Govorov A.O., Carmeli I. (2007) Hybrid structures composed of photosynthetic system and metal nanoparticles: plasmon enhancement effect. *Nano Lett* 7(**3**): 620–625

Harborne J.B., Baxter H. and Moss G.P. (1999). Phytochemical Dictionary-A hand book of bioactive compounds from plants, *2 nd Edition, Taylor and Francis Ltd, London*, p.528.

Helaly M.N., El-Metwally M.A., El-Hoseiny H., Omar S.A., El-Sheery N.I. (2014) Effect of nanoparticles on biological contamination of in vitro cultures and organogenic regeneration of banana. *Aust J Crop Sci* 8: 612–624

Himejima M., Hobson K.R., Otsuka T., Wood D.L., and Kubo I. (1992). Antimicrobial terpenes from oleoresin of ponderosa pine tree Pinus ponderosa: a defense mechanism against microbial invasion. *Journal of Chemistry and Ecology*, 18: 1809-1818.

HongE.-H., SongJ.-H., ShimA., LeeB.-R., KwonB.-E., SongH.-H., KimY.-J., ChangS.- Y., JeongH.G., KimJ.G.(2015) Coadministration of Hedera helix L. extract enabled mice to overcome insufficient protection against influenza A/PR/8 virus infection under suboptimal treatment with oseltamivir, *PLoS One* 10 e0131089.

Hong F., Zhou J., Liu C., Yang F., Wu C., Zheng L., Yang P. (2005a) Effect of nano-TiO2 on photochemical reaction of chloroplasts of spinach. *Biol Trace Elem Res*, 105(**1–3**): 269–279

HootonT., BlairA., TurckM., CountsG. (1984) Synergism at clinically attainable concentrations of aminoglycoside and beta-lactam antibiotics, *Antimicrob. AgentsChemother.* 26: 535–538.

Hoult J.R.S., and Paya M. (1996). Pharmacological and biochemical actions of simple coumarins: natural products with therapeutic potential. *General Pharmacology*, 27: 713-722.

Iwu M.M., Unaeze N.C., Okunji C.O., Corley D.J., Sanson D.R. and Tempesta M.S. (1991). Antibacterial aromatic isothiocyanates from the essential oil of Hippocrateawelwitschii roots. *International Journal of Pharmacognosy*, 29: 154-158.

Jha A.K., Prasad K., Kumar V., Prasad K.(2009). Biosynthesis of silver nanoparticles using Eclipta leaf. *Biotechnol. Prog.* 25, 1475-1477.

Khodakovskaya M.V., de Silva K., Biris A.S., Dervishi E., Villagarcia H. (2012) Carbon nanotubes induce growth enhancement of tobacco cells. *ACS Nano* 6(3): 2128–2135

Kirschbaum MUF. (2011) Does enhanced photosynthesis enhance growth? Lessons learned from CO2 enrichment studies. *Plant Physiol* 155: 117–124

Krishnaraj C., Jagan E.G., Ramachandran R., Abirami S.M., Mohan N., Kalaichelvan P.T. (2012) Effect of biologically synthesized silver nanoparticles on Bacopa monnieri (Linn.) Wettst. Plant growth metabolism. *Process Biochem*, 47(**4**): 651–658

Ku P., Rani S., Rana V., Pal GK. (2020) Aquatic plants as a natural source of antimicrobial and functional ingredients, Editor(s): Bhanu Prakash, Functional and Preservative Properties of Phytochemicals, Academic Press. *Elsevier*93-118

Kubo I., Muroi H. and Himejima M. (1993). Combination effects of antifungal agilactones against Candida albicans and two other fungi with phenylpropanoids. *Journal of NaturalProducts*, 56: 220-226.

Kumar V., Guleria P., Kumar V., Yadav S.K. (2013) Gold nanoparticle exposure induces growth and yield enhancement in Arabidopsis thaliana. *Sci Total Environ* 461: 462–468

Kumar V., Yadav S.C., Yadav S.K.(2010)Syzygiumcumini leaf and seed extract mediated biosyn- thesis of silver nanoparticles and their characterization. *J Chem TechnolBiotechnol*,85: 1301–9.

Kunjiappan S., Chowdhury R., BhattacharjeeC.(2014) A green chemistry approach for the synthesis and characterization of bioactive gold nanoparticles using Azolla microphylla methanol extract. *Front. Mater. Sci.*, 8, 123–135.

Levin H., Hazenfrantz R., Friedman J. and Perl M. (1988). Partial purification and some properties of the antibacterial compounds from Aloe vera. *Phytotherapy Research*,1: 1-3.

Li S., Shen Y., Xie A., Yu X., Qui L., Zhang L., *et al.* (2007a)Green synthesis of silver nanoparticlesusing Capsicum annum L. extract,*Green Chem*, 9: 852–858.

Lin M.T., Occhialini A., Andralojc P.J., Parry M.A.J., Hanson M.R. (2014) A faster Rubisco with potential to increase photosynthesis in crops. *Nature.* doi: 10.1038/nature13776

Makarov V.V., Love A.J., Sinitsyna O.V., Makarova S.S., Yaminsky I.V.,Taliansky M.E., KalininaN.O.(2014) "Green" nanotechnologies: Synthesis of metal nanoparticles dsing plants. *Acta Nat.*, 6, 35–44.

Marslin G., Siram K., Maqbool Q., Selvakesavan R.K., Kruszka D., Kachlicki P., Franklin G. (2018) Secondary Metabolites in the Green Synthesis of Metallic Nanoparticles. *Materials* 3; 11(**6**): 940.

McMahon J.B., Currens M.J., Gulakowski R.J., Buckheit R.W.J., Lackman-Smith C., Hallock Y.F., and Boyd M.R. (1995). Michellamine B, a novel plant alkaloid, inhibits human immunode?ciency virus-induced cell killing by at least two distinct mechanisms. *Antimicrobial Agents Chemotherapy*, 39: 484-488.

Meyers M.A., Mishra A., Benson D.J.(2006) Mechanical properties of nanocrystalline materials.*Prog Mater Sci*,51: 427–556.

Michael A Grusak,(2002). Phytochemicals in plants: genomics-assisted plant improvement for nutritional and health benefits, *Current Opinion in Biotechnology*, 13(5); 508-511.

Mishra V., Mishra R.K., Dikshit A., Pandey A.C. (2014) Interactions of nanoparticles with plants: an emerging prospective in the agriculture industry. In: Ahmad P, Rasool S (eds) *Emerging technologies and management of crop stress tolerance: biological techniques*, 1, 159–180

MovahedE., TanG.M.Y., MunusamyK., YeowT.C., TayS.T., WongW.F., LooiC.Y. (2016). Triclosan demonstrates synergic effect with amphotericin B and fluconazole and induces apoptosis-like cell death in Cryptococcus neoformans, *Front. Microbiol.* 7

Muhammad A., Farhat U., Abdul S., Farman U., Muhammad O., Jawad A., Hari Prasad D.(2019) Synergistic interactions of phytochemicals with antimicrobial agents: Potential strategy to counteract drug resistance, *Chemico-Biological Interactions*,308: 294-303.

Mukherjee P., Ahmad A., Mandal D., Senapati S., Sainkar S.R., Khan MI.(2001) Fungus-mediated synthesis of silver nanoparticles and their immobilization in the mycelial matrix: a novel biological approach to nanoparticle synthesis. *Nano Lett*,1(**10**): 515–519.

Nethathe B.B., Ndip R.N.(2011) Bioactivity of Hydnoraafricana on selected bacterial pathogens: Preliminary phytochemical screening. *Afr J Microbiol Res.*; 5(**18**): 2820–2826

NgoH.X., Garneau-TsodikovaS., GreenK.D. (2016) A complex game of hide and seek: the search for new antifungals, *Med Chem Comm* 7: 1285–1306.

O'Kennedy R., and Thornes R.D. (1997). Coumarins: biology, applications and mode of action. *John Wiley and Sons, Inc., New York, N.Y.*

Omulokoli E., Khan B., and Chhabra S.C. (1997). Antiplasmodial activity of four Kenyan medicinal plants. *Journal of Ethno-pharmacology*, 56: 133-137.

Pengsuparp T., Cai L., Fong H.H.S., Kinghorn A.D., Pezzuto J.M., WaniM.C. and Wall M.E. (1994). Pentacyclic triterpenes derived from Maprouneaafricana are potent inhibitors of HIV-1 reverse transcriptase. *Journal of Natural Products*, 57: 415-418.

Prasad R., Bhattacharyya A. NguyenQ.D.(2017) Nanotechnology in sustainable agriculture: Recent developments, challenges, and perspectives. *Front. Microbiol.*, 8, 1014.

Rucker G., Kehrbaum S., SakulasH., Lawong, B., and Goeltenboth F. (1992). Acetylenic glucosides from Microglossapyrifolia. *Planta Medica*, 58: 266-269.

Salama H.M.H. (2012) Effects of silver nanoparticles in some crop plants, common bean (Phaseolus vulgaris L.) and corn (Zea mays L.). *Int Res J Biotech*,3(**10**): 190–197

SalihK.A. (2016)Synergistic effects of plant extracts and antifungal drugs on C. Albicans, *J. Dev.* Drugs 5

SanhuezaL., MeloR., MonteroR., MaiseyK., MendozaL., WilkensM. (2017) Synergistic interactions between phenolic compounds identified in grape pomace extract with antibiotics of different classes against Staphylococcus aureus and Escherichia coli, *PLoS One* 12 e0172273.

Sanna V., Pala N., Dessi G., Manconi P., Mariani A., Dedola S., Rassu M., Crosio C., Iaccarino C., Sechi M.(2014) Single-step green synthesis and characterization of gold-conjugated polyphenol nanoparticles with antioxidant and biological activities. *Int. J.Nanomed.*, 9, 4935–4951.

Scalbert A. (1991). Antimicrobial properties of tannins. *Phytochemistry*, 30: 3875-3883.

Sepeur S. (2008) Nanotechnology: technical basics and applications. Hannover: Vincentz;

Siddiqui Dr. Manzer H. and Al-whaibi, Mohamed and Mohammad, Firoz and Al-Khaishany, Mutahhar. (2015). Role of Nanoparticles in Plants,*Nanotechnology and Plant Sciences* 10.1007/978-3-319-14502-0_2.

Srinivasan C., Saraswathi R. (2010) Nano-agriculture-carbon nanotubes enhance tomato seed germination and plant growth. *Curr Sci*, 99: 273–275

StermitzF.R., LorenzP., TawaraJ.N., ZenewiczL.A., LewisK.(2000) Synergy in a medicinal plant: antimicrobial action of berberine potentiated by 52 -methoxyhydnocarpin, a multidrug pump inhibitor, *Proc. Natl. Acad. Sci. Unit. States* Am. 97: 1433–1437.

Sun H.D., Qiu S.X., Lin L.Z., Wang Z.Y., Lin Z.W., Pengsuparp T., Pezzuto J.M., Fong H.H., Cordell G.A. and Farnsworth N.R. (1996). Nigranoic acid, a triterpenoid from*Schisandra sphaerandra* that inhibits HIV-1 reverse transcriptase.*Journal of Natural Products,* 59: 525-527.

Suriyaprabha R., Karunakaran G., Yuvakkumar R., Rajendran V., Kannan N. (2012) Silica nanoparticles for increased silica availability in maize (Zea mays L) seeds under hydroponic conditions. *Curr Nano sci* 8: 902–908

VanderJagt T. J., Ghattas R., VanderJagt D. J., Crossey M., and Glew R. H. (2002). Comparison of the total antioxidant content of 30 widely used medicinal plants of New Mexico. *Life Sciences,* 70(**9**), 1035-1040.

Yang F., Hong F., You W., Liu C., Gao F., Wu C., Yang P. (2006) Influence of nano-anatase TiO2 on the nitrogen metabolism of growing spinach. *Biol Trace Elem Res*,110(**2**): 179–190.

Chapter 17

DNA Sequencing: It's Generations and Interpretations A Review

Harinder Singh[1], Sunita Khatak[1] and Manjit Kaur[2]*

[1]Department of Biotechnology, University Institute of Engineering and Technology, Kurukshetra University, Kurukshetra – 136 119, Haryana
[2]CSIR-CBRI, Roorkee – 247 667, Uttarakhand
**e-mail: sunitakhatak2019@gmail.com*

Introduction

In China discovery of Covid-19 from late December 2019, the expanding worldwide infectious COVID-19 disease arose from the new Severe Acute Respiratory Syndrome Coronavirus 2 (SARS-CoV-2) poses a serious danger to worldwide health and in turn effecting the world economy (Azizzadeh *et al.*, 2019). Ever since, works were going on to identify the genome sequence of this virus, its unique order of nucleotide, the genetic building block that make it up. In result, found it to be a 30,000 base pair long (Paden *et al.*, 2020). Likewise, through a worldwide consortium of Telomere-to-Telomere (T2T) complete genome sequencing achieved for Human Genome Project by May-2021, which includes sequencing of all 22 autosomes and X chromosome (Sergey *et al.*, 2021).Numerous genomic databases and genomic variation maps are been drawn for plants and animals worldwide (Cuiping *et al.*, 2021). So, one can figure out the importance of Genome Sequencing as a matter of fact that, many countries and its numerous scientists, technologists are working together. They are developing database of genome sequences from human to animals, plants to microbes. The pursue of this review study is to figure out the importance of genome sequencing, its generations of development, analysis and interpretation.

Generations of Sequencing

It has been more than three decades since the first generation of sequencing of DNA technology came out in 1977 (James *et al.,* 2014).With the improvement of technology which can decide the (bp) base-pair series of (deoxyribonucleic acid) DNA, the capacity to series the genes have provided a lot to technological know-how and medicinal. Although, it endured a tremendously high priced and exhausting procedure, hampering the use as an ordinary biomedicine tool. In these days are seeing speedy trends in this domain, each with inside of availability of narrative platforms of sequencing, as helping technology concerned in record and analysis (Alice *et al.,* 2020). From the first generation to the third generation and even the fourth generation, the sequencing generation has experienced long-term and short-term exploration. The fourth sequencing technology developed unexpectedly during the course (Alice *et al.,* 2020).

By 1952, Hershey and Chase discovered the T2 phage infection of microorganism experiment, identifying DNA as a genetic material (Boyle *et al.,* 1965). In 1953, Crick and Watson defined DNA version with "double helix" (Kantha 2013). In 1958, Francis Crick proposed the genetic critical dogma, which become reiterated in Nature in 1970 (Olby, 1970; Maxam and Gilbert, 1977) initially installed a DNA fragment series approach, which is also known as Maxam-Gilbert chemical degradation approach. Currently, this chemical degradation approach and enzymatic approach (dideoxy chain termination approach) proposed through Sanger are rapid sequencing techniques(Allan 1980). In 1983, PCR (polymerase chain reaction) become evolved *via* way of means of Dr. Kary B.Mullis. It is a molecular biology method and used to increase unique DNA fragments, which may be appeared because the unique DNA replication *in vitro* (Kaltenboeck *et al.,* 2005). In 1986, the American Pe Abi Company developed the first automatic sequencer apriorism 310 gene analysers. Hood and Smith subsequently used fluorescently labelled dNTPs for electrophoresis. In 2005 Roche Technology developed 454 (Genome Sequencing System 20), an ultra-high-throughput genome sequencing system that Nature recognized as a breakthrough in sequencing technology development (Momiao *et al.,* 2010).

First Generation of Sequencing

The sequencing technology of Sanger is classified as the 1st-Generation Sequencing Technology (Thudi *et al.,* 2012; Liu *et al.,* 2012). First published in 1977, creating a precedent for DNA sequencing.

Sanger Sequencing

In the 1970s, with the use of enzyme DNA polymerase (Sanger *et al.,* 1977), Sanger and their co-worker developed Sanger sequencing which is also known as enzymatic DNA sequencing. They published their DNA sequencing method as "plus and minus" and bacteriophage sequence. After two years they published the

'Chain Termination Method', a less complicated version on the 'Plus and Minus' Method (Smith *et al.*, 1985) which utilised radioactively or fluorescently labelled dideoxynucleosides performing as chain terminators (Atkinson *et al.*, 1969).

Maxam and Gilbert Sequencing

In 70s, Maxam and Gilbert evolved a DNA sequencing technique which is non-enzymatic that belongs to the first technology of sequencing called the chemical degradation technique (Maxam *et al.*, 1977). Samples to be sequenced need few complicated practice than in Sanger Sequencing. Relies at the cleaving of nucleotides with the aid of using compounds (chemical) and is maximum reactive with small nucleotides polymers, but reads are shorter (most one hundred bp). This results in a series of selected fragments which may be segregated in step according to their length by the aid of using electrophoresis (Masoudi *et al.*, 2013). The system itself is technically complicated and involves using unsafe chemical compounds (Maxam *et al.*, 1977).

Second Generation of Sequencing

The basic principle of all second-generation DNA sequencing technologies is the "massive parallelization", which exponentially increases capacity and reduces costs. The sequencing techniques of this generation are also known as "clonal sequencing" and "next generation sequencing (NGS)" (Pettersson *et al.*, 2009). All the chemical processes of NGS have one thing in common, that is, spatially separated clonal amplification amplicons are read in parallel cycles (Mardis *et al.*, 2008).

Chemistry of 2nd Generation of Sequencing

Pyrosequencing

Supporting platforms for this chemistry are Genome Sequencer Junior (*GS Junior*), Genome Sequencer FLX and Roche (Moorthie *et al.*, 2011). Here the synthesis is carried out for Sequencing and depends upon detection of pyrophosphate (PPi) released (Margulies *et al.*, 2005). In a series of reaction that involved enzymes, the light produced (easily visible) proportionally identifies the nucleotides number present in said DNA template.

Likewise, in Roche, is pyrosequencing-based platform primarily depends on the pyrophosphate detection launched after every incorporation of nucleotide inside the formed artificial strand of DNA (Heo *et al.*, 2015)

Reversible Termination

Here also sequencing is carried out through sequencing. With the use of terminator nucleotide (*fluorescent lab bled*) –The Process of chain termination is reversed (Bentley *et al.*, 2008). Platforms using this technology are Illumina/ Solexa, Genome Analyzer 11e and 11x (GA11e, GA11x), HiSeq 1000 and 2000, Mi-Seq (Moorthie *et al.*, 2011).

Likewise, in Illumina/Solexa Genome Analyzer is capable of giving very quick reads ~35 basepairs and they had a bonus in the production of paired-end (PE) quick reads, wherein series at each ends of every Deoxyribonucleic acid cluster is recorded. Loading control is important while using this sequencing technique as overlapping of sequence may occur (Shendure *et al.,* 2008; Balasubramanian *et al.,* 2015).

Sequencing by Ligation

Unlike prior methods, this one relies on the ligation of hybridisation probes that is labelled by florescence rather than polymerase-based DNA synthesis (Shendure *et al.,* 2005). Likewise, Life Technologies/Applied Biosystems for Ligation and Detection, commercialised SOLiD (Sequencing by Oligonucleotide Detection) (Moorthie *et al.,* 2011).

Table 17.1: Summarizes the Existing Second-generation Sequencing Techniques

Platform	*Roche/454*	*Life Technologies/ SOLiD*	*Illumina/Solexa,MiSeq*
Sequencing (Enzyme)	DNA Polymerase	DNA Ligase	DNA Polymerase
Method	Pyrosequencing	Ligation Sequencing	Transmission (Reversible)
Parallelisation	In high density plate beads are loaded.	In high density glass slide beads are immobilized.	Bridge Amplification generated by Surface PCR.
Loading	On beads adapters on template DNA binds.	On beads adapters on template DNA binds.	High density primers across surface of the slide.
Amplification	Through Emulsion PCR for creating cluster on beads.	Through Emulsion PCR for creating cluster on beads.	Surface PCR for generating bridge amplification

Source: Moorthie *et al.*, 2011.

Second Generation Sequencing: Target Enrichment Methods

A few ways to deal with target enhancement have been created, which can contrast concerning a few boundaries: 1) consistency, for example changeability in grouping inclusion across target districts 2) explicitness, or level of the all-out grouping yield that guides to the objective districts 3) ease of utilization 4) the measure of DNA required 5) affectability, or level of bases addressed 6) reproducibility, or then again how well results relate between reproduces.

PCR Enrichment

Because of its capacity to intensify a solitary DNA particle have giant explicitness, PCR is the most utilized strategy for preparing tests for sequencing (Saiki *et al.*, 1988). Sequencing of Sanger depends upon perusing of single amplicons. For its implementation to NGS it is fundamental to expand PCR.

Targeted enrichment through PCR, unlike other enrichment methods like as probe-hybridization, does not need a good amount of DNA used for sequencing areas. Several hundred kilobases are possible (Mamanova *et al.* 2010). It has a high enrichment ratio and only a few off-target readings. It is also true in the case of tiny amplicons (Mertes *et al.*, 2011).

Circularisation Enrichment

DNA circularisation comprises in the utilization of atomic reversal tests (MIP) (Porreca *et al.*, 2010) or selector tests (Dahl *et al.*, 2005). It depends upon the utilization of Deoxyribonucleic acid tests, which are oligonucleotides (single-abandoned) comprising of typical linker flanked by target-explicit groupings. In MIPs the tests are toughened to the underlying DNA test, going through circularisation all the while. The hybridized finishes of the test flank the objective district. The hole between the two finishes, non-circularized DNA at that point filled is eliminated by an enzyme exonucleases (Johansson *et al.*, 2011) (Emily *et al.*, 2009).

Hybridization Capture

Enhancement by hybridisation depends on information procured through microarray examination (Schena *et al.*, 1988) and applied to the advancement of tests for NGS (Albert *et al.*, 2007). Vague half and halves are eliminated by washing and the designated DNA acquired is developed (Okou *et al.*, 2007; Hodges *et al.*, 2007).

Second Generation Sequencing: Clonal Amplification

Following the preparation of library, NGS technologies must amplify the sample before sequencing it by cloning, as the signal created must be amplified in order to be detected during the sequencing procedure. This procedure seeks to make many duplicates of each DNA library fragment while keeping these segregated to generate a signal(monoclonal) that is identifiable (Mardis *et al.*, 2008; Shendure *et al.*, 2008).

Emulsion PCR (emPCR)

Emulsion PCR is a variation of the traditional amplification procedure in which every DNA fragment is separated in its own microreactor (aqueous) encircled by an oil phase (Shendure *et al.*, 2005; Nakano *et al.*, 2003) and the templates that are intensify on primer-coated beads (Dressman *et al.*, 2003). The emulsion is broken down using organic solvents after the process, and the beads are separated from extracting the aqueous phase.

Bridge PCR

Using a solid oligonucleotide coated surface (also called as a "flow cell"), bridge amplification permits clonal amplification of a high no. of DNA fragments at the same time (Adessi *et al.,* 2000). The ds DNA that results is covalently bonded to the surface. The single strand of this ds DNA is then denatured and hybridised to neighbouring primers, producing a bridge.

Third Generation of Sequencing

In comparison to the 1st generation of sequencing technologies, 2nd generation of sequencing technologies (SGC) has revolutionised DNA analysis and is the most frequently utilised. However, most SGS methods require a PCR amplification phase, which is a time-consuming operation with a high sequencing cost. Furthermore, it became evident that genomes are extremely complicated, with many repeated regions, and that SGS technologies are unable to address them, and that genome assembly is made more difficult by the comparatively short reads (Levene *et al.,* 2003). Researchers have created a updated version of sequencing dubbed "third generation sequencing" to address the issues presented by SGS technology. These third ages of sequencing may provide a low sequencing cost and simple test readiness excluded the need PCR enhancement in an execution time fundamentally quicker than SGS advances (Schadt *et al.,* 2010).

A wide number of firms are now developing this form of sequencing, each utilising different methods and chemistry. Based on methods, it can be categorized into three variants as Electronic, Fluorescence and Atomic based.

Single Molecule Real Time Sequencing (SMRTS)

Development of SMRTS is done by Pacific Biosciences (McCoy *et al.,* 2014). The technique is established on a single DNA polymerase detecting natural DNA production. Phosphate-labelled nucleotides (also known as phosphor linked nucleotides) are introduced at the same time, allowing for continuous and control polymerisation (Rhoads *et al.,* 2015).

Pacific Bioscience technique provides numerous benefits to SGS (Liu *et al.,* 2012). The sample preparation is incredibly quick, taking only 4 to 6 hours instead of days (Chin *et al.,* 2016). The error rate is around 13 per cent. Individually, very lengthy readings can be as long as 60 kbp, well beyond the capabilities of any SGS technology (Koren *et al.,* 2012).

DNA Sequencing using Nanopores

GridION, Oxford Nanopore Technologies' nanopore sequencing technology, was unveiled, along with MinION (Mikheyev *et al.,* 2014), a disposable nanopore sequencing device (Laehnemann *et al.,* 2015). It is a portable single-molecule Nanopore sequencing device that is 4 inches in length and connects to a laptop computer through a USB 3.0 connector (Laver *et al.,* 2015).

A hairpin connects the first strand of a DNA molecule to its corresponding strand in this sequencing technique. A protein nanopore (a nanopore is a tiny hole formed of proteins or synthetic materials) is used to transmit the DNA fragment (Jain *et al.,* 2016). When a motor protein linked to the pore translates the DNA fragment through the pore, due to variations in the moving nucleotides filling the pore, it causes an ionic current fluctuation. This ionic fluctuation is progressively listed on a graphic model and then analysed to determine it (Lu *et al.,* 2016; Heo *et al.,* 2015).

DNA Direct Imaging using Electron Microscope

Halcyon Molecular and ZS Genetics developed this method. It allows direct imaging and chemical detection of atoms to identify nucleotides (Lundquist *et al.,* 2008). Its ability to detect atoms in a planar surface by annular dark-field imaging in an aberration-corrected scanning (TEM) has recently been demonstrated as proof-of-its working (Krivanek *et al.,* 2010; Mardis *et al.,* 2008).

DNA Sequencing using Transistor Mediation

This nanostructured sequencing technology is in development by IBM to specific bases in a single molecule of deoxyribonucleic acid electrically (Krems *et al.,* 2009). Alternating layers of metal with pores and dielectric elements make up the structure. As deoxyribonucleic acid travels along the pores, the gadget recognises every base based on how it interacts with the device. This technique has a very high throughput rate (millions of bases/sec). Because of its label, it is inexpensive with no-optics feature (Luan *et al.,* 2010).

Table 17.2: Summarizes the Three Generations of Sequencing Techniques

Parameter	*First Generation Sequencing*	*Second Generation Sequencing*	*Third Generation Sequencing*
Sequencing Principle	Chain Termination Method	Pyrosequencing	Chain Termination Method
RNA-Sequencing Method	c-DNA Sequencing	c-DNA Sequencing	Direct-RNA Sequencing
Data Analysis	Routine	Complex because of large data volume.	Complex because of large data volume.
Sample Preparation	Moderately complex, PCR Amplification	Complex, PCR Amplification	
Throughput	Low	High	Moderate
Primary Result	Quality values. High Accuracy	Quality values. Amplified template are generated during sequencing.	Quality values, Potential and base information.

Analysis of DNA Sequence

The preliminary assessment of every novel DNA sequence begins with a reference list to determine whether coding regions of this gene (protein) exist and, if so, what is known about them. Apparently, seeking for the exact identical sequence is a simple task. Simply take very first character of the template strand, look for it in the database for the first time, and then see if the second letter of the doubt appears in the topic. If they are the same, the software may verify the third letter, then the fourth, and so on until the query is completed. If the subject's second letter differs from the query's second letter, the computer should look for another occurrence of the first letter, and so on. This will provide a list of all database sequences that are identical to the query sequence (Miller *et al.,* 2001). Though this method is rudimentary in terms of computing, and there are sophisticated text finding techniques that are much more efficient (Binstock *et al.,* 1995). But before coming to computing and text finding, targeted DNA sequence needs to be shotgun sequenced (inportions) aligned for analysis.

Shotgun Sequencing

Shotgun sequencing is a basic method that scientists employ to analyse a single gene in its entirety. A succession of shorter overlapping sequences makes up the lengthy DNA sequence (Motahari *et al.,* 2013). Machines, known as sequencing machines are utilized to extricate short arbitrary DNA arrangements from a specific genome we wish to decide (target genome). These short pieces are called reads. Programmes are utilized to collect these reads as per how they cross-over, to produce persistent strings called contigs (Mohajer *et al.,* 2013). These contigs can be simply the entire objective genome, or portions of the genome(Christoph *et al.,* 2004).

Sequence Alignment

Sequence alignmentis a technique for identifying areas of similarity in DNA, RNA, or protein sequences (Rajko *et al.,* 2004; Biswanath *et al.,* 2017). The similarity seen between the sequences might be due to functional, structural, or evolutionary connections. Pairwise sequence alignment is when two sequences are compared. Multiple sequence alignment occurs when we compare more than two sequences (Kazutaka *et al.,* 2019). For computer assembly-analysis, certain software/tools are available.

Functional Assignment of Genes

Branch points, transcription terminators, splice sites, binding sites of ribosomes, topoisomerase II binding and cleavage sites, start and stop codons are among the DNA regions that researchers look for (Gelfand 1995; Krogh 2000). These functionally active sites are named as signal and ways-means to identify these signals are known to be signal sensors. Introns and Exons are named as contents. Whereas method to identify them is termed as content sensor. Sensors content exon utilise statistical models of nucleotide frequencies and dependencies found in codon structures to distinguish coding from noncoding regions in eukaryotes (Sherrif *et al.*, 2001). Markov models are the most widely used statistical models. In exon detectors, neural networks are utilised to integrate multiple coding clues with signal sensors for the flanking splice locations (Staden1984).

Non-Coding Assessment of Genomic Region

The ability to interpret noncoding genomic areas important in gene regulation (based only on DNA sequence) is a long-standing problem in developmental genetics (Aerts *et al.*, 2012; Alonso *et al.*, 2009).While there are very easy principles for predicting which genomic areas serve significant non-coding activities, it has been recognised that there are few rigorous rules for predicting the genomic regions serve as non-coding functions (Davidson *et al.*, 2015). A popular technique for identifying potential functional parts of non-coding sequences is to examine conservation of sequence, a procedure called as phylogenetic foot printing (Tagle *et al.*, 1988). This method takes use of the fact that key regulatory sequence functions are typically restricted, come out in the conservation of crucial binding sites with surrounding sequences change (Pennacchio *et al.*, 2001). Numerous examples show that phylogenetic foot printing is effective in discovering highly conserved binding sites that perform a variety of regulatory roles (Hardison *et al.*, 2000; Bulyk *et al.*, 2003).

Data Interpretations and Analysis

While genome sequencing generates a huge number of databases and the interpretation-analysis of which needs to have accurate database along with computational powers. This process of interpretation-analysis can be categorized into three steps as –

a) Base Calling – This is known to be the primary analysis. Likewise, the fluorescence or electronic or atomic signal intensities of nucleotide are converted into readable formats (Moorthie *et al.*, 2010). By this time quality score is poor and raw.

b) Variant and Alignment Calling – or Secondary Analysis. Here in our raw data of sequencing is referred to the databanks or other references (Flicek and Birney, 2009). Similarity and variations are checked through the help of software/algorithm package. Sequence data is explained with biological, structural and functional information and shown using genome browser or graphical interface or in the last step of the alignment procedure (Magi *et al.*, 2010; Ding *et al.*, 2010).

c) Interpretations – or Territory Analysis – Here in the functional impacts of resultant DNA sequence is estimated through filtering variations and through accessing repetends, unique or novelty and inheritance (Kuhlenbnumer *et al.*, 2010). In case of variation found in tested DNA sequence to its reference database, estimations and interpretations are made on pathogenicity poses by that mutation or aberration (Kuntzer *et al.*, 2010; Huang *et al.*, 2010). Certain software packages interpret it. In clinical estimations this analysis is worked from pathogenicity to benign.

Conclusion

The study of biomolecules has been transformed by recent advances. Molecular markers, amplification, and nucleic acid sequencing are among them. There are three generations of the latter. The 1st permits tiny DNA fragments to be sequenced. The 2nd boosts throughput while lowering turn around and cost, making it easier to sequence whole genomes. The 3rd generation is pushing of technology to its boundaries, allowing single molecules to be sequenced excluding prior development, which was previously not possible. Furthermore, this is a new revolutionary era, for the researchers to sequence RNA directly without the need for retro transcription. Biotechnology, agronomy, medicine abd ecology are few of the fields where these technologies are having a big influence. Biomolecular research is also yielding fascinating evolutionary facts. Decrypting all that makes humans human, including occurrences like non-coding RNA growth, is part of this. All of this is changing the way we think about genes and transcripts. New genome editing technologies, like as CRISPR, have made basic analysis and applications easier. All of these advancements, particularly nucleic-acid sequencing, are ushering in a new age of biomolecule analysis and applications, such as customised medicine and disease detection and prevention for humans and other animals.

REFERENCES

Adessi C. *et al.* (2000) Solid phase DNA amplification: characterisation of primer attachment and amplification mechanisms, *Nucleic Acids Res.* 28 E87.

Aerts S. (2012) Computational strategies for the genome-wide identification of coregulatory elements and transcriptional targets. *Curr. Top. Dev. Biol.* 98, 121–145.

Albert T. (2007) Direct selection of human genomic loci by microarray hybridization, *Nat. Methods* 4 903–905.

Albert T., *et al.,* (2007) Direct selection of human genomic loci by microarray hybridization, *Nat. Methods* 4: 903–905.

Alice M.G. (2020) Long walk to genomics: History and current approaches to genome sequencing and assembly,*Computational and Structural Biotechnology Journal*: 18, 9-19.

Allan M. M, Walter G. (1980) Sequencing end-labelled DNA with base-specific chemical cleavages,Methods in Enzymology, Elsevier, 65, 499-560.

Alonso M.E., Pernaute B., Crespo M., Gomez-Skarmeta J., Manzanares M.(2009). Understanding the regulatory genome. *Int. J. Dev. Biol.* 53, 1367–1378.

Atkinson M.R., Deutscher M.P., Kornberg A., Russell A.F., Moffatt J.G. (1969) Enzymatic synthesis of deoxyribonucleic acid. XXXIV. Termination of chain growth by a 22 ,32 -dideoxyribonucleotide, *Biochemistry* 8: 4897–4904.

Azizzadeh K., Hamdan U.S., Salehi P.P. (2020) Effect of Coronavirus Disease and Pandemics on Global Surgical Outreach. *JAMA Otolaryngol Head Neck Surg.*; 146(**9**)783–784

Balasubramanian S. (2015) Solexa sequencing: Decoding genomes on a population scale. *Clin Chem* 61: 21-24.

Bentley D.R.,et.al. (2008) Accurate whole human genome sequencing using reversible terminator chemistry. *Nature* 456: 53–59.

Binstock A., Rex J. (1995) Practical algorithms for programmers. *Addiison-WesleyPublishing Co., Reading, MA.*

Biswanath C., Gautam G.(2017) A review on multiple sequence alignment from the perspective of genetic algorithm,Genomics, 109(**5–6**): 419-431.

Boyle J.M., Ritchie D.A., Symonds N. (1965). Density-gradient separation of host-modified bacteriophage T2, *Journal of Molecular Biology*,14, (**1**) 48-54

Bulyk M.L., McClay D., Hood L., Sarig O., Ziv Y., Barkai N., Smith H., Yandell M., Evans C., Holt R.(2003). Computational prediction of transcription-factor binding site locations. *Genome Biol.* 5, 201.

Chin CS, Peluso P, Sedlazeck FJ, Nattestad M, Concepcion GT, *et al.* (2016) Phased diploid genome assembly with single molecule real- time sequencing. BioRxiv 13: 1050-1054.

Christoph A. (2004) Information theory in molecular biology, *Physics of LifeReviews*,1(1): 3-22.

Cuiping Li, *et al.* (2021). Genome Variation Map: a worldwide collection of genome variations across multiple species, *Nucleic Acids Research*, 49(**D1**) D1186–D1191

Dahl F., Gullberg M., Stenberg J., Landegren U., Nilsson M. (2005) Multiplex amplification enabled by selective circularization of large sets of genomic DNA fragments, *Nucleic Acids Res.* 33 e71.

Davidson E.H., Peter I.S. (2015). Genomic Control Process, Genomic Control Process. *Elsevier.*

Ding L., Wendl M.C., Koboldt D.C., Mardis E.R. (2010) Analysis of next generation genomic data in cancer: accomplishments and challenges. *Hum Mol Genet* 19(**R2**): R188–R196

Dressman D. *et al.,* (2003) Transforming single DNA molecules into fluorescent magnetic particles for detection and enumeration of genetic variations, *Proc. Natl. Acad. Sci. USA* 100 8817–8822.

E. Mardis (2008) The impact of next-generation sequencing technology on genetics, *Trends Genet.* 24: 133–141.

Emily H., Choli L., Sarah B., Deborah A., Shendure J. (2009) Massively parallel exon capture and library-free resequencing across 16 individuals, *Nat. Methods* 16: 315–316.

Flicek P., Birney E. (2009) Sense from sequence reads: methods for alignment and assembly. *Nat Methods*6: S6–S12

Gelfand M.S. (1995) Prediction of function in DNA sequence analysis. *J ComputBiol* 2: 87–115

Hardison R.C. (2000). Conserved noncoding sequences are reliable guides to regulatory elements. *Trends Genet.* 16, 369–372.

Heo and Yun (2015) Improving quality of high-throughput sequencing reads,*Ideals*

Hodges E. (2007) Genome-wide in situ exon capture for selective resequencing, *Nat.Genet.* 39: 1522–1527.

Huang N., Lee I., Marcotte E.M., Hurles M.E. (2010) Characterising and predicting haploinsufficiency in the human genome. *PLoS Genet* 6: e1001154

Jain M, Hugh EO, Paten B, Akeson M (2016) The oxford nanopore MinION: Delivery of nanopore sequencing to the genomics community.*Genome Biology* 17: 239.

James M. Heather, Benjamin Chain (2016). The sequence of sequencers: The history of sequencing DNA,*Genomics-Elsevier*: 107, (1)1-8.

Johansson H. *et al.* (2011). Targeted resequencing of candidate genes using selector probes, *Nucleic Acids Res.* 3 e8.

Kaltenboeck B., and Wang C. (2005). Advances in real-time PCR: application to clinical laboratory diagnostics. *Advances in clinical chemistry*, 40, 219–259.

Kantha S. (2013). Citation patterns of the four seminal DNA double-helix model papers by Watson and Crick in 1953–54. *CURRENT SCIENCE,* 104(9) 1237-1239.

Kazutaka K., John R., Kazunori D.Y. (2019) MAFFT online service: multiple sequence alignment, interactive sequence choice and visualization, *Briefings in Bioinformatics,* 20(**4**) 1160–1166.

Koren S, Schatz M, Walenz B, Martin J, Howard J, *et al.* (2012) Hybrid error correction and de novo assembly of single-molecule sequencing reads. *Nature Biotechnology* 30: 693-700.

Krems M., Zwolak M., Pershin Y.V., Di Ventra M. (2009) Effect of noise on DNA sequencing via transverse electronic transport, *Biophys. J.* 97 1990–1996.

Krivanek O.L *et al.,* (2010) Atom-by-atom structural and chemical analysis by annular dark-field electron microscopy, *Nature* 464: 571–574.

Krogh A. (2000) Using database matches with for HMMGene for automated gene detection in Drosophila. *Genome Res* 10: 523–528

Kuhlenbnumer G., Hullmann J, Appenzellerm S. (2010) Novel genomic techniques open new avenues in the analysis of monogenic disorders, *Human Mutation*,32(**2**): 144-151

Kuntzer J., Eggle D., Klostermann S., Burtscher H. (2010) Human variation databases. *Database.*, Article ID baq015

Laehnemann D, Borkhardt A, McHardy AC (2015) Denoising DNA deep sequencing data—high-throughput sequencing errors and their correction. *Brief Bioinformatics* 17: 154-79.

Laver T, Harrisona J, O'Neill PA, Moorea K, Farbosa A, *et al.* (2015) Studholme. Assessing the performance of the oxford nanopore technologies MinION. *Biomolecular Detection and Quantification* 3: 1-8.

Levene M.J., *et al.,* (2003) Zero-mode waveguides for single-molecule analysis at high concentrations, *Science* 299: 682–686.

Liu L, Li Y, Li S, Hu N, He Y, *et al.* (2012) Comparison of next-generation sequencing Systems. *Journal of Biomedicine and Biotechnology*, 251-364.

Lora B. (2013) DNA Sequence Nomenclature and Variant Interpretation, *American association of clinical chemistry*, Web page

Lu H, Giordano F, Ning Z. (2016) Oxford nanopore MinION sequencing and genome assembly. *Genomics Proteomics Bioinformatics* 14: 265-279.

Luan B., Peng H., Polonsky S., Rossnagel S., Stolovitzky G., Martyna G. (2010) Base-by-base ratcheting of single stranded DNA through a solid-state nanopore, *Phys. Rev. Lett.* 104

Lundquist P.M. *et al.,* (2008) Parallelconfocal detection of single molecules in real time, *Opt.Lett.* 33: 1026–1028.

MacArthur D.G., Tyler-Smith C. (2010) Loss-of-function variants in the genomes of healthy humans. *Hum Mol Genet* 19: R125–R130

Magi A., Benelli M., Gozzini A., Girolami F., Torricelli F., Brandi M. (2010) Bioinformatics for next generation sequencing data. *Genes* 1: 294–307

Mamanova L., Coffey A., Scott C., Kozarewa I., Turner E., Kumar A., Howard E., Shendure J., Turner D. (2010) Target-enrichment strategies for next-generation sequencing, *Nat. Methods* 7: 111–118.

Mardis E. (2008) Next-generation DNA sequencing methods, *Annu. Rev. GenomicsHum. Genet.* 9: 387–402.

Margulies M, Egholm M, Altman WE, Attiya S, Bader JS, Bemben LA, Berka J, Braverman MS, Begley RF, Rothberg JM (2005) Genome sequencing in microfabricated high-density picolitre reactors. *Nature* 437: 376–380.

Masoudi-Nejad A, Narimani Z, Hosseinkhan N. (2013) Next generation sequencing and sequence assembly. Methodologies and algorithms. *Springer.*

Maxam A.M., Gilbert W.A. (1977) A new method for sequencing DNA. *Proc Natl AcadSci* 74: 560-564.

McCoy R.C., *et al.* (2014) Illumina TruSeq synthetic long-reads empower de novo assembly and resolve complex, highly-repetitive transposable elements. *PLoS ONE*9: e106689.

Mertes F., Elsharawy A., Sauer S., Van Helvoort J., Van D.Z., Franke A., Nilsson M., Lehrach H., Brookes A. (2011) Targeted enrichment of genomic DNA regions for next-generation sequencing, *Brief. Funct. Genomics* 10: 374–386.

Mikheyev A.S., Tin M.M.Y. (2014) A first look at the oxford nanopore MinION sequencer. *Molecular Ecology Resources* 14: 1097–1102.

Miller W. (2001) Comparison of genomic DNA sequences: solved and unsolved problems. *Bioinformatics,*17: 391–397.

Momiao X. (2010) "Next-Generation Sequencing", BioMed Research International, Article ID 370710, 2 pages,

Moorthie S., Mattocks C. J., and Wright C. F. (2011). Review of massively parallel DNA sequencing technologies. *The HUGO journal*, 5(**1-4**), 1–12.

Moorthie S., Mattocks C.J., Wright C.F. (2011) Review of massively parallel DNA sequencing technologies. *Hugo J.*; 5(1-4): 1-12.

Mostafa R.(2001) Pyrosequencing Sheds Light on DNA Sequencing, *Genome Res, ColdSpring Harbor Laboratory Press*, 11: 3-11

Motahari A. S., Bresler G. and Tse D. N. C. (2013), "Information Theory of DNA Shotgun Sequencing," in IEEE Transactions on Information Theory, 59(10): 6273-6289.

Motahari A. S., Bresler G. and Tse D. N. C. (2013), "Reference-based DNA shotgun sequencing: Information theoretic limits," *2013 IEEE International Symposium on Information Theory,* 1635-1639.

Nakano M.et.al. (2003) Single molecule PCR using water-in-oil emulsion, *J. Biotechnol.* 117–124.

Ng P.C., Henikoff S. (2006) Predicting the effects of amino acid substitutions on protein function. *Annu Rev Genomics Hum Genet* 7: 61–80

Okou D. (2007) Microarray-based genomic selection for high-throughput resequencing, *Nat. Methods* 4: 907–909.

Olby R. (1970). Francis Crick, DNA, and the Central Dogma. *Daedalus*, 99(**4**), 938-987.

Paden C. R., Tao Y., Queen K., Zhang J., Li Y., Uehara,A., and Tong, S. (2020). Rapid, Sensitive, Full-Genome Sequencing of Severe Acute Respiratory Syndrome Coronavirus Emerging *infectious diseases*; 26(**10**) 2401–2405.

Pennacchio L.A., Rubin E.M. (2001). Genomic strategies to identify mammalian regulatory sequences. *Nat. Rev. Genet.* 2, 100–109.

Pettersson E., Lundeberg J., Ahmadian A. (2009) Generations of sequencing technologies. *Genomics* 93: 105–111

Porreca G., Zhang K., Li J., Xie B., Austin D., Vassallo S., Leproust E., Peck B., Emig, F. Dahl C., Gao Y., Church G., Shendure J. (2007) Multiplex amplification of large sets of human exons, *Nat. Methods* 4: 931–936.

Rajko S. and Aluru S. (2004.) "Space and time optimal parallel sequence alignments," in *IEEE Transactions on Parallel and Distributed Systems*, 15(**12**)1070-1081

Rhoads A., Au K.F. (2015) PacBio sequencing and its applications. *Genomics, Proteomics and Bioinformatics* 13: 178-289.

Saiki R., Gelfand D., Stoffel S., Scharf S., Higuchi R., Horn G., Mullis K., Erlich H. (1988) Primer-directed enzymatic amplification of DNA with a thermostable DNA polymerase, *Science* 239 487–491.

Sanger F., Nicklen S., Coulson A.R. (1977) DNA sequencing with chain-terminating inhibitors, *Proc. Natl. Acad. Sci.* U.S.A. 74: 5463–5467.

Schadt E.E.,Turner, S., Kasarskis A. *et al.* (2010). "A window into third-generation sequencing". *Human Molecular Genetics.* 19 (**R2**): 227-40.

Schena M. *et al.,* (1998) Microarrays: biotechnology's discovery platform for functional genomics, *Trends Biotechnol.* 17: 301–306.

Sergey N., *et al.*, (2021). The complete sequence of a human genome *bioRxiv*.05.26.445798.

Shendure J, Ji H (2008) Next-generation DNA sequencing. *Nature Biotechnology* 26: 135–1145.

Shendure J. (2005) Accurate multiplex polony sequencing of an evolved bacterial genome, *Science* 309: 1728–1732.

Shendure J., Ji H. (2008) Next-generation DNA sequencing, *Nat. Biotechnol.* 26: 1135–1145.

Shendure J., Porreca G.J., Reppas N.B., Lin X., McCutcheon J.P., Rosenbaum A.M., Wang M.D., Zhang K., Mitra R.D., Church G.M. (2005) Accurate multiplex polony sequencing of an evolved bacterial genome. *Science* 309: 1728–1732.

Sherriff A., Ott J. (2001) Applications of neural networks for gene finding. *Adv Genet* 42: 287–297

Smith L.M., Fung S., Hunkapiller M.W., Hunkapiller T.J., Hood L.E. (1985) The synthesis of oligonucleotides containing an aliphatic amino group at the 52 terminus: synthesis of fluorescent DNA primers for use in DNA sequence analysis, *Nucleic Acids Res.* 13: 2399–2412.

Soto J.et al (2015). The impact of next-generation sequencing on the DNA-methylation based translational cancer research. *Translational Research.*169, (**10**)1016

Staden R. (1984) Computer methods to locate signals in nucleic acid sequences. *NucleicAcids Res* 12: 505–519.

Tagle D.A., Koop B.F., Goodman M., Slightom J.L., Hess D.L., Jones R.T. (1988). Embryonic epsilon and gamma globin genes of a prosimian primate (Galago crassicaudatus). Nucleotide and amino acid sequences, developmental regulation and phylogenetic footprints. *J. Mol. Biol.* 203, 439–455.

Thudi M, Li Y, Jackson SA, May GD, Varshney RK (2012) Current stateof- art of sequencing technologies for plant genomics research. *Brief Funct Genomics* 11, 3-11.

Chapter 18

Abiotic Stress Alleviation by Plant Growth Promoting Rhizobacteria (PGPR) for Environmental Sustainability

Swati Sindhu[1], Basanti Brar[2] and Monika Jangra[3]*

[1]Senior Research Fellow, Department of Veterinary Microbiology, LUVAS, Hisar, Haryana
[2]Chief Scientific Consultant, HABITAT, GIPPCL Biofertilizer Production Lab, CCSHAU, Hisar, Haryana
[3]Assistant Professor, Department of Zoology, Chaudhary Bansi Lal University, Bhiwani, Haryana
**Corresponding author*

ABSTRACT

Rigorous exercise of chemically synthesized fertilizers in agriculture seeks increased crop yields at the expense of environmental degradation and decreased soil fertility. Besides this, abiotic stresses such as high temperatures, high salt concentrations and drought negatively affects our ecosystem which ultimately limits crop production in field conditions. These environmental issues led us to a sustainable and ecofriendly approach, where plant growth promoting rhizobacteria have been used and highly recommended under abiotic stressed conditions in the soil system. These PGPR promotes overall expansion, development and plant yield under various environmental stress circumstances, through direct and indirect methods. The PGPR used as bioinoculants for many agricultural crops as cheap, eco-friendly and easily accessible sources to mitigate variuos biotic and abiotic stresses. By implementing PGPR, we can reduce the utilization of chemical fertilizers, which disrupt our ecosystem and also resulted in improvement of agricultural yield and productivity.

Keywords: *Abiotic stress, PGPR, Temperature, Drought, Salinity, Environmental sustainability.*

Introduction

In the 21st century, climate change and global warming creates alarming situations and several research studies have documented the fact that environmental stresses threaten the food security on a global level (Battisti and Naylor, 2009). A consequence of inevitable population growth, world's population is projected to grow from approximately 7 billion to 8.9 billion by 2050 (Singh *et al.*, 2011). Due to the indefensible nature of our conventional agriculture system, most of the world's food will be insufficient and it is become very difficult to fulfil the nutritional requirements of a population worldwide (Masciarelli *et al.*, 2014). On the other way, agricultural practices that indiscriminately apply chemicals such as fertilizers, pesticides, weedicides *etc.* cause a serious loss of useful soil microbes. Various abiotic factors are the major limitation in the path of agricultural production and responsible for the reduction in crop yield. There are a number of abiotic stresses that have a high negative impact on survival, biomass production and growth of food crops by as much as 70 per cent, resulting in threats to food security worldwide. Crop plants require some affective and intrinsic physiological and biological mechanisms to fight with adverse impact produced by abiotic stresses. Microbes works as the natural mitigator of abiotic stress for plants as a source of enormous metabolic capacities in diversified or extreme environments (Figure 18.1). As a result of their interactions with plants, they evoke various kinds of local and systemic responses that allows the plants to be more resilient against different abiotic stresses (Nguyen *et al.*, 2016).

PGPR also known as enhancer of plant growth as they can profoundly increase their growth and yield of cereals and valuable agronomic crop plants (Santoyo *et al.*, 2016). Various kind of PGPRs found in rhizospheric region of soil promote crop yield by many direct or indirect processes, as suppressor of many pathogens and circuitous production of a variety of plant growth regulators, organic matter putrefiction and improvement in available sources for various inorganic nutrients such as phosphorus, iron and zinc (Valencia-Cantero *et al.*, 2007). PGPR resides in or on the rhizospheric soil mainly belonging to strains of *Bacillus, Burkholderia, Klebsiella, Enterobacter, Azotobacter, Pseudomonas, Pantoea, Serratia, Paenibacillus, Achromobacter, Arthrobacter, Azospirillum, Streptomyces etc.* They are cheap, ecofriendly and easily available resources which are generally used as bioinoculants for the important agricultural crops in various harsh environmental circumstances. PGPR usually improve plant growth promotion (increased plant biomass, plant height, root and shoot growth) mainly through active production of phytohormones, antioxidants and manufacture of growth regulators like siderophore, bacteriocin and many other known and unknown factors (Yang *et al.*, 2009, Kumar and Verma, 2017, Mohanty *et al.*, 2021, Harkousse *et al.*, 2021).

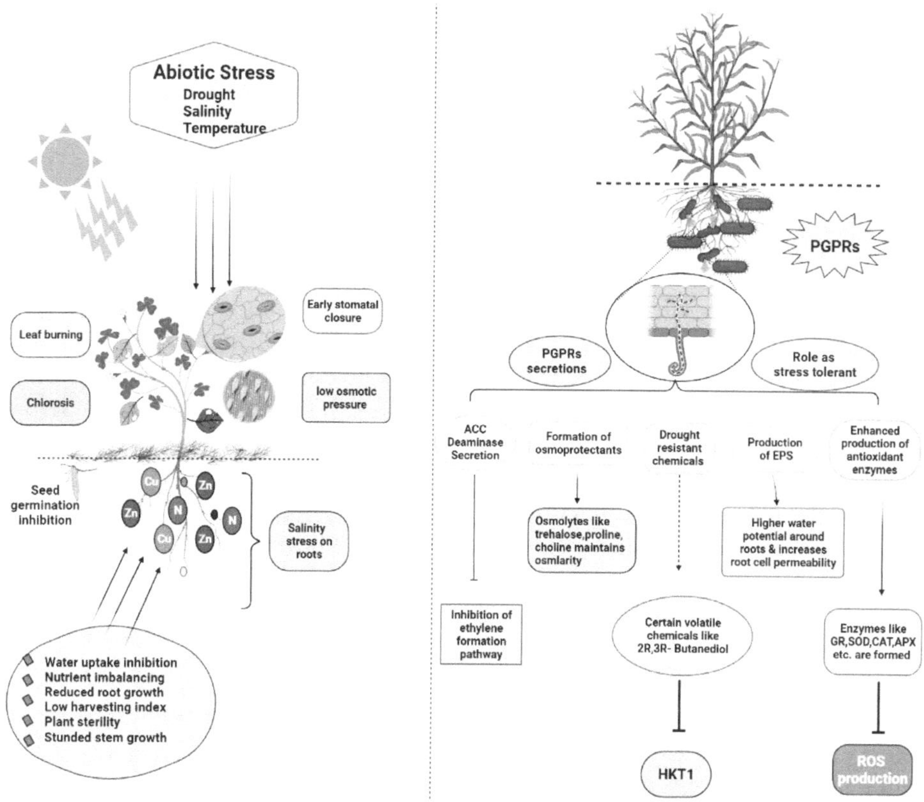

Figure 18.1: Negative Impact of Major Abiotic Factors on Plant Growth (Left side), Improved Growth of Plants (Right side) by Inoculation of Abiotic Stress Tolerant (Temperature, drought and salinity) Microbes

High Salinity Effects on Plant

Saline conditions with high salt concentrations negatively affects the physico-chemical characteristics of soil, which ultimately reduce the growth and population of living organisms resides in the soil vicinity like microbes, plants and nematodes directly. Salinity stress has a major influence on the agro-economic area of a nation, as high salinity stress severely decrease imitation and also reduce plant yield which negatively affects the fertile lands. Saline stress negatively affect approximately 50 per cent of the soil region and according to an estimation, to raise upto 7 per cent worldwide (Riadh *et al.*, 2010; Ruan *et al.*, 2010). Globally, the amount of saline soil is varied in arid, semi-arid and irrigated regions, in which 40 per cent of irrigated and 15 per cent of the arid and semi-arid area were observed as salt prone. Presence of different inorganic minerals in the rhizospheric soil such as magnesium, bicarbonate, sodium, potassium, sulphate, chloride, calcium and carbonate is called salinization. Among all these, one mineral sodium chloride is the most responsible component which is responsible for salinization as their ions are found to be very lethal for plant growth at high levels above the threshold salt concentrations. The accumulation of salts in the plant parts hinder the process of photosynthesis and also increase the reactive oxygen species (ROS) production which resulted in activation of many toxic reactions resulted in protein denaturation, cell membrane damage and mutations in the genetic material (Sudhir and Murthy, 2004; Islam *et al.*, 2015).

Salinity not only reduces blooming but also decrease crop yields by disrupting important processes like pollination which ultimately reduces grain yield and seed set (Gill, 1979). Many agriculturally important crops, including cotton, barley, wheat, rice, and bean have lower yields in a saline-stressed environment, Keating and Fisher (1985). Salinization of the rhizosphere region of soil might have a negative impact on growth of plants as the roots are unable to take water and nutrients from salt-caked places and a high quantity of salt ions in the water is another key issue that inhibits their growth. Plants expend the majority of their energy in such salt-stressed conditions to restore osmotic equilibrium by producing suitable solutes and enzymes (Robinson and Jones, 1986; Kempf and Bremer, 1998). It has also been demonstrated that some beneficial bacteria serve a key role in reducing salt stress in plants, resulting in higher crop yields.

PGPR Strains Mediate Salt Tolerance

Plants are born with stress tolerance capacities that allow them to respond to different types of stress and have unique methods for dealing with salt stress. Some plant growth promoters, in addition to the plant's intrinsic systems, some plant growth promoting rhizobacteria (PGPR) have specific mechanisms that are important for salt stress tolerance and plant growth and yield promotion (Cherif-Silini *et al.*, 2021). Through a mechanism known as systemic induced resistance (SIR), PGPR also serves as bioprotectants. These bacteria cause plants to produce auxin and other plant growth hormones to ameliorate stress prone conditions (Morcillo *et al.*, 2021). In salt-prone zones of barley and oats, *Acinetobacter* spp. and *Pseudomonas* produce ACC deaminase and indole acetic acid (IAA) to enhance plant growth (Chang *et al.*, 2014). In barley (*Hordeum vulgare* L.), *Hartmannibacter diazotrophicus* E19 inoculation resulted in higher root and shoot dry weight, ACC deaminase activity, and decreased ethylene concentration (Suarez *et al.*, 2015). Only *Pseudomonas chlororaphis* TSAU13 and *Pseudomonas extremorientalis* TSAU20 were shown to boost plant growth and act as a biocontrol for tomato foot and root rot disease among the five strains examined. Under 200 mM concentrated salt, the majority of the researched bacterial strains promoted plant development, and the results revealed that the investigated bacterial strains, EN1 (*Bacillus sp.*), EN3 (*Z. halophilus*), and EN4 (*Z. halophilus*), all promoted plant growth (Egamberdieva *et al.*,2017). *Planococcus rifietoensis*, a halophilic bacteria with IAA generating, phosphate solubilizing, and ACC deaminase activity, enhanced the growth and production of *Triticum aestivum* under salinity stress (Rajput *et al.*, 2013).

Furthermore, halophilic bacteria can create halotolerant and thermotolerant chitinases, which aid in the decomposition of chitin-based organic matter and thereby used as biocontrol agents. Beneficial bacteria have been shown to lower salt stress in maize and wheat by about 50 per cent. Previous research on plants grown under salt stress (generally 100 mM NaCl) has demonstrated that plants injected with bacterial strains develop faster than plants that are not inoculated. Inoculated peanuts (*Arachis hypogaea*) with *Brachybacterium saurashtrense* and *Brevibacterium casei* and *Haererohalobacter* in comparison to control plants boosted plant growth (Kotuby-Amazher *et al.*, 2000). Similar findings have been observed for radish inoculated with *Staphylococcus kloosii* and *Kocuria erythromyxa* (Yildirim *et al.*, 2008), lettuce infected with *Bacillus subtilis, Bacillus atrophaeus, Bacillus spharicus, Spharicus, S. kloosii*, and *K. erythromyxa* (Karlidag *et al.*, 2011) and in salt stress prone conditions, strawberry inoculated with *B. subtilis, B. atrophaeus, B. spharicus, S. kloosii*, and *K. erythromyxa* (Yildirim *et al.*, 2011).

Drought Tolerance Mediated by PGPRs

High drought level is accepted as a major constraint among different abiotic stresses for loss of agricultural production (Yadav *et al.*, 2018). Approximately 9-10 per cent decrease was observed in total cereal production of country due to drought prone conditions (Lesk *et al.*, 2016). It was also observed that water deficit conditions (drought) affected more than half of universal land area (Cramer *et al.*, 2011). According to the American Meteorological Society (2004), drought term are of four kinds: first is meteorological drought, known as a dearth of precipitation for a certain time; second is hydrological drought, termed as paucity water resources as well as adequate surface during water resource management strategies; third is running shortage of water in the water resources to mitigate the water demands and fourth one is agricultural drought, is a period where soil moisture is decrease, hence outcoming in crop lapse. The Science provides a variety of sustainable elucidations to overcome these hurdles by the use of plant growth promoting rhizobacteria (PGPR) in various agriculture crops which will become the trend setters for the future agricultural production. There are some known or tested strains of PGPRs that can facilitate the plant to tolerate drought stress (Table 18.1). PGPRs colonize the plants rhizospheric region and contribute to the drought tolerance by exopolysaccharides (EPS) production, which plays an essential in influencing soil structure by stimulating the soil-water binding capacity and plays a role in water as well as nutrient supply regulation to roots. Exopolysaccharides permanently binds the microbial species to the surface of roots with the help of network of fibrillar material. EPS production has been investigated in various species such as *Bacillus subtilis, Streptococcus mutans* and *Pseudomonas aeruginosa* (Vimala and Lalitha kumar, 2003).

Table 18.1: Effects of Inoculation of PGPRs in Drought Stress Mitigation in different Crop Plants

Sl. No.	*Microbial Inoculation*	*Experimented Plant*	*Tolerance Strategy*	*References*
1.	*GA-P45-Pseudomonas putida*	*Helianthus annuus*	Exopolysaccharides (EPS) production	Sandhya *et al.*, 2009
2.	*K11 Bacillus licheformis strain*	*Capsicum annum*	Protein and genes related with stress	Lim and Kim, 2013
3.	AR156 *Bacillus cereus*, SM21 *Bacillus subtilis* and *XY21 Serratia Sp.*	*Cucumis sativa*	Stress tolerant gene expression due the production of antioxidant enzymes, proline and monodehydro ascorbate	Wang *et al.*, 2012
4.	*Alcaligenes faecalis* (AF3)	Maize	Increased root length (10 per cent)	Naseem and Bano, 2014
5.	*Burkholderia phytofirmans* strain PsJN	Mazurka and Kaleo cultivars, respectively	Considerably enhanced root biomass by 70 and 58 per cent (alterations in root architecture) led to improved water uptake	Naseem and Bano, 2014
6.	BOFC15-*Bacillus megaterium*	Arabidopsis plantsnts	Abassic acid and polyamines levels increase, improved root system architecture as well as enhanced photosynthetic capacity, enhanced plant biomass	Zhou *et al.*, 2016
7.	H-2-3 *pseudomonas putida*	Soybean	Gibberellins secretion	Sang-Mo *et al.*, 2014
8.	*Bacillis thuringenesis*	Lavandula dentate	IAA production resulted in increased K content and reduced glutathione reductase and ascorbate peroxidase	Armada *et al.*, 2014
9.	*Rhizobium leguminosarum* (LR-30), *Mesorhizobium ciceri* (CR-39) and *Rhizobium phaseoli* (*MR-2*)	Wheat	IAA production improved the plant biomass, growth drought tolerance capacity	Hussain *et al.*, 2014
10.	*Rhizophagus irregularis*	Tomato	Photosynthetic rate increase, decrease lipid oxidation and enhanced root water conductivity as well as oxidative phosphorylation in plants	Calvo- Polanco *et al.*, 2016

Sl. No.	Microbial Inoculation	Experimented Plant	Tolerance Strategy	References
11.	*Az19-Azospirillum spp.*	Maize	Enhanced productivity and plant growth	Garcia *et al.*, 2017
12.	*Piriformspora indica*	Wheat	Increased adaptation of plants by enhancing water and nutrient absorption, increased biomass, root growth sand chlorophyll content	Hosseini *et al.*, 2017

Bacterial isolates can be enhanced the resistance to drought by intracellular accumulation of osmoprotectants like trehalose, glycine, betaine. *Bacillus* inoculation also help in reduction of different antioxidant enzymes like catalyse, glutathione peroxidise, ascorbate peroxidase and also decreased electrolyte leakage at minimum available water potential and enhance the sugar, free amino acids and proline content in maize seedling (Vardharajulaa *et al.*, 2011). *Bacillus megaterium* BOFC15 inoculated in *Arabidopsis* secrete spermidine (Spd), a type of ascorbic acid and polyamines content during drought stress. The treated plant had a markedly higher viability rate of 83.7 per cent than control/untreated plants (18.6 per cent) (Zhou *et al.*, 2016). *Phyllobacterium brassicacearum* STM196 strain extracted from *Brassica napus* increased osmotic stress tolerance in treated *Arabidopsis* plant by increasing the ABA amount which helps in reducing leaf transpiration (Bresson *et al.*, 2013). *Platycladus orientalis* seedling inoculated with *Bacillus subtilis* increase the ABA levels in shoots and enhanced the stomatal conductance which showed the drought resistance in treated plants. It was also reported that in spite of water supply amount water, production of organic acids, amino acids and sugars increased considerably because of *Bacillus subtilis* treatment (Liu *et al.*, 2013). In comparison with control plants, cytokinin levels in the leaves of *Sambucus williamsii* enhanced (91.17 per cent) in inoculated seedling with drought-tolerant PGPR (*Acinetobacter calcoaceticus* X128) in severe drought conditions (Liu *et al.*, 2019).

Temperature Tolerance

Plants are greatly sensitive to climatic and environmental fluctuations in temperature. Heat stress results in serious damage problems to the cell structure by interference in the function of cell differentiation, cell elongation, cell expression, cell permeability and microtubular organisation. It may results in carbon flux variation of the chloroplast stroma, rubisco enzymes, thylokoid membrane which ultimately lead to failure of the plant production and crop yield (Bita and Gerats, 2013). High soil temperatures also play a major problem in microbial colonization (Cheng, 2014). It was well documented in some scientific studies in Africa and Asia that a temperature rise of 3-40°C can cause a decrease in crop production up to 15-35 per cent (Ortiz *et al.*, 2008). Plants require an optimum temperature for the proper functioning as a threshold temperature that influences various stages of plant growth and production like emergence, flowering, seedling, productive phase, grain yield, grain filling which are highly dependent on temperature. However, determination of a reliable upper threshold temperature is very difficult because the plant behaviour may show variations according to the different environmental conditions (Miller *et al.,* 2011).

Different crop plants require different temperature conditions for their proper growth and maintenance (Table 18.2). A small fluctuation in temperature conditions (exposure to low and high temperature) can cause enzymes inactivation, accumulation of protein and reactive species in wheat crops resulting in oxidative stress. Wang and cowrkers (2004) reported that there are some controlling factors at molecular and physiological levels which concerned in signalling pathways of transcriptional control for temperature resistance like chaperones, proteins, free radical scavengers, osmoprotectants, ion transportes. In a recent study, *Bacillus cereus* SA1 act as thermotolerant bacterium in amelioration of heat stress loss in soyabean crop (Khan *et al.,* 2020).

Table 18.2: Optimum Temperature Range Required for different Crop Plants at Various Growth Stages

Sl.No.	*Crop Species*	*Optimum Temperature (°C)*	*Growth Stage*	*References*
1.	Wheat	26	After anthesis	Stone and Nicolas, 1994
2.	Rice	34	Grain yield	Morita *et al.,* 2004
3.	Cotton	45	Reproduction	Rehman *et al.,* 2004
4.	Corn	38	Grain maturing	Thompson, 1986
5.	Pearl millet	35	Seedling	Ashraf and Hafeez, 2004
6.	Cowpea	41	Florescence	Patel and Hall, 1990
7.	Groundnut	34	Pollen synthesis	Vara Prasad *et al.,* 2000
8.	Brassica	29	Florescence	Morrison and Stewart, 2002
9.	Tomato	30	Initial growth	Camejo *et al.,* 2005

High and Low Temperature Tolerance

Many PGPRs have the capability to produce multifunctional polysaccharides that plays a major part in root colonisation and biofilm formation microbial association with roots or other parts that protect the plants from desiccation and high temperature (Qurashi and Sabri, 2012). Formation of bacterial biofilms is one of the chief mechanisms to tolerance of high temperature in plants (Vurukonda *et al.,* 2016). Bacterial biofilms can produce infinite range of macromolecules like sugars, oligo and polysaccharides which plays important role in microbe-plant associations such as improvement in water retention in roots and other parts for proper functioning of plant growth. At high temperature, pre-inoculation of crop plants with heat tolerance *Bacillus* sp. enospores and *Burkholderia phytofirmans* results in enhanced plant growth. It has been observed that the *Bacillus* sp. is an ideal candidate for development of biological products due to their capability to produce heat tolerance endospores (Yanez-Mendizabal *et al.*, 2012). *Pseudomonas aerugnosa* 2CpS1 strain has been investigated to produce even at a high temperature up to 42°C and seed inoculation of wheat with *Pseudomonas aerugnosa* reported considerable increase in leaf area, root length, total chlorophyll content, relative water content, plant height and decrease in cell membrane damage due to high environmental temperature (Meena *et al.*, 2015). *Pseudomonas aerugenosa* AMK-P6 strain (thermo-tolerant) was extracted from sorghum seedlings in semi-arid regions that induced thermo tolerance by formation of protein with high molecular weight in plant leaves and increase biomass of plant along with their biochemical characteristics such as sugar, amino acids, proline and chlorophyll content (Ali *et al.*, 2009). Similarly, Srivastva and coworkers (2008) extracted a thermo-tolerant *Psseudomonas putida* strain NBRI0987 from drought exaggerated rhizospheric region of chickpea crop plant.

Cold resistant SRM strain of *Serratia marscescens* and 1A strain of *Pantoea dispersa* showing plant growth promoting features such as P-solubilization, HCN production, IAA and siderophore production in inoculated wheat plants at 4°C and 15°C. Microbial treated seeds grown under cold temperature showed significantly increased biomass and nutrient absorption of wheat seedlings than control (Selvakumar *et al.*, 2007a, b). Among 19 PGPR strains extracted from nodules of pea plants, on the basis of their morphological, biochemical tests and differential colony formation characterstics, four isolates showed better growth as compared to P09 reference strain at low temperature. *Rhizobium leguminosarum sp.* extracted from *L. Pratensis and L. Japonicas* has been reported for low temperature resistance and used as a bioinoculant in crop production under cold areas (Manivannan *et al.*, 2012). *Serratia nematodiphila*, a gibberllin formation of PGPR, increased *Capsicum annum* growth under low temperature conditions and treated pepper plants showed high gibberellins and ABA production and reduce amount of jasmonate and salicylate (Kang *et al.*, 2015). Grapevine (*Vitis vinifera*) plantlets inoculated with *Burkholderia phytofirmans* PsJN, modulated the carbohydrate metabolism that reduce low temperature damage during plantlets exposure to low temperature stress and supports the plant growth (Fernandez *et al.*, 2012). *Solanum lycopercsicum* plants inoculated with OB155 strain of *Pseudomonas vancoverensis* and OS 261 strain of *P. frederiks bergensis* under low temperature conditions showed enhanced expression of genes related with cold acclimation and antioxidant activity in leaf parts of plants (Subramanian *et al.*, 2015). It was observed that cold SRM strain of *Serratia marcescens* was found efficient in reducing cold stress injury due to induction of plants with cold tolerance ability (Selvakumar *et al.*, 2008).

Barka and coworkers (2006) investigated that seed treatment with PGPR has an intial effect on plants shoot as well as root growth by modifying the plant hormones including gibberlins, that play an essential role in seed germination.

This concluded that *B. phytofirmans* strain PsJN inoculation of *Vitis vinifera L.* cv. enhances grapevine root growth, plantlet biomass and increase its capability to withstand at low temperature prone environment. Some of the research reported that AAC deaminase potential in producing bacterial strains help in proper plant growth when exposed to drastic temperatures by reducing the accelerated ethylene content induced due to temperature stress. It has been also observed that a *Pseudomonas putida* UW4- a psychrotolerant deaminase bacterium enhanced the canola plant growth at lower temperature (Cheng *et al.*, 2007).

Conclusion

The growth of both plants and plant growth promoting rhizobacteria is negatively affected by different abiotic stress consisting high and low temperature, drought and salinity. Abiotic stresses main problem is faced during field conditions in agroecosystems that makes bio-inoculants survival a problematic issue. A significant difference was found in efficacy of microbial bioinoculants from laboratory to field level because of different abiotic stresses faced at field level. Microbial inoculants has been highly recommended under abiotic stressed conditions which promotes growth and yield of crop plants. The PGPR used as bioinoculants for many agricultural crops as an ecological and sustainable option for the mitigation of different abiotic stresses. Efficacy of PGPR strains can be enhanced under field conditions by screening the abiotic stress tolerant microbial isolates. Stress-tolerant microbes could offer more advantageous effects in pot house and field conditions leading to improved crop production in agriculture.

REFERENCES

American Meteorological Society (AMS), 2004. Statement on meteorological drought. *Bull. Am. Meteorol. Soc.* 85: 771–773.

Armada, E., Roldan, A. and Azcon, R. 2014. Differential activity of autochthonous bacteria in controlling drought stress in native Lavandula and Salvia plants species under drought conditions in natural arid soil. *Microbiol. Ecol.* 67: 410–420.

Ashraf, M. and Hafeez, M. 2004.Thermotolerance of pearl millet and maize at early growth stages: growth and nutrient relations. *Biol. Plant.* 48: 81–86.

Battisti, D.S. and Naylor, R.L. 2009. Historical warnings of future food insecurity with unprecedented seasonal heat. Science 323: 240–244.

Bita, C.E. and Gerats, T. 2013. Plant tolerance to high temperature in a changing environment: scientific fundamentals and production of heat stress-tolerant crops. *Front. Plant Sci.* 4: 273, 1–18.

Bresson, J., Varoquaux, F., Bontpart, T., Touraine, B. and Vile, D. 2013. The PGPR strain *Phyllobacterium brassicacearum* STM196 induces a reproductive delay and physiological changes that result in improved drought tolerance in Arabidopsis. *New Phytol.* 200: 558–569.

Calvo-Polanco, M., Sanchez-Romera, B., Aroca, R., Asins, M.J., Declerck, S., Dodd, I.C., Martinez-Andujar, C., Albacete, A. and Ruiz-Lozano, J.M. 2016. Exploring the use of recombinant inbred lines in combination with beneficial microbial inoculants (AM fungus and PGPR) to improve drought stress tolerance in tomato. *Environ. Exp. Bot.* 131: 47–57.

Camejo, D., Rodrý́guez, P., Angeles Morales, M., Miguel Dell'Amico, J., Torrecillas, A. and Alarcon, J.J. 2005. High temperature effects on photosynthetic activity of two tomato cultivars with different heat susceptibility. *J. Plant Physiol.* 162(3): 281–289.

Chang, P., Gerhardt, K.E., Huang, X.D., Yu, X.M., Glick, B.R., Gerwing, P.D. and Greenberg, B.M. 2014. Plant growth-promoting bacteria facilitate the growth of barley and oats in salt-impacted soil: implications for phytoremediation of saline soils. *Int. J. Phytoremedation.* 16(7-12): 1133–1147.

Chang, W.S., van de Mortel, M., Nielsen, L., Nino de Guzman, G., and Li, X. *et al.*, 2007. Alginate production by *Pseudomonas putida* creates a hydrated microenvironment and contributes to biofilm architecture and stress tolerance under water limiting conditions. *J. Bacteriol.* 189: 8290–8299.

Cheng, Z., Park, E. and Glick, B. R. 2007. 1-Aminocyclopropane-1-carboxylate (ACC) deaminase from Pseudomonas putida UW4 facilitates the growth of canola in the presence of salt. *Can. J. Microbiol.* 53: 912–918.

Cherif-Silini, H., Silini, A., Chenari Bouket, A., Alenezi, F.N., Luptakova, L., Bouremani, N., Nowakowska, J.A., Oszako, T. and Belbahri, L. 2021.Tailoring Next Generation Plant Growth Promoting Microorganisms as Versatile Tools beyond Soil Desalinization: A Road Map towards Field Application. *Sustainability.* 13: 1–22.

Cramer, G.R., Urano, K., Delrot, S., Pezzotti, M. and Shinozaki, K. 2011. Effects of abiotic stress on plants: a systems biology perspective. *BMC Plant Biol.* 11: 163. DOI 10.1186/1471-2229-11-163.

Egamberdieva, D., Davranov, K., Wirth, S., Hashem, A. and AbdAllah, E.F. 2017. Impact of soil salinity on the plant growth promoting and biological control abilities of root associated bacteria. *Saudi J. Biol.* Sci. 24(7): 1601–1608.

Garcia, J.E., Maroniche, G., Creus, C., Suarez-Rodriguez, R., Ramirez-Trujillo, J.A. and Groppa, M.D. 2017. In vitro PGPR properties and osmotic tolerance of different Azospirillum native strains and their effects on growth of maize under drought stress. *Microbiol. Res.* 202: 21–29.

Gill, K. 1979. Effect of soil salinity on grain filling and grain development in barley. *Biol. Plant.* 21(4): 241–244.

Harkousse, O., Slimani, A., Jadrane, I., Aitboulahsen, M., Mazri, M.A., Zouahri, A., Ouahmane, L., Koussa, T. and Al Feddy, M.N. 2021. Role of Local Biofertilizer in Enhancing the Oxidative Stress Defence Systems of Date Palm Seedling (*Phoenix dactylifera*) against Abiotic Stress. *Hindawi Appl. Environ. Soil Sci.* Article ID 6628544. pp.1–13.

Hosseini, F., Mosaddeghi, M.R. and Dexter, A.R. 2017. Effect of the fungus *Piriformospora indica* on physiological characteristics and root morphology of wheat under combined drought and mechanical stresses. *Plant Physiol. Biochem.* 118: 107–120.

Hussain, M.B., Zahir, Z.A., Asghar, H.N. and Asghar, M. 2014. Exopolysaccharides producing rhizobia ameliorate drought stress in wheat. *Int. J. Agric. Biol.* 16: 3–13.

Islam, F., Yasmeen, T., Ali, S., Ali, B., Farooq, M.A. and Gill, R.A. 2015. Priming-induced antioxidative responses in two wheat cultivars under saline stress. *Acta Physiol. Plant.* 37 (8): 1–12.

Jha, Y. And Subramanian, R.B. 2014. PGPR regulate caspase-like activity, programmed celldeath, and antioxidant enzyme activity in paddy under salinity. *Physiol. Mol. Biol. Plant.* 20: 201–207.

Karlidag, H., Esitken, A., Yildirim, E., Donmez, M.F. and Turan, M. 2011. Effects of plant growth promoting bacteria (PGPB) on yield, growth, leaf water content, membrane permeability and ionic composition of strawberry under saline conditions. *J. Plant Nutr.* 34: 34–45.

Keating, B. and Fisher, M. 1985. Comparative tolerance of tropical grain legumes to salinity. *Crop Pasture Sci.* 36(3): 373–383.

Kempf, B. and Bremer, E. 1998. Uptake and synthesis of compatible solutes as microbialstress responses to high-osmolality environments. *Arch. Microbiol.* 170(5): 319–330.

Khan, M.A., Asaf, S., Khan, A.L., Jan, R., Kang, S.M., Kim, K.M. and Lee, I. J. 2020. Thermotolerance effect of plant growth-promoting Bacillus cereus SA1 on soybean during heat stress. *BMC Microbiol.* 20(175): 1–14.

Kotuby-Amacher, J., Koenig, K. and Kitchen, B. 2000. Salinity and Plant Tolerance. Available at https: //extension.usu.edu/files/publications/publication/AG-SO-03.pdf.

Kumar, A. and Verma, J.P. 2017. Does plant-microbe interaction confer stress tolerance in plants: a review? *Microbiol. Res.* 207: 41–52.

Lesk, C., Rowhani, P. And Ramankutty, N. 2016. Influence of extreme weather disasters on global crop production. *Nature.* 529.7584, 84–87.

Lim, J.H. and Kim, S.D. 2013. Induction of drought stress resistance by multi-functional PGPR *Bacillus licheniformis* K11 in Pepper. *Plant Pathol. J.* 29: 201–208.

Liu, F., Ma, H., Du, Z., Ma, B., Liu, X., Peng, L. and Zhang, W. 2019. Physiological response of North China red elder container seedlings to inoculation with plant growth-promoting rhizobacteria under drought stress. *PLoS One* 14(12): 1–14. e0226624.

Liu, F., Xing, S., Ma, H., Du, Z. and Ma, B. 2013.Cytokinin producing, plant growth promoting rhizobacteria that confer resistance to drought stress in *Platycladus orientalis* container seedlings. *Appl. Microbiol. Biotechnol.* 97: 9155–9164.

Masciarelli, O., Llanes, A. and Luna, V. 2014. A new PGPR co-inoculated with *Bradyrhizobium japonicum* enhances soybean nodulation. *Microbiol. Res.* 169: 609–615.

Miller, P., Lanier, W. and Brandt, S. 2001. Using Growing Degree Days to Predict Plant Stages. Ag/Extension Communications Coordinator, Communications Services, Montana State University-Bozeman, Bozeman, MO.

Mohanty, P., Singh, P.K., Chakraborty, D., Mishra, S. and Pattnaik, R. 2021. Insight Into the Role of PGPR in Sustainable Agriculture and Environment. *Front. Sustain. Food Syst.* 5: 1–12.

Morcillo, R.J.L., Manzanera, M. 2021. The Effects of Plant-Associated Bacterial Exopolysaccharides on Plant Abiotic Stress Tolerance. *Metabolites.* 11: 1–19.

Morita, S., Siratsuchi, H., Takanashi, J. and Fujita, K. 2004. Effect of high temperature on ripening in rice plant. Analysis of the effect of high night and high day temperature applied to the panicle in other parts of the plant. *Jpn. J. Crop Sci.* 73: 77–83.

Morrison, M.J. and Stewart, D.W. 2002. Heat stress during flowering in summer brassica. *Crop Sci.* 42: 797–803.

Naseem, H. and Bano, A. 2014. Role of plant growth-promoting rhizobacteria and their exopolysaccharide in drought tolerance of maize. *J. Plant Interact.* 9: 689–701.

Nguyen, D., Rieu, I., Mariani, C. and van Dam, N.M. 2016. How plants handle multiple stresses: hormonal interactions underlaying responses to abiotic stress and insect herbivory. *Plant Mol. Biol.* 91: 72–740.

Ortiz, R., Braun, J., Crossa, J., Crouch, J.H., Davenport, G., Dixon, J., *et al.,* 2008. Wheat genetic resources enhancement by the International Maize and Wheat Improvement Center (CIMMYT). *Genet. Resour. Crop Evol.* 55: 1095–1140.

Patel, P.N. and Hall, A.E. 1990. Genotypic variation and classification of cowpea for reproductive responses to high temperatures under long photoperiods. *Crop Sci.* 30: 614–621.

Qurashi, A.W. and Sabri, A.N. 2012. Bacterial exopolysaccharide and biofilm formation stimulate chickpea growth and soil aggregation under salt stress. *Braz. J Microbiol.* 43(3): 1183–91.

Rajput, L., Imran, A., Mubeen, F. and Hafeez, F.Y. 2013. Salt tolerant PGPR strain Planococcus rifietoensis promotes the growth and yield of wheat (*Triticum aestivum*) cultivated in saline soil. *Pak. J. Bot.* 45: 1955–1962.

Rehman, H., Malik, S.A. and Saleem, M., 2004. Heat tolerance of upland cotton during the fruiting stage evaluated during cellular membrane thermostability. *Field Crops Res.* 85: 149–158.

Riadh, K., Wided, M., Hans-Werner, K. and Chedly, A. 2010. Responses of halophytes to environmental stresses with special emphasis to salinity. *Adv. Bot. Res.* 53: 117–145.

Robinson, S. and Jones, G. 1986. Accumulation of glycinebetaine in chloroplasts provides osmotic adjustment during salt stress. *Funct. Plant Biol.* 13(5): 659–668.

Ruan, C.J., da Silva, J.A.T., Mopper, S., Qin, P. and Lutts, S. 2010. Halophyte improvement for a salinized world. *Crit. Rev. Plant Sci.* 29(6): 329–359.

Sandhya, V., Ali, S.k.Z., Grover, M., Reddy, G. and Venkateswarlu, B. 2009. Alleviation of drought stress effects in sunflower seedlings by the exopolysaccharides producing *Pseudomonas putida* strain GAP-P45. *Biol. Fertil. Soils.* 46: 17–26.

Sang-Mo, K., Radhakrishnan, R., Khan, A.L., Min-Ji, K., Jae-Man, P., Bo-Ra, K., Dong-Hyun, S. and In-Jung, L. 2014. Gibberellin secreting rhizobacterium, *Pseudomonas putida* H-2-3 modulates the hormonal and stress physiology of soybean to improve the plant growth under saline and drought conditions. *Plant Physiol. Biochem.* 84: 115–124.

Santoyo, G., Moreno-Hagelsieb, G., del Carmen Orozco-Mosqueda, M. and Glick, B.R. 2016. Plant growth-promoting bacterial endophytes. *Microbiol. Res.* 183: 92–99.

Selvakumar, G., Mohan, M., Kundu, S., Gupta, A. D., Joshi, P., Nazim, S. and Gupta, H. S. 2008. Cold tolerance and plant growth promotion potential of *Serratia marcescens* strain SRM (MTCC8708) isolated from flowers of summer squash (cucurbita pepo). *Lett. Appl. Microbiol.* 46: 171175.

Shrivasta, U.P. and Kumar, A. 2014. Characterization and optimization of 1-aminocyclopropane-1-carboxylate deaminase (ACCD) activity in different rhizospheric PGPR along with Microbacterium sp. Strain ECI-12A. *Int. J. Appl. Sci. Biotechnol.* 1: 11–15.

Singh, J.S., Pandey, V.C. and Singh, D.P. 2011. Efficient soil microorganisms: a new dimension for sustainable agriculture and environmental development. *Agric. Ecosyst. Environ.* 140: 339–353.

Stone, P.J. and Nicol´as, M.E. 1994. Wheat cultivars vary widely in their responses of grain yield and quality to short periods of post-anthesis heat stress. *Aust. J. Plant Physiol.* 21: 887–900.

Suarez, C., Cardinale, M., Ratering, S., Steffens, D., Jung, S., Montoya, A.M. Z., Geissler-Plaum, R. and Schnell, S. 2015. Plant growth-promoting effects of Hartmannibacter diazotrophic on summer barley (*Hordeum vulgare L.*) under salt stress. *Appl. Soil Ecol.* 95: 23–30.

Sudhir, P. and Murthy, S. 2004. Effect of salt stress on basic processes of photosynthesis. *Photosynthetica*. 42(4): 481–486.

Thompson, L.M. 1986. Climatic change, weather variability and corn production. *Agron. J.* 78: 649–653.

Valencia-Cantero, E., Hernández-Calderón, E., Velázquez-Becerra, C., López-Meza, J.E., Alfaro-Cuevas, R. and López-Bucio, J. 2007. Role of dissimilatory fermentative iron-reducing bacteria in Fe uptake by common bean (*Phaseolus vulgaris L.*) plants grown inalkaline soil. *Plant Soil.* 291(1-2): 263–273.

Vara Prasad, P.V., Craufurd, P.Q., Summerfield, R.J. and Wheeler, T.R. 2000. Effect of short epidosed of heat stress on flower production and fruit set of groundnut (*Arachis hypogea L.*). *J. Exp. Bot.* 51: 777–784.

Vardharajula, S., Zulfikar, A.S., Grover, M., Reddy, G. and Bandi, V. 2011. Drought-tolerant plant growth promoting Bacillus spp., effect on growth, osmolytes and antioxidant status of maize under drought stress. *J. Plant Inter.* 6: 1–14.

Vimala, P. and Lalithakumari, D. 2003. Characterization of exopolysaccharide (EPS) produced by Leuconostoc sp. V 41. *Asian J. Microbiol. Biotechnol. Environ. Sci.* 5(2): 161–165.

Vurukonda, S.S.K.P., Vardharajula, S., Shrivastava, M. and Ali, SkZ. 2016. Enhancement of drought stress tolerance in crops by plant growth promoting rhizobacteria. *Microbiol. Res.* 184: 13–24.

Wang, C.J., Yang, W., Wang, C., Gu, C., Niu, D.D., Liu, H.X., Wang, Y.P., Guo, J.H. 2012. Induction of drought tolerance in cucumber plants by a consortium of three plant growth-promoting rhizobacterium strains. *PLoS One*. 7: 1–10.

Yadav, A.N., et al., 2018. Microbiome in Crops: Diversity, distribution and potential role in crops improvements. In: Prasad, R., Gill, S.S. and Tuteja, N. (Eds.). Crop Improvement through Microbial Biotechnology, pp. 305–332.

Yanez, M.V., Vinas, I., Usali, J., Canamas, T. and Teixido, N. 2012. Endospore production allows use of spray-drying as a possible formulation system of the biocontrol agent Bacillus subtilis CPA-8. *Biotechnol. Lett.* 34: 729–735.

Yang, J., Kloepper, J.W. and Ryu, C.M. 2009. Rhizosphere bacteria help plants tolerate abiotic stress. *Trends Plant Sci.* 14(1): 1–4.

Yildirim, E., Turan, M. and Donmez, M.F. 2008. Mitigation of salt stress in radish (Raphanus sativus) by plant growth promoting rhizobacteria. Romanian *Biotechnol. Lett.* 13: 3933–3943.

Yildirim, E., Turan, M., Ekinci, M., Dursun, A. and Cakmakci, R. 2011. Plant growth promoting rhizobacteria ameliorate deleterious effect of salt stress on lettuce. *Sci. Res. Essays*. 6: 4389–4396.

Zhou, C., Ma, Z., Zhu, L., Xiao, X., Xie, Y., Zhu, J. and Wang, J. 2016. Rhizobacterial strain Bacillus megaterium BOFC15 induces cellular polyamine changes that improve plant growth and drought resistance. *Int. J. Mol. Sci.* 17(6): 976.

Index

C

D

E

F

G

H

I

K

L

M

N

9 789354 614941

Printed at Repro India Ltd.